Ford Aerostar Mini-vans Automotive Repair Manual

by Larry Warren, Mark Christman and John H Haynes
Member of the Guild of Motoring Writers

Models covered:
All 2WD Ford Aerostar mini-vans
1986 through 1997

(3L18 - 36004)
(1476)

ABCDE
FGHI

2

Haynes Publishing Group
Sparkford Nr Yeovil
Somerset BA22 7JJ England

Haynes North America, Inc
861 Lawrence Drive
Newbury Park
California 91320 USA

© **Haynes North America, Inc.** **1988, 1990, 1992, 1994,**
1996, 2000, 2004

With permission from J.H. Haynes & Co. Ltd.

A book in the Haynes Automotive Repair Manual Series

Printed in the U.S.A.

ISBN 1 56392 374 2

Library of Congress Control Number 00-100920

While every attempt is made to ensure that the information in this manual is correct, no liability can be accepted by the authors or publishers
for loss, damage or injury caused by any errors in, or omissions from,
the information given.

04-272

Contents

Haynes photographer, mechanic and author with an Aerostar

About this manual

Its purpose

The purpose of this manual is to help you get the best value from your vehicle. It can do so in several ways. It can help you decide what work must be done, even if you choose to have it done by a dealer service department or a repair shop; it provides information and procedures for routine maintenance and servicing; and it offers diagnostic and repair procedures to follow when trouble occurs.

We hope you use the manual to tackle the work yourself. For many simpler jobs, doing it yourself may be quicker than arranging an appointment to get the vehicle into a shop and making the trips to leave it and pick it up. More importantly, a lot of money can be saved by avoiding the expense the shop must pass on to you to cover its labor and overhead costs. An added benefit is the sense of satisfaction and accomplishment that you feel after doing the job yourself.

Using the manual

The manual is divided into Chapters. Each Chapter is divided into numbered Sections, which are headed in bold type between horizontal lines. Each Section consists of consecutively numbered paragraphs.

At the beginning of each numbered Section you will be referred to any illustrations which apply to the procedures in that Section. The reference numbers used in illustration captions pinpoint the pertinent Section and the Step within that Section. That is, illustration 3.2 means the illustration refers to Section 3 and Step (or paragraph) 2 within that Section.

Procedures, once described in the text, are not normally repeated. When it's necessary to refer to another Chapter, the reference will be given as Chapter and Section number. Cross references given without use of the word "Chapter" apply to Sections and/or paragraphs in the same Chapter. For example, "see Section 8" means in the same Chapter.

References to the left or right side of the vehicle assume you are sitting in the driver's seat, facing forward.

Even though we have prepared this manual with extreme care, neither the publisher nor the author can accept responsibility for any errors in, or omissions from, the information given.

NOTE

A **Note** provides information necessary to properly complete a procedure or information which will make the procedure easier to understand.

CAUTION

A **Caution** provides a special procedure or special steps which must be taken while completing the procedure where the Caution is found. Not heeding a Caution can result in damage to the assembly being worked on.

WARNING

A **Warning** provides a special procedure or special steps which must be taken while completing the procedure where the Warning is found. Not heeding a Warning can result in personal injury.

Introduction

The Aerostar mini-van is a conventional front engine, rear drive vehicle available in passenger and cargo van configurations. Engines available include a 2.3 liter overhead cam inline four cylinder (early models only), a 2.8 liter V6 (1986 only), a 3.0 liter V6 and a 4.0 liter V6. All models other than those with the 2.8 liter V6 are equipped with port-type electronic fuel injection.

Transmissions installed in the Aerostar include a 5-speed manual overdrive and a 4-speed automatic overdrive. The 4-speed automatic transmission is equipped with a torque converter clutch, which eliminates torque converter slippage at high speeds and results in greater fuel economy.

The front suspension consists of upper and lower control arms (A-arms) and coil springs. A shock absorber is mounted inside each coil spring. The rear suspension is composed of longitudinally mounted control arms and coil springs attached to the outer ends of the solid rear axle.

The brakes are power assisted disc-type at the front and drum at the rear.

Vehicle identification numbers

Modifications are a continuing and unpublicized part of automotive manufacturing. Since spare parts manuals and lists are compiled on a numerical basis, individual vehicle numbers are essential to correctly identify new parts that may be needed.

The VIN number is attached to the dash on the driver's side of the vehicle

Vehicle Identification Number (VIN)

A seventeen digit combination of numbers and letters forms the VIN. It's stamped on a metal plate attached to the dashboard near the windshield on the driver's side **(see illustration)**. The VIN also appears on the Vehicle Certificate of Title and Registration and the Safety Compliance Certification label. It contains valuable information such as vehicle make and type, GVWR, brake system installed, model, body type, engine installed, etc. **(see illustration)**.

Safety Compliance Certification label

The English Safety Compliance Certification label is attached to the door latch edge on the driver's side door. The French (Quebec) label is attached in the same location on the passenger's side door. The label contains the month and year of production,

the certification statement and the VIN. It also includes GVWR, wheel and tire information and codes for additional vehicle data **(see illustration)**.

Build date stamp

The build date is stamped on the front surface of the radiator support on the passenger side of the vehicle.

Engine identification number

The number eight digit in the VIN denotes the type engine originally installed in the vehicle **(see illustration)**. The engine identification number (serial number) is located on a tag, which is usually attached to the rocker arm cover **(see illustration)**.

Vehicle Emissions Control Information label

The Vehicle Emissions Control Information label is attached to the underside of the hood (see Chapter 6 for an illustration).

Sample Safety Compliance Certification label

The engine serial number is located on a tag that's attached to the rocker arm cover

Buying parts

Replacement parts are available from many sources, which generally fall into one of two categories - authorized dealer parts departments and independent retail auto parts stores. Our advice concerning these parts is as follows:

Retail auto parts stores: Good auto parts stores will stock frequently needed components which wear out relatively fast, such as clutch components, exhaust systems, brake parts, tune-up parts, etc. These stores often supply new or reconditioned parts on an exchange basis, which can save a considerable amount of money. Discount auto parts stores are often very good places to buy materials and parts needed for general vehicle maintenance such as oil, grease, filters, spark plugs, belts, touch-up paint, bulbs, etc. They also usually sell tools and general accessories, have convenient hours, charge lower prices and can often be found not far from home.

Authorized dealer parts department: This is the best source for parts which are unique to the vehicle and not generally available elsewhere (such as major engine parts, transmission parts, trim pieces, etc.).

Warranty information: If the vehicle is still covered under warranty, be sure that any replacement parts purchased - regardless of the source - do not invalidate the warranty!

To be sure of obtaining the correct parts, have engine and chassis numbers available and, if possible, take the old parts along for positive identification.

Maintenance techniques, tools and working facilities

Maintenance techniques

There are a number of techniques involved in maintenance and repair that will be referred to throughout this manual. Application of these techniques will enable the home mechanic to be more efficient, better organized and capable of performing the various tasks properly, which will ensure that the repair job is thorough and complete.

Fasteners

Fasteners are nuts, bolts, studs and screws used to hold two or more parts together. There are a few things to keep in mind when working with fasteners. Almost all of them use a locking device of some type, either a lockwasher, locknut, locking tab or thread adhesive. All threaded fasteners should be clean and straight, with undamaged threads and undamaged corners on the hex head where the wrench fits. Develop the habit of replacing all damaged nuts and bolts with new ones. Special locknuts with nylon or fiber inserts can only be used once. If they are removed, they lose their locking ability and must be replaced with new ones.

Rusted nuts and bolts should be treated with a penetrating fluid to ease removal and prevent breakage. Some mechanics use turpentine in a spout-type oil can, which works quite well. After applying the rust penetrant, let it work for a few minutes before trying to loosen the nut or bolt. Badly rusted fasteners may have to be chiseled or sawed off or removed with a special nut breaker, available at tool stores.

If a bolt or stud breaks off in an assembly, it can be drilled and removed with a special tool commonly available for this purpose. Most automotive machine shops can perform this task, as well as other repair procedures,

such as the repair of threaded holes that have been stripped out.

Flat washers and lockwashers, when removed from an assembly, should always be replaced exactly as removed. Replace any damaged washers with new ones. Never use a lockwasher on any soft metal surface (such as aluminum), thin sheet metal or plastic.

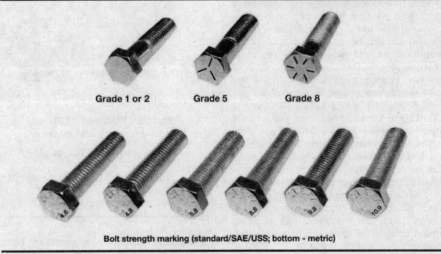

Grade 1 or 2   Grade 5   Grade 8

Bolt strength marking (standard/SAE/USS; bottom - metric)

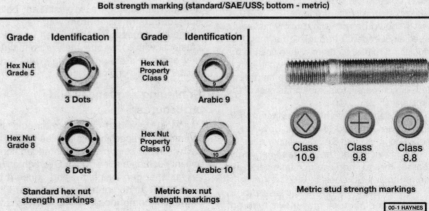

Grade	Identification
Hex Nut Grade 5	3 Dots
Hex Nut Grade 8	6 Dots

Standard hex nut strength markings

Grade	Identification
Hex Nut Property Class 9	Arabic 9
Hex Nut Property Class 10	Arabic 10

Metric hex nut strength markings

Class 10.9   Class 9.8   Class 8.8

Metric stud strength markings

00-1 HAYNES

Fastener sizes

For a number of reasons, automobile manufacturers are making wider and wider use of metric fasteners. Therefore, it is important to be able to tell the difference between standard (sometimes called U.S. or SAE) and metric hardware, since they cannot be interchanged.

All bolts, whether standard or metric, are sized according to diameter, thread pitch and length. For example, a standard 1/2 - 13 x 1 bolt is 1/2 inch in diameter, has 13 threads per inch and is 1 inch long. An M12 - 1.75 x 25 metric bolt is 12 mm in diameter, has a thread pitch of 1.75 mm (the distance between threads) and is 25 mm long. The two bolts are nearly identical, and easily confused, but they are not interchangeable.

In addition to the differences in diameter, thread pitch and length, metric and standard bolts can also be distinguished by examining the bolt heads. To begin with, the distance across the flats on a standard bolt head is measured in inches, while the same dimension on a metric bolt is sized in millimeters (the same is true for nuts). As a result, a standard wrench should not be used on a metric bolt and a metric wrench should not be used on a standard bolt. Also, most standard bolts have slashes radiating out from the center of the head to denote the grade or strength of the bolt, which is an indication of the amount of torque that can be applied to it. The greater the number of slashes, the greater the strength of the bolt. Grades 0 through 5 are commonly used on automobiles. Metric bolts have a property class (grade) number, rather than a slash, molded into their heads to indicate bolt strength. In this case, the higher the number, the stronger the bolt. Property class numbers 8.8, 9.8 and 10.9 are commonly used on automobiles.

Strength markings can also be used to distinguish standard hex nuts from metric hex nuts. Many standard nuts have dots stamped into one side, while metric nuts are marked with a number. The greater the number of dots, or the higher the number, the greater the strength of the nut.

Metric studs are also marked on their ends according to property class (grade). Larger studs are numbered (the same as metric bolts), while smaller studs carry a geometric code to denote grade.

It should be noted that many fasteners, especially Grades 0 through 2, have no distinguishing marks on them. When such is the case, the only way to determine whether it is standard or metric is to measure the thread pitch or compare it to a known fastener of the same size.

Standard fasteners are often referred to as SAE, as opposed to metric. However, it should be noted that SAE technically refers to a non-metric fine thread fastener only. Coarse thread non-metric fasteners are referred to as USS sizes.

Since fasteners of the same size (both standard and metric) may have different

Metric thread sizes	Ft-lbs	Nm
M-6	6 to 9	9 to 12
M-8	14 to 21	19 to 28
M-10	28 to 40	38 to 54
M-12	50 to 71	68 to 96
M-14	80 to 140	109 to 154

Pipe thread sizes		
1/8	5 to 8	7 to 10
1/4	12 to 18	17 to 24
3/8	22 to 33	30 to 44
1/2	25 to 35	34 to 47

U.S. thread sizes		
1/4 - 20	6 to 9	9 to 12
5/16 - 18	12 to 18	17 to 24
5/16 - 24	14 to 20	19 to 27
3/8 - 16	22 to 32	30 to 43
3/8 - 24	27 to 38	37 to 51
7/16 - 14	40 to 55	55 to 74
7/16 - 20	40 to 60	55 to 81
1/2 - 13	55 to 80	75 to 108

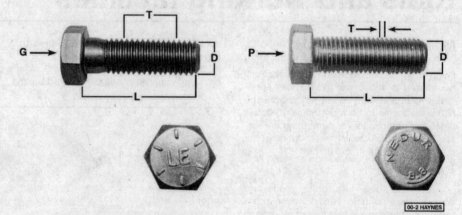

Standard (SAE and USS) bolt dimensions/grade marks

G   Grade marks (bolt strength)
L   Length (in inches)
T   Thread pitch (number of threads per inch)
D   Nominal diameter (in inches)

Metric bolt dimensions/grade marks

P   Property class (bolt strength)
L   Length (in millimeters)
T   Thread pitch (distance between threads in millimeters)
D   Diameter

strength ratings, be sure to reinstall any bolts, studs or nuts removed from your vehicle in their original locations. Also, when replacing a fastener with a new one, make sure that the new one has a strength rating equal to or greater than the original.

Tightening sequences and procedures

Most threaded fasteners should be tightened to a specific torque value (torque is the twisting force applied to a threaded component such as a nut or bolt). Overtightening the fastener can weaken it and cause it to break, while undertightening can cause it to eventually come loose. Bolts, screws and studs, depending on the material they are

made of and their thread diameters, have specific torque values, many of which are noted in the Specifications at the beginning of each Chapter. Be sure to follow the torque recommendations closely. For fasteners not assigned a specific torque, a general torque value chart is presented here as a guide. These torque values are for dry (unlubricated) fasteners threaded into steel or cast iron (not aluminum). As was previously mentioned, the size and grade of a fastener determine the amount of torque that can safely be applied to it. The figures listed here are approximate for Grade 2 and Grade 3 fasteners. Higher grades can tolerate higher torque values.

Fasteners laid out in a pattern, such as cylinder head bolts, oil pan bolts, differential cover bolts, etc., must be loosened or tightened in sequence to avoid warping the component. This sequence will normally be shown in the appropriate Chapter. If a specific pattern is not given, the following procedures can be used to prevent warping.

Initially, the bolts or nuts should be assembled finger-tight only. Next, they should be tightened one full turn each, in a criss-cross or diagonal pattern. After each one has been tightened one full turn, return to the first one and tighten them all one-half turn, following the same pattern. Finally, tighten each of them one-quarter turn at a time until each fastener has been tightened to the proper torque. To loosen and remove the fasteners, the procedure would be reversed.

Component disassembly

Component disassembly should be done with care and purpose to help ensure that the parts go back together properly. Always keep track of the sequence in which parts are removed. Make note of special characteristics or marks on parts that can be installed more than one way, such as a grooved thrust washer on a shaft. It is a good idea to lay the disassembled parts out on a clean surface in the order that they were removed. It may also be helpful to make sketches or take instant photos of components before removal.

When removing fasteners from a component, keep track of their locations. Sometimes threading a bolt back in a part, or putting the washers and nut back on a stud, can prevent mix-ups later. If nuts and bolts cannot be returned to their original locations, they should be kept in a compartmented box or a series of small boxes. A cupcake or muffin tin is ideal for this purpose, since each cavity can hold the bolts and nuts from a particular area (i.e. oil pan bolts, valve cover bolts, engine mount bolts, etc.). A pan of this type is especially helpful when working on assemblies with very small parts, such as the carburetor, alternator, valve train or interior dash and trim pieces. The cavities can be marked with paint or tape to identify the contents.

Whenever wiring looms, harnesses or connectors are separated, it is a good idea to identify the two halves with numbered pieces of masking tape so they can be easily reconnected.

Gasket sealing surfaces

Throughout any vehicle, gaskets are used to seal the mating surfaces between two parts and keep lubricants, fluids, vacuum or pressure contained in an assembly.

Many times these gaskets are coated with a liquid or paste-type gasket sealing compound before assembly. Age, heat and pressure can sometimes cause the two parts to stick together so tightly that they are very difficult to separate. Often, the assembly can be loosened by striking it with a soft-face

hammer near the mating surfaces. A regular hammer can be used if a block of wood is placed between the hammer and the part. Do not hammer on cast parts or parts that could be easily damaged. With any particularly stubborn part, always recheck to make sure that every fastener has been removed.

Avoid using a screwdriver or bar to pry apart an assembly, as they can easily mar the gasket sealing surfaces of the parts, which must remain smooth. If prying is absolutely necessary, use an old broom handle, but keep in mind that extra clean up will be necessary if the wood splinters.

After the parts are separated, the old gasket must be carefully scraped off and the gasket surfaces cleaned. Stubborn gasket material can be soaked with rust penetrant or treated with a special chemical to soften it so it can be easily scraped off. A scraper can be fashioned from a piece of copper tubing by flattening and sharpening one end. Copper is recommended because it is usually softer than the surfaces to be scraped, which reduces the chance of gouging the part. Some gaskets can be removed with a wire brush, but regardless of the method used, the mating surfaces must be left clean and smooth. If for some reason the gasket surface is gouged, then a gasket sealer thick enough to fill scratches will have to be used during reassembly of the components. For most applications, a non-drying (or semi-drying) gasket sealer should be used.

Hose removal tips

Warning: *If the vehicle is equipped with air conditioning, do not disconnect any of the A/C hoses without first having the system depressurized by a dealer service department or a service station.*

Hose removal precautions closely parallel gasket removal precautions. Avoid scratching or gouging the surface that the hose mates against or the connection may leak. This is especially true for radiator hoses. Because of various chemical reactions, the rubber in hoses can bond itself to the metal spigot that the hose fits over. To remove a hose, first loosen the hose clamps that secure it to the spigot. Then, with slip-joint pliers, grab the hose at the clamp and rotate it around the spigot. Work it back and forth until it is completely free, then pull it off. Silicone or other lubricants will ease removal if they can be applied between the hose and the outside of the spigot. Apply the same lubricant to the inside of the hose and the outside of the spigot to simplify installation.

As a last resort (and if the hose is to be replaced with a new one anyway), the rubber can be slit with a knife and the hose peeled from the spigot. If this must be done, be careful that the metal connection is not damaged.

If a hose clamp is broken or damaged, do not reuse it. Wire-type clamps usually weaken with age, so it is a good idea to replace them with screw-type clamps whenever a hose is removed.

Tools

A selection of good tools is a basic requirement for anyone who plans to maintain and repair his or her own vehicle. For the owner who has few tools, the initial investment might seem high, but when compared to the spiraling costs of professional auto maintenance and repair, it is a wise one.

To help the owner decide which tools are needed to perform the tasks detailed in this manual, the following tool lists are offered: *Maintenance and minor repair, Repair/overhaul* and *Special.*

The newcomer to practical mechanics should start off with the *maintenance and minor repair* tool kit, which is adequate for the simpler jobs performed on a vehicle. Then, as confidence and experience grow, the owner can tackle more difficult tasks, buying additional tools as they are needed. Eventually the basic kit will be expanded into the *repair and overhaul* tool set. Over a period of time, the experienced do-it-yourselfer will assemble a tool set complete enough for most repair and overhaul procedures and will add tools from the special category when it is felt that the expense is justified by the frequency of use.

Maintenance and minor repair tool kit

The tools in this list should be considered the minimum required for performance of routine maintenance, servicing and minor repair work. We recommend the purchase of combination wrenches (box-end and open-end combined in one wrench). While more expensive than open end wrenches, they offer the advantages of both types of wrench.

*Combination wrench set (1/4-inch to
    1 inch or 6 mm to 19 mm)*
Adjustable wrench, 8 inch
Spark plug wrench with rubber insert
Spark plug gap adjusting tool
Feeler gauge set
Brake bleeder wrench
*Standard screwdriver (5/16-inch x
    6 inch)*
Phillips screwdriver (No. 2 x 6 inch)
Combination pliers - 6 inch
Hacksaw and assortment of blades
Tire pressure gauge
Grease gun
Oil can
Fine emery cloth
Wire brush
Battery post and cable cleaning tool
Oil filter wrench
Funnel (medium size)
Safety goggles
Jackstands (2)
Drain pan

Note: *If basic tune-ups are going to be part of routine maintenance, it will be necessary to purchase a good quality stroboscopic timing light and combination tachometer/dwell meter. Although they are included in the list of special tools, it is mentioned here because they are absolutely necessary for tuning most vehicles properly.*

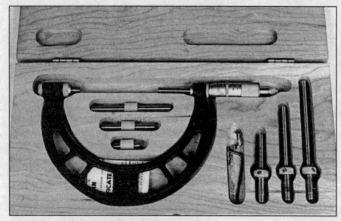

Micrometer set

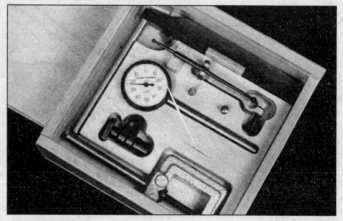

Dial indicator set

Repair and overhaul tool set

These tools are essential for anyone who plans to perform major repairs and are in addition to those in the maintenance and minor repair tool kit. Included is a comprehensive set of sockets which, though expensive, are invaluable because of their versatility, especially when various extensions and drives are available. We recommend the 1/2-inch drive over the 3/8-inch drive. Although the larger drive is bulky and more expensive, it has the capacity of accepting a very wide range of large sockets. Ideally, however, the mechanic should have a 3/8-inch drive set and a 1/2-inch drive set.

> *Socket set(s)*
> *Reversible ratchet*
> *Extension - 10 inch*
> *Universal joint*
> *Torque wrench (same size drive as sockets)*
> *Ball peen hammer - 8 ounce*
> *Soft-face hammer (plastic/rubber)*
> *Standard screwdriver (1/4-inch x 6 inch)*
> *Standard screwdriver (stubby - 5/16-inch)*
> *Phillips screwdriver (No. 3 x 8 inch)*
> *Phillips screwdriver (stubby - No. 2)*
> *Pliers - vise grip*
> *Pliers - lineman's*
> *Pliers - needle nose*
> *Pliers - snap-ring (internal and external)*
> *Cold chisel - 1/2-inch*
> *Scribe*
> *Scraper (made from flattened copper tubing)*
> *Centerpunch*
> *Pin punches (1/16, 1/8, 3/16-inch)*
> *Steel rule/straightedge - 12 inch*
> *Allen wrench set (1/8 to 3/8-inch or 4 mm to 10 mm)*
> *A selection of files*
> *Wire brush (large)*
> *Jackstands (second set)*
> *Jack (scissor or hydraulic type)*

Note: *Another tool which is often useful is an electric drill with a chuck capacity of 3/8-inch and a set of good quality drill bits.*

Special tools

The tools in this list include those which are not used regularly, are expensive to buy, or which need to be used in accordance with their manufacturer's instructions. Unless these tools will be used frequently, it is not very economical to purchase many of them. A consideration would be to split the cost and use between yourself and a friend or friends. In addition, most of these tools can be obtained from a tool rental shop on a temporary basis.

This list primarily contains only those tools and instruments widely available to the public, and not those special tools produced by the vehicle manufacturer for distribution to dealer service departments. Occasionally, references to the manufacturer's special tools are included in the text of this manual. Generally, an alternative method of doing the job without the special tool is offered. However, sometimes there is no alternative to their use. Where this is the case, and the tool cannot be purchased or borrowed, the work should be turned over to the dealer service department or an automotive repair shop.

> *Valve spring compressor*
> *Piston ring groove cleaning tool*
> *Piston ring compressor*
> *Piston ring installation tool*
> *Cylinder compression gauge*
> *Cylinder ridge reamer*
> *Cylinder surfacing hone*
> *Cylinder bore gauge*
> *Micrometers and/or dial calipers*
> *Hydraulic lifter removal tool*
> *Balljoint separator*
> *Universal-type puller*
> *Impact screwdriver*
> *Dial indicator set*
> *Stroboscopic timing light (inductive pick-up)*
> *Hand operated vacuum/pressure pump*
> *Tachometer/dwell meter*
> *Universal electrical multimeter*
> *Cable hoist*
> *Brake spring removal and installation tools*
> *Floor jack*

Buying tools

For the do-it-yourselfer who is just starting to get involved in vehicle maintenance and repair, there are a number of options available when purchasing tools. If maintenance and minor repair is the extent of the work to be done, the purchase of individual tools is satisfactory. If, on the other hand, extensive work is planned, it would be a good idea to purchase a modest tool set from one of the large retail chain stores. A set can usually be bought at a substantial savings over the individual tool prices, and they often come with a tool box. As additional tools are needed, add-on sets, individual tools and a larger tool box can be purchased to expand the tool selection. Building a tool set gradually allows the cost of the tools to be spread over a longer period of time and gives the mechanic the freedom to choose only those tools that will actually be used.

Tool stores will often be the only source of some of the special tools that are needed, but regardless of where tools are bought, try to avoid cheap ones, especially when buying screwdrivers and sockets, because they won't last very long. The expense involved in replacing cheap tools will eventually be greater than the initial cost of quality tools.

Care and maintenance of tools

Good tools are expensive, so it makes sense to treat them with respect. Keep them clean and in usable condition and store them properly when not in use. Always wipe off any dirt, grease or metal chips before putting them away. Never leave tools lying around in the work area. Upon completion of a job, always check closely under the hood for tools that may have been left there so they won't get lost during a test drive.

Some tools, such as screwdrivers, pliers, wrenches and sockets, can be hung on a panel mounted on the garage or workshop wall, while others should be kept in a tool box or tray. Measuring instruments, gauges, meters, etc. must be carefully stored where they cannot be damaged by weather or impact from other tools.

When tools are used with care and

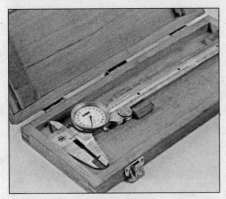

Dial caliper

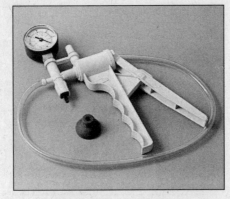

Hand-operated vacuum pump

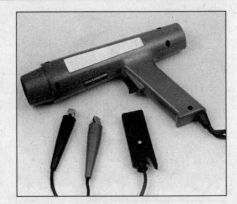

Timing light

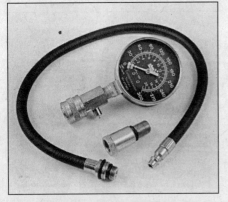

Compression gauge with spark plug hole adapter

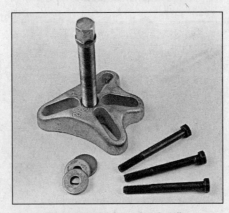

Damper/steering wheel puller

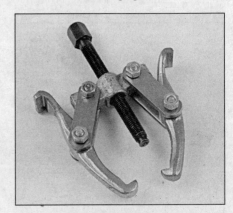

General purpose puller

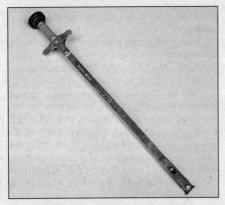

Hydraulic lifter removal tool

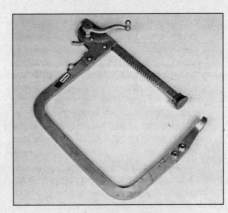

Valve spring compressor

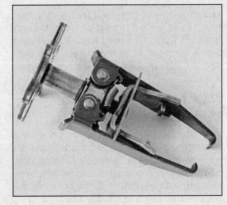

Valve spring compressor

Ridge reamer

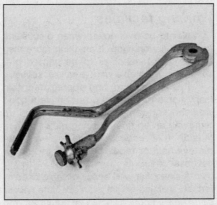

Piston ring groove cleaning tool

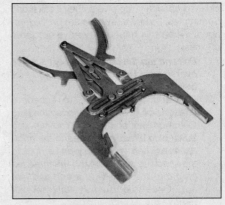

Ring removal/installation tool

Ring compressor

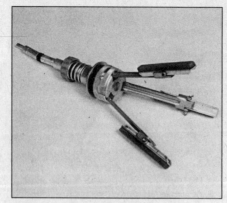

Cylinder hone

Brake hold-down spring tool

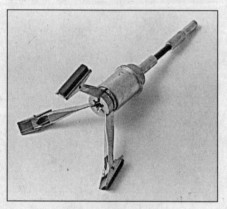

Brake cylinder hone

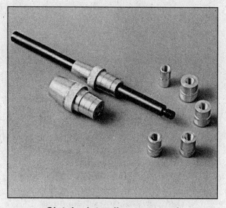

Clutch plate alignment tool

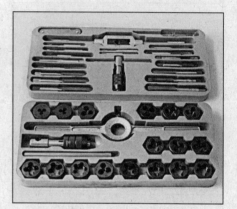

Tap and die set

stored properly, they will last a very long time. Even with the best of care, though, tools will wear out if used frequently. When a tool is damaged or worn out, replace it. Subsequent jobs will be safer and more enjoyable if you do.

How to repair damaged threads

Sometimes, the internal threads of a nut or bolt hole can become stripped, usually from overtightening. Stripping threads is an all-too-common occurrence, especially when working with aluminum parts, because aluminum is so soft that it easily strips out.

Usually, external or internal threads are only partially stripped. After they've been cleaned up with a tap or die, they'll still work. Sometimes, however, threads are badly damaged. When this happens, you've got three choices:

1) *Drill and tap the hole to the next suitable oversize and install a larger diameter bolt, screw or stud.*
2) *Drill and tap the hole to accept a threaded plug, then drill and tap the plug to the original screw size. You can also buy a plug already threaded to the original size. Then you simply drill a hole to the specified size, then run the threaded plug into the hole with a bolt and jam nut. Once the plug is fully seated, remove the jam nut and bolt.*

3) *The third method uses a patented thread repair kit like Heli-Coil or Slimsert. These easy-to-use kits are designed to repair damaged threads in straight-through holes and blind holes. Both are available as kits which can handle a variety of sizes and thread patterns. Drill the hole, then tap it with the special included tap. Install the Heli-Coil and the hole is back to its original diameter and thread pitch.*

Regardless of which method you use, be sure to proceed calmly and carefully. A little impatience or carelessness during one of these relatively simple procedures can ruin your whole day's work and cost you a bundle if you wreck an expensive part.

Working facilities

Not to be overlooked when discussing tools is the workshop. If anything more than routine maintenance is to be carried out, some sort of suitable work area is essential.

It is understood, and appreciated, that many home mechanics do not have a good workshop or garage available, and end up removing an engine or doing major repairs outside. It is recommended, however, that the overhaul or repair be completed under the cover of a roof.

A clean, flat workbench or table of comfortable working height is an absolute necessity. The workbench should be equipped with

a vise that has a jaw opening of at least four inches.

As mentioned previously, some clean, dry storage space is also required for tools, as well as the lubricants, fluids, cleaning solvents, etc. which soon become necessary.

Sometimes waste oil and fluids, drained from the engine or cooling system during normal maintenance or repairs, present a disposal problem. To avoid pouring them on the ground or into a sewage system, pour the used fluids into large containers, seal them with caps and take them to an authorized disposal site or recycling center. Plastic jugs, such as old antifreeze containers, are ideal for this purpose.

Always keep a supply of old newspapers and clean rags available. Old towels are excellent for mopping up spills. Many mechanics use rolls of paper towels for most work because they are readily available and disposable. To help keep the area under the vehicle clean, a large cardboard box can be cut open and flattened to protect the garage or shop floor.

Whenever working over a painted surface, such as when leaning over a fender to service something under the hood, always cover it with an old blanket or bedspread to protect the finish. Vinyl covered pads, made especially for this purpose, are available at auto parts stores.

Booster battery (jump) starting

Observe these precautions when using a booster battery to start a vehicle:

a) Before connecting the booster battery, make sure the ignition switch is in the Off position.
b) Turn off the lights, heater and other electrical loads.
c) Your eyes should be shielded. Safety goggles are a good idea.
d) Make sure the booster battery is the same voltage as the dead one in the vehicle.
e) The two vehicles MUST NOT TOUCH each other!
f) Make sure the transaxle is in Neutral (manual) or Park (automatic).
g) If the booster battery is not a maintenance-free type, remove the vent caps and lay a cloth over the vent holes.

Connect the red jumper cable to the positive (+) terminals of each battery (see illustration).

Connect one end of the black jumper cable to the negative (-) terminal of the booster battery. The other end of this cable should be connected to a good ground on the vehicle to be started, such as a bolt or bracket on the body.

Start the engine using the booster battery, then, with the engine running at idle speed, disconnect the jumper cables in the reverse order of connection.

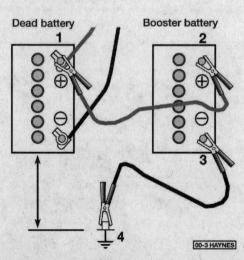

Make the booster battery cable connections in the numerical order shown (note that the negative cable of the booster battery is NOT attached to the negative terminal of the dead battery)

Jacking and towing

Jacking

The jack supplied with the vehicle should only be used for changing a tire or placing jackstands under the frame. **Warning:** *Never work under the vehicle or start the engine while this jack is being used as the only means of support!* If a hoist is available, be sure to raise the vehicle only at the points shown in the accompanying illustrations.

The vehicle should be on level ground with the wheels blocked and the transmission in Park (automatic) or Reverse (manual). If a tire must be changed, remove the hub cap or wheel cover (if equipped). Loosen the lug nuts one-half turn and leave them in place until the wheel is raised off the ground. **Caution:** *On vehicles equipped with an under chassis mounted spare tire, remove the spare before raising the vehicle. Once it's in the air, removal of the spare may cause the vehicle to shift and fall.*

Position the jack as shown in the accompanying illustrations.

When using a floor jack, position it under the axles or under the frame lift points, as close to the wheels as possible. Refer to the jacking decal and the vehicle's owner's guide if necessary.

Operate the jack with a slow, smooth motion until the wheel is raised off the ground. Remove the tire and install the spare.

Tighten the lug nuts until they're snug, but wait until the vehicle is lowered to use the wrench.

Lower the vehicle, remove the jack and tighten the lug nuts (if loosened or removed) in a criss-cross pattern.

Towing

As a general rule, the vehicle should be towed with the rear (drive) wheels off the ground. If they can't be raised, either place them on a dolly or disconnect the driveshaft from the differential. When a vehicle is towed with the rear wheels raised, the steering wheel must be clamped in the straight ahead position with a special device designed for use during towing. **Caution:** *Don't use the steering column lock to keep the front wheels pointed straight ahead.*

Vehicles can be towed with all four wheels on the ground, provided that speeds don't exceed 35 mph and the distance is not over 50 miles. When the rear wheels are off the ground there's no distance limitation, but don't exceed 50 mph.

Equipment specifically designed for towing should be used. It should be attached to the main structural members of the vehicle, not the bumpers or brackets.

Safety is a major consideration when towing and all applicable state and local laws must be obeyed. A safety chain system must be used at all times.

While towing, the parking brake should be released and the transmission must be in Neutral. The steering must be unlocked (ignition switch in the Off position). Remember that power steering and power brakes will not work with the engine off.

Front jacking point

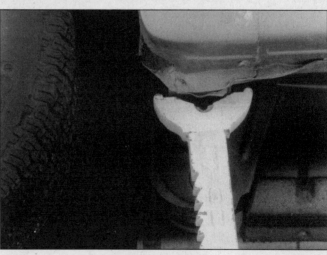

Rear jacking point

Automotive chemicals and lubricants

A number of automotive chemicals and lubricants are available for use during vehicle maintenance and repair. They include a wide variety of products ranging from cleaning solvents and degreasers to lubricants and protective sprays for rubber, plastic and vinyl.

Cleaners

Carburetor cleaner and choke cleaner is a strong solvent for gum, varnish and carbon. Most carburetor cleaners leave a dry-type lubricant film which will not harden or gum up. Because of this film it is not recommended for use on electrical components.

Brake system cleaner is used to remove brake dust, grease and brake fluid from the brake system, where clean surfaces are absolutely necessary. It leaves no residue and often eliminates brake squeal caused by contaminants.

Electrical cleaner removes oxidation, corrosion and carbon deposits from electrical contacts, restoring full current flow. It can also be used to clean spark plugs, carburetor jets, voltage regulators and other parts where an oil-free surface is desired.

Demoisturants remove water and moisture from electrical components such as alternators, voltage regulators, electrical connectors and fuse blocks. They are non-conductive and non-corrosive.

Degreasers are heavy-duty solvents used to remove grease from the outside of the engine and from chassis components. They can be sprayed or brushed on and, depending on the type, are rinsed off either with water or solvent.

Lubricants

Motor oil is the lubricant formulated for use in engines. It normally contains a wide variety of additives to prevent corrosion and reduce foaming and wear. Motor oil comes in various weights (viscosity ratings) from 0 to 50. The recommended weight of the oil depends on the season, temperature and the demands on the engine. Light oil is used in cold climates and under light load conditions. Heavy oil is used in hot climates and where high loads are encountered. Multi-viscosity oils are designed to have characteristics of both light and heavy oils and are available in a number of weights from 5W-20 to 20W-50.

Gear oil is designed to be used in differentials, manual transmissions and other areas where high-temperature lubrication is required.

Chassis and wheel bearing grease is a heavy grease used where increased loads and friction are encountered, such as for wheel bearings, balljoints, tie-rod ends and universal joints.

High-temperature wheel bearing grease is designed to withstand the extreme temperatures encountered by wheel bearings in disc brake equipped vehicles. It usually contains molybdenum disulfide (moly), which is a dry-type lubricant.

White grease is a heavy grease for metal-to-metal applications where water is a problem. White grease stays soft under both low and high temperatures (usually from -100 to +190-degrees F), and will not wash off or dilute in the presence of water.

Assembly lube is a special extreme pressure lubricant, usually containing moly, used to lubricate high-load parts (such as main and rod bearings and cam lobes) for initial start-up of a new engine. The assembly lube lubricates the parts without being squeezed out or washed away until the engine oiling system begins to function.

Silicone lubricants are used to protect rubber, plastic, vinyl and nylon parts.

Graphite lubricants are used where oils cannot be used due to contamination problems, such as in locks. The dry graphite will lubricate metal parts while remaining uncontaminated by dirt, water, oil or acids. It is electrically conductive and will not foul electrical contacts in locks such as the ignition switch.

Moly penetrants loosen and lubricate frozen, rusted and corroded fasteners and prevent future rusting or freezing.

Heat-sink grease is a special electrically non-conductive grease that is used for mounting electronic ignition modules where it is essential that heat is transferred away from the module.

Sealants

RTV sealant is one of the most widely used gasket compounds. Made from silicone, RTV is air curing, it seals, bonds, waterproofs, fills surface irregularities, remains flexible, doesn't shrink, is relatively easy to remove, and is used as a supplementary sealer with almost all low and medium temperature gaskets.

Anaerobic sealant is much like RTV in that it can be used either to seal gaskets or to form gaskets by itself. It remains flexible, is solvent resistant and fills surface imperfections. The difference between an anaerobic sealant and an RTV-type sealant is in the curing. RTV cures when exposed to air, while an anaerobic sealant cures only in the absence of air. This means that an anaerobic sealant cures only after the assembly of parts, sealing them together.

Thread and pipe sealant is used for sealing hydraulic and pneumatic fittings and vacuum lines. It is usually made from a Teflon compound, and comes in a spray, a paint-on liquid and as a wrap-around tape.

Chemicals

Anti-seize compound prevents seizing, galling, cold welding, rust and corrosion in fasteners. High-temperature ant-seize, usually made with copper and graphite lubricants, is used for exhaust system and exhaust manifold bolts.

Anaerobic locking compounds are used to keep fasteners from vibrating or working loose and cure only after installation, in the absence of air. Medium strength locking compound is used for small nuts, bolts and screws that may be removed later. High-strength locking compound is for large nuts, bolts and studs which aren't removed on a regular basis.

Oil additives range from viscosity index improvers to chemical treatments that claim to reduce internal engine friction. It should be noted that most oil manufacturers caution against using additives with their oils.

Gas additives perform several functions, depending on their chemical makeup. They usually contain solvents that help dissolve gum and varnish that build up on carburetor, fuel injection and intake parts. They also serve to break down carbon deposits that form on the inside surfaces of the combustion chambers. Some additives contain upper cylinder lubricants for valves and piston rings, and others contain chemicals to remove condensation from the gas tank.

Miscellaneous

Brake fluid is specially formulated hydraulic fluid that can withstand the heat and pressure encountered in brake systems. Care must be taken so this fluid does not come in contact with painted surfaces or plastics. An opened container should always be resealed to prevent contamination by water or dirt.

Weatherstrip adhesive is used to bond weatherstripping around doors, windows and trunk lids. It is sometimes used to attach trim pieces.

Undercoating is a petroleum-based, tar-like substance that is designed to protect metal surfaces on the underside of the vehicle from corrosion. It also acts as a sound-deadening agent by insulating the bottom of the vehicle.

Waxes and polishes are used to help protect painted and plated surfaces from the weather. Different types of paint may require the use of different types of wax and polish. Some polishes utilize a chemical or abrasive cleaner to help remove the top layer of oxidized (dull) paint on older vehicles. In recent years many non-wax polishes that contain a wide variety of chemicals such as polymers and silicones have been introduced. These non-wax polishes are usually easier to apply and last longer than conventional waxes and polishes.

Conversion factors

Length (distance)

Inches (in)	X	25.4	= Millimeters (mm)	X 0.0394	= Inches (in)
Feet (ft)	X	0.305	= Meters (m)	X 3.281	= Feet (ft)
Miles	X	1.609	= Kilometers (km)	X 0.621	= Miles

Volume (capacity)

Cubic inches (cu in; in^3)	X	16.387	= Cubic centimeters (cc; cm^3)	X 0.061	= Cubic inches (cu in; in^3)
Imperial pints (Imp pt)	X	0.568	= Liters (l)	X 1.76	= Imperial pints (Imp pt)
Imperial quarts (Imp qt)	X	1.137	= Liters (l)	X 0.88	= Imperial quarts (Imp qt)
Imperial quarts (Imp qt)	X	1.201	= US quarts (US qt)	X 0.833	= Imperial quarts (Imp qt)
US quarts (US qt)	X	0.946	= Liters (l)	X 1.057	= US quarts (US qt)
Imperial gallons (Imp gal)	X	4.546	= Liters (l)	X 0.22	= Imperial gallons (Imp gal)
Imperial gallons (Imp gal)	X	1.201	= US gallons (US gal)	X 0.833	= Imperial gallons (Imp gal)
US gallons (US gal)	X	3.785	= Liters (l)	X 0.264	= US gallons (US gal)

Mass (weight)

Ounces (oz)	X	28.35	= Grams (g)	X 0.035	= Ounces (oz)
Pounds (lb)	X	0.454	= Kilograms (kg)	X 2.205	= Pounds (lb)

Force

Ounces-force (ozf; oz)	X	0.278	= Newtons (N)	X 3.6	= Ounces-force (ozf; oz)
Pounds-force (lbf; lb)	X	4.448	= Newtons (N)	X 0.225	= Pounds-force (lbf; lb)
Newtons (N)	X	0.1	= Kilograms-force (kgf; kg)	X 9.81	= Newtons (N)

Pressure

Pounds-force per square inch (psi; lbf/in^2; lb/in^2)	X	0.070	= Kilograms-force per square centimeter (kgf/cm^2; kg/cm^2)	X 14.223	= Pounds-force per square inch (psi; lbf/in^2; lb/in^2)
Pounds-force per square inch (psi; lbf/in^2; lb/in^2)	X	0.068	= Atmospheres (atm)	X 14.696	= Pounds-force per square inch (psi; lbf/in^2; lb/in^2)
Pounds-force per square inch (psi; lbf/in^2; lb/in^2)	X	0.069	= Bars	X 14.5	= Pounds-force per square inch (psi; lbf/in^2; lb/in^2)
Pounds-force per square inch (psi; lbf/in^2; lb/in^2)	X	6.895	= Kilopascals (kPa)	X 0.145	= Pounds-force per square inch (psi; lbf/in^2; lb/in^2)
Kilopascals (kPa)	X	0.01	= Kilograms-force per square centimeter (kgf/cm^2; kg/cm^2)	X 98.1	= Kilopascals (kPa)

Torque (moment of force)

Pounds-force inches (lbf in; lb in)	X	1.152	= Kilograms-force centimeter (kgf cm; kg cm)	X 0.868	= Pounds-force inches (lbf in; lb in)
Pounds-force inches (lbf in; lb in)	X	0.113	= Newton meters (Nm)	X 8.85	= Pounds-force inches (lbf in; lb in)
Pounds-force inches (lbf in; lb in)	X	0.083	= Pounds-force feet (lbf ft; lb ft)	X 12	= Pounds-force inches (lbf in; lb in)
Pounds-force feet (lbf ft; lb ft)	X	0.138	= Kilograms-force meters (kgf m; kg m)	X 7.233	= Pounds-force feet (lbf ft; lb ft)
Pounds-force feet (lbf ft; lb ft)	X	1.356	= Newton meters (Nm)	X 0.738	= Pounds-force feet (lbf ft; lb ft)
Newton meters (Nm)	X	0.102	= Kilograms-force meters (kgf m; kg m)	X 9.804	= Newton meters (Nm)

Vacuum

Inches mercury (in. Hg)	X	3.377	= Kilopascals (kPa)	X 0.2961	= Inches mercury
Inches mercury (in. Hg)	X	25.4	= Millimeters mercury (mm Hg)	X 0.0394	= Inches mercury

Power

Horsepower (hp)	X	745.7	= Watts (W)	X 0.0013	= Horsepower (hp)

Velocity (speed)

Miles per hour (miles/hr; mph)	X	1.609	= Kilometers per hour (km/hr; kph)	X 0.621	= Miles per hour (miles/hr; mph)

Fuel consumption*

Miles per gallon, Imperial (mpg)	X	0.354	= Kilometers per liter (km/l)	X 2.825	= Miles per gallon, Imperial (mpg)
Miles per gallon, US (mpg)	X	0.425	= Kilometers per liter (km/l)	X 2.352	= Miles per gallon, US (mpg)

Temperature

Degrees Fahrenheit = (°C x 1.8) + 32

Degrees Celsius (Degrees Centigrade; °C) = (°F - 32) x 0.56

*It is common practice to convert from miles per gallon (mpg) to liters/100 kilometers (l/100km), where mpg (Imperial) x l/100 km = 282 and mpg (US) x l/100 km = 235

DECIMALS to MILLIMETERS

Decimal	mm	Decimal	mm
0.001	0.0254	0.500	12.7000
0.002	0.0508	0.510	12.9540
0.003	0.0762	0.520	13.2080
0.004	0.1016	0.530	13.4620
0.005	0.1270	0.540	13.7160
0.006	0.1524	0.550	13.9700
0.007	0.1778	0.560	14.2240
0.008	0.2032	0.570	14.4780
0.009	0.2286	0.580	14.7320
		0.590	14.9860
0.010	0.2540		
0.020	0.5080		
0.030	0.7620		
0.040	1.0160	0.600	15.2400
0.050	1.2700	0.610	15.4940
0.060	1.5240	0.620	15.7480
0.070	1.7780	0.630	16.0020
0.080	2.0320	0.640	16.2560
0.090	2.2860	0.650	16.5100
		0.660	16.7640
0.100	2.5400	0.670	17.0180
0.110	2.7940	0.680	17.2720
0.120	3.0480	0.690	17.5260
0.130	3.3020		
0.140	3.5560		
0.150	3.8100		
0.160	4.0640	0.700	17.7800
0.170	4.3180	0.710	18.0340
0.180	4.5720	0.720	18.2880
0.190	4.8260	0.730	18.5420
		0.740	18.7960
0.200	5.0800	0.750	19.0500
0.210	5.3340	0.760	19.3040
0.220	5.5880	0.770	19.5580
0.230	5.8420	0.780	19.8120
0.240	6.0960	0.790	20.0660
0.250	6.3500		
0.260	6.6040		
0.270	6.8580	0.800	20.3200
0.280	7.1120	0.810	20.5740
0.290	7.3660	0.820	21.8280
		0.830	21.0820
0.300	7.6200	0.840	21.3360
0.310	7.8740	0.850	21.5900
0.320	8.1280	0.860	21.8440
0.330	8.3820	0.870	22.0980
0.340	8.6360	0.880	22.3520
0.350	8.8900	0.890	22.6060
0.360	9.1440		
0.370	9.3980		
0.380	9.6520		
0.390	9.9060	0.900	22.8600
0.400	10.1600	0.910	23.1140
0.410	10.4140	0.920	23.3680
0.420	10.6680	0.930	23.6220
0.430	10.9220	0.940	23.8760
0.440	11.1760	0.950	24.1300
0.450	11.4300	0.960	24.3840
0.460	11.6840	0.970	24.6380
0.470	11.9380	0.980	24.8920
0.480	12.1920	0.990	25.1460
0.490	12.4460	1.000	25.4000

FRACTIONS to DECIMALS to MILLIMETERS

Fraction	Decimal	mm	Fraction	Decimal	mm
1/64	0.0156	0.3969	33/64	0.5156	13.0969
1/32	0.0312	0.7938	17/32	0.5312	13.4938
3/64	0.0469	1.1906	35/64	0.5469	13.8906
1/16	0.0625	1.5875	9/16	0.5625	14.2875
5/64	0.0781	1.9844	37/64	0.5781	14.6844
3/32	0.0938	2.3812	19/32	0.5938	15.0812
7/64	0.1094	2.7781	39/64	0.6094	15.4781
1/8	0.1250	3.1750	5/8	0.6250	15.8750
9/64	0.1406	3.5719	41/64	0.6406	16.2719
5/32	0.1562	3.9688	21/32	0.6562	16.6688
11/64	0.1719	4.3656	43/64	0.6719	17.0656
3/16	0.1875	4.7625	11/16	0.6875	17.4625
13/64	0.2031	5.1594	45/64	0.7031	17.8594
7/32	0.2188	5.5562	23/32	0.7188	18.2562
15/64	0.2344	5.9531	47/64	0.7344	18.6531
1/4	0.2500	6.3500	3/4	0.7500	19.0500
17/64	0.2656	6.7469	49/64	0.7656	19.4469
9/32	0.2812	7.1438	25/32	0.7812	19.8438
19/64	0.2969	7.5406	51/64	0.7969	20.2406
5/16	0.3125	7.9375	13/16	0.8125	20.6375
21/64	0.3281	8.3344	53/64	0.8281	21.0344
11/32	0.3438	8.7312	27/32	0.8438	21.4312
23/64	0.3594	9.1281	55/64	0.8594	21.8281
3/8	0.3750	9.5250	7/8	0.8750	22.2250
25/64	0.3906	9.9219	57/64	0.8906	22.6219
13/32	0.4062	10.3188	29/32	0.9062	23.0188
27/64	0.4219	10.7156	59/64	0.9219	23.4156
7/16	0.4375	11.1125	15/16	0.9375	23.8125
29/64	0.4531	11.5094	61/64	0.9531	24.2094
15/32	0.4688	11.9062	31/32	0.9688	24.6062
31/64	0.4844	12.3031	63/64	0.9844	25.0031
1/2	0.5000	12.7000	1	1.0000	25.4000

Safety first!

Regardless of how enthusiastic you may be about getting on with the job at hand, take the time to ensure that your safety is not jeopardized. A moment's lack of attention can result in an accident, as can failure to observe certain simple safety precautions. The possibility of an accident will always exist, and the following points should not be considered a comprehensive list of all dangers. Rather, they are intended to make you aware of the risks and to encourage a safety conscious approach to all work you carry out on your vehicle.

Essential DOs and DON'Ts

DON'T rely on a jack when working under the vehicle. Always use approved jackstands to support the weight of the vehicle and place them under the recommended lift or support points.

DON'T attempt to loosen extremely tight fasteners (i.e. wheel lug nuts) while the vehicle is on a jack - it may fall.

DON'T start the engine without first making sure that the transmission is in Neutral (or Park where applicable) and the parking brake is set.

DON'T remove the radiator cap from a hot cooling system - let it cool or cover it with a cloth and release the pressure gradually.

DON'T attempt to drain the engine oil until you are sure it has cooled to the point that it will not burn you.

DON'T touch any part of the engine or exhaust system until it has cooled sufficiently to avoid burns.

DON'T siphon toxic liquids such as gasoline, antifreeze and brake fluid by mouth, or allow them to remain on your skin.

DON'T inhale brake lining dust - it is potentially hazardous (see *Asbestos* below).

DON'T allow spilled oil or grease to remain on the floor - wipe it up before someone slips on it.

DON'T use loose fitting wrenches or other tools which may slip and cause injury.

DON'T push on wrenches when loosening or tightening nuts or bolts. Always try to pull the wrench toward you. If the situation calls for pushing the wrench away, push with an open hand to avoid scraped knuckles if the wrench should slip.

DON'T attempt to lift a heavy component alone - get someone to help you.

DON'T rush or take unsafe shortcuts to finish a job.

DON'T allow children or animals in or around the vehicle while you are working on it.

DO wear eye protection when using power tools such as a drill, sander, bench grinder, etc. and when working under a vehicle.

DO keep loose clothing and long hair well out of the way of moving parts.

DO make sure that any hoist used has a safe working load rating adequate for the job.

DO get someone to check on you periodically when working alone on a vehicle.

DO carry out work in a logical sequence and make sure that everything is correctly assembled and tightened.

DO keep chemicals and fluids tightly capped and out of the reach of children and pets.

DO remember that your vehicle's safety affects that of yourself and others. If in doubt on any point, get professional advice.

Asbestos

Certain friction, insulating, sealing, and other products - such as brake linings, brake bands, clutch linings, torque converters, gaskets, etc. - may contain asbestos. Extreme care must be taken to avoid inhalation of dust from such products, since it is hazardous to health. If in doubt, assume that they do contain asbestos.

Fire

Remember at all times that gasoline is highly flammable. Never smoke or have any kind of open flame around when working on a vehicle. But the risk does not end there. A spark caused by an electrical short circuit, by two metal surfaces contacting each other, or even by static electricity built up in your body under certain conditions, can ignite gasoline vapors, which in a confined space are highly explosive. Do not, under any circumstances, use gasoline for cleaning parts. Use an approved safety solvent.

Always disconnect the battery ground (-) cable at the battery before working on any part of the fuel system or electrical system. Never risk spilling fuel on a hot engine or exhaust component. It is strongly recommended that a fire extinguisher suitable for use on fuel and electrical fires be kept handy in the garage or workshop at all times. Never try to extinguish a fuel or electrical fire with water.

Fumes

Certain fumes are highly toxic and can quickly cause unconsciousness and even death if inhaled to any extent. Gasoline vapor falls into this category, as do the vapors from some cleaning solvents. Any draining or pouring of such volatile fluids should be done in a well ventilated area.

When using cleaning fluids and solvents, read the instructions on the container carefully. Never use materials from unmarked containers.

Never run the engine in an enclosed space, such as a garage. Exhaust fumes contain carbon monoxide, which is extremely poisonous. If you need to run the engine, always do so in the open air, or at least have the rear of the vehicle outside the work area.

If you are fortunate enough to have the use of an inspection pit, never drain or pour gasoline and never run the engine while the vehicle is over the pit. The fumes, being heavier than air, will concentrate in the pit with possibly lethal results.

The battery

Never create a spark or allow a bare light bulb near a battery. They normally give off a certain amount of hydrogen gas, which is highly explosive.

Always disconnect the battery ground (-) cable at the battery before working on the fuel or electrical systems.

If possible, loosen the filler caps or cover when charging the battery from an external source (this does not apply to sealed or maintenance-free batteries). Do not charge at an excessive rate or the battery may burst.

Take care when adding water to a non maintenance-free battery and when carrying a battery. The electrolyte, even when diluted, is very corrosive and should not be allowed to contact clothing or skin.

Always wear eye protection when cleaning the battery to prevent the caustic deposits from entering your eyes.

Household current

When using an electric power tool, inspection light, etc., which operates on household current, always make sure that the tool is correctly connected to its plug and that, where necessary, it is properly grounded. Do not use such items in damp conditions and, again, do not create a spark or apply excessive heat in the vicinity of fuel or fuel vapor.

Secondary ignition system voltage

A severe electric shock can result from touching certain parts of the ignition system (such as the spark plug wires) when the engine is running or being cranked, particularly if components are damp or the insulation is defective. In the case of an electronic ignition system, the secondary system voltage is much higher and could prove fatal.

Troubleshooting

Contents

This Section provides an easy reference guide to the more common problems that may occur during the operation of your vehicle. Various symptoms and their probable causes are grouped under headings denoting components or systems, such as Engine, Cooling system, etc. They also refer to the Chapter and/or Section that deals with the problem.

Remember that successful troubleshooting isn't a mysterious "black art' practiced only by professional mechanics, it's simply the result of knowledge combined with an intelligent, systematic approach to a problem. Always use a process of elimination starting with the simplest solution and working through to the most complex - and never overlook the obvious. Anyone can run the gas tank dry or leave the lights on overnight, so don't assume that you're exempt from such oversights.

Finally, always establish a clear idea why a problem has occurred and take steps to ensure that it doesn't happen again. If the electrical system fails because of a poor connection, check all other connections in the system to make sure they don't fail as well. If a particular fuse continues to blow, find out why - don't just go on replacing fuses. Remember, failure of a small component can often be indicative of potential failure or incorrect functioning of a more important component or system.

Engine

1 Engine will not rotate when attempting to start

1 Battery terminal connections loose or corroded. Check the cable terminals at the battery. Tighten the cable or remove corrosion as necessary.
2 Battery discharged or faulty. If the cable connections are clean and tight on the battery posts, turn the key to the On position and switch on the headlights and/or windshield wipers. If they fail to function, the battery is discharged. On 1986 models with a 3.0 liter V6 engine and auxiliary A/C, a discharged battery may be due to excess battery demand, especially if it occurs when both A/C blowers and other accessories have been operating.
3 Automatic transmission not completely engaged in Park or clutch not completely depressed.
4 Broken, loose or disconnected wiring in the starting circuit. Inspect all wiring and connectors at the battery, starter solenoid and ignition switch.
5 Starter motor pinion jammed in flywheel ring gear. If equipped with a manual transmission, place the transmission in gear and rock the vehicle to manually turn the engine. Remove the starter and inspect the pinion and flywheel at earliest convenience.
6 Starter solenoid faulty (Chapter 5).
7 Starter motor faulty (Chapter 5).
8 Ignition switch faulty (Chapter 12).
9 On 1986 models with a 3.0 liter V6 engine and automatic transmission, a no start condition may be caused by a damaged Neutral start switch wire (especially if it's accompanied by no back-up lights or turn signals).
10 On vehicles equipped with a 2.3 liter engine, built between 8/6/86 and 9/1/86, a no start condition may be caused by a damaged engine control wire harness.

2 Engine rotates but will not start

1 Fuel tank empty.
2 Battery discharged (engine rotates slowly). Check the operation of electrical components as described in previous Section.
3 Battery terminal connections loose or corroded. See previous Section.
4 Carburetor flooded and/or fuel level in carburetor incorrect. This will usually be accompanied by a strong fuel odor from under the engine cover. Wait a few minutes, depress the accelerator pedal all the way to the floor and attempt to start the engine.
5 Choke control inoperative (Chapter 4).
6 Fuel not reaching carburetor or fuel injector(s). With the ignition switch in the Off position, remove the engine cover, remove the top plate of the air cleaner assembly and observe the top of the carburetor (manually move the choke plate back if necessary). Depress the accelerator pedal and check that fuel spurts into the carburetor. If not, check

the fuel filter (Chapter 1), fuel lines and fuel pump (Chapter 4).
7 Fuel injector(s), fuel pump or fuel pump relay faulty (fuel injected vehicles) (Chapter 4).
8 No power to fuel pump (Chapter 4).
9 Worn, faulty or incorrectly gapped spark plugs (Chapter 1).
10 Broken, loose or disconnected wiring in the starting circuit (see previous Section).
11 Distributor loose, causing ignition timing to change. Turn the distributor as necessary to start the engine, then set the ignition timing as soon as possible (Chapter 1).
12 Broken, loose or disconnected wires at the ignition coil or faulty coil (Chapter 5).

3 Starter motor operates without rotating engine

1 Starter pinion sticking. Remove the starter (Chapter 5) and inspect.
2 Starter pinion or flywheel teeth worn or broken. Remove the cover at the rear of the engine and inspect.

4 Engine hard to start when cold

1 Battery discharged or low. Check as described in Section 1.
2 Choke control inoperative or out of adjustment (Chapter 4).
3 Carburetor flooded (see Section 2).
4 Fuel supply not reaching the carburetor (see Section 2).
5 Carburetor/fuel injection system in need of overhaul (Chapter 4).
6 Distributor rotor carbon tracked and/or damaged (Chapter 1).
7 Fuel injection malfunction (Chapter 4).
8 On 1986 models, hard starting when cold may be caused by sludge in the throttle body (2.3 liter engine).
9 On vehicles equipped with a 2.3 liter engine, built between 8/6/86 and 9/1/86, hard starting and/or stalling may be caused by a damaged engine control wire harness.

5 Engine hard to start when hot

1 Air filter clogged (Chapter 1).
2 Fuel not reaching the injector(s) (see Section 2).
3 Corroded electrical leads at the battery (Chapter 1).
4 Bad engine ground (Chapter 12).
5 Starter worn (Chapter 5).
6 Corroded electrical leads at the fuel injector (Chapter 4).
7 See Paragraph 9 in Section 4.

6 Starter motor noisy or excessively rough in engagement

1 Pinion or flywheel gear teeth worn or broken. Remove the cover at the rear of the engine (if so equipped) and inspect.

2 Starter motor mounting bolts loose or missing.

7 Engine starts but stops immediately

1 Loose or faulty electrical connections at distributor, coil or alternator.
2 Insufficient fuel reaching the carburetor or fuel injector(s). Disconnect the fuel line. Place a container under the disconnected fuel line and observe the flow of fuel from the line. If little or none at all, check for blockage in the lines and/or replace the fuel pump (Chapter 4).
3 Vacuum leak at the gasket surfaces of the carburetor or fuel injection unit. Make sure that all mounting bolts/nuts are tightened securely and that all vacuum hoses connected to the carburetor or fuel injection unit and manifold are positioned properly and in good condition.

8 Engine lopes while idling or idles erratically

1 Vacuum leakage. Check mounting bolts/nuts at the carburetor/fuel injection unit and intake manifold for tightness. Make sure that all vacuum hoses are connected and in good condition. Use a stethoscope or a length of fuel hose held against your ear to listen for vacuum leaks while the engine is running. A hissing sound will be heard. Check the carburetor/fuel injector and intake manifold gasket surfaces.
2 Leaking EGR valve or plugged PCV valve (see Chapters 1 and 6).
3 Air filter clogged (Chapter 1).
4 Fuel pump not delivering sufficient fuel to the carburetor/fuel injector (see Section 7).
5 Carburetor out of adjustment (Chapter 4).
6 Leaking head gasket. If this is suspected, take the vehicle to a repair shop or dealer where the engine can be pressure checked.
7 Timing chain and/or gears worn (Chapter 2).
8 Camshaft lobes worn (Chapter 2).

9 Engine misses at idle speed

1 Spark plugs worn or not gapped properly (Chapter 1).
2 Faulty spark plug wires (Chapter 1).
3 Choke not operating properly (Chapter 1).
4 Sticking or faulty emissions system components (Chapter 6).
5 Clogged fuel filter and/or foreign matter in fuel. Remove the fuel filter (Chapter 1) and inspect.
6 Vacuum leaks at the intake manifold or at hose connections. Check as described in Section 8.

7 Incorrect idle speed or idle mixture (Chapter 1).
8 Incorrect ignition timing (Chapter 1).
9 Uneven or low cylinder compression. Check compression as described in Chapter 2.
10 On 1987 models, rough idle may be caused by loose wire terminals in the fuel injector connectors.

10 Engine misses throughout driving speed range

1 Fuel filter clogged and/or impurities in the fuel system (Chapter 1). Also check fuel output at the carburetor/fuel injector (see Section 7).
2 Faulty or incorrectly gapped spark plugs (Chapter 1).
3 Incorrect ignition timing (Chapter 1).
4 Check for cracked distributor cap, disconnected distributor wires and damaged distributor components (Chapter 1).
5 Leaking spark plug wires (Chapter 1).
6 Faulty emissions system components (Chapter 6).
7 Low or uneven cylinder compression pressures. Remove the spark plugs and test the compression with gauge (Chapter 2).
8 Weak or faulty ignition system (Chapter 5).
9 Vacuum leaks at the carburetor/fuel injection unit or vacuum hoses (see Section 8).

11 Engine stalls

1 Idle speed incorrect (Chapter 1).
2 Fuel filter clogged and/or water and impurities in the fuel system (Chapter 1).
3 Choke improperly adjusted or sticking (Chapter 4).
4 Distributor components damp or damaged (Chapter 5).
5 Faulty emissions system components (Chapter 6).
6 Faulty or incorrectly gapped spark plugs (Chapter 1). Also check spark plug wires (Chapter 1).
7 Vacuum leak at the carburetor/fuel injection unit or vacuum hoses. Check as described in Section 8.
8 On 1986 models, stalling during idle or deceleration may be caused by sludge in the throttle body (2.3 liter engine).
9 Refer to Paragraph 9 in Section 4.
10 On 1986 and 1987 models with a 2.3 liter engine, hesitation or stalling during acceleration may be caused by carbon build-up on the intake valves.

12 Engine lacks power

1 Incorrect ignition timing (Chapter 1).
2 Excessive play in distributor shaft. At the same time, check for worn rotor, faulty distributor cap, wires, etc. (Chapters 1 and 5).

3 Faulty or incorrectly gapped spark plugs (Chapter 1).
4 Fuel injection unit not adjusted properly or excessively worn (Chapter 4).
5 Faulty coil (Chapter 5).
6 Brakes binding (Chapter 1).
7 Automatic transmission fluid level incorrect (Chapter 1).
8 Clutch slipping (Chapter 8).
9 Fuel filter clogged and/or impurities in the fuel system (Chapter 1).
10 Emissions control system not functioning properly (Chapter 6).
11 Use of substandard fuel. Fill tank with proper octane fuel.
12 Low or uneven cylinder compression pressures. Test with compression tester, which will detect leaking valves and/or blown head gasket (Chapter 2).

13 Engine backfires

1 Emissions system not functioning properly (Chapter 6).
2 Ignition timing incorrect (Chapter 1).
3 Faulty secondary ignition system (cracked spark plug insulator, faulty plug wires, distributor cap and/or rotor) (Chapters 1 and 5).
4 Carburetor/fuel injection unit in need of adjustment or worn excessively (Chapter 4).
5 Vacuum leak at the fuel injection unit or vacuum hoses. Check as described in Section 8.
6 Valves sticking (Chapter 2).
7 Crossed plug wires (Chapter 1).

14 Pinging or knocking engine sounds during acceleration or uphill

1 Incorrect grade of fuel. Fill tank with fuel of the proper octane rating.
2 Ignition timing incorrect (Chapter 1).
3 Carburetor/fuel injection unit in need of adjustment (Chapter 4).
4 Improper spark plugs. Check plug type against Emissions Control Information label located under hood. Also check plugs and wires for damage (Chapter 1).
5 Worn or damaged distributor components (Chapter 5).
6 Faulty emissions system (Chapter 6).
7 Vacuum leak. Check as described in Section 8.

15 Engine diesels (continues to run) after switching off

1 Idle speed too high (Chapter 1).
2 Electrical solenoid at side of carburetor not functioning properly (not all models, see Chapter 4).
3 Ignition timing incorrectly adjusted (Chapter 1).

4 Thermo-controlled air cleaner heat valve not operating properly (Chapter 1).
5 Excessive engine operating temperature. Probable causes of this are malfunctioning thermostat, clogged radiator, faulty water pump (Chapter 3).

Engine electrical system

16 Battery will not hold a charge

1 Alternator drivebelt defective or not adjusted properly (Chapter 1).
2 Electrolyte level low or battery discharged (Chapter 1).
3 Battery terminals loose or corroded (Chapter 1).
4 Alternator not charging properly (Chapter 5).
5 Loose, broken or faulty wiring in the charging circuit (Chapter 5).
6 Short in vehicle wiring causing a continual drain on battery.
7 Battery defective internally.

17 Alternator light stays on

1 Fault in alternator or charging circuit (Chapter 5).
2 Alternator drivebelt defective or not properly adjusted (Chapter 1).

18 Alternator light fails to come on when key is turned on

1 Warning light bulb defective (Chapter 12).
2 Alternator faulty (Chapter 5).
3 Fault in the printed circuit, dash wiring or bulb holder (Chapter 12).

19 "Check engine' light comes on

Emission system malfunction. Take the vehicle to a dealer service department.

Fuel system

20 Excessive fuel consumption

1 Dirty or clogged air filter element (Chapter 1).
2 Incorrectly set ignition timing (Chapter 1).
3 Choke sticking or improperly adjusted (Chapter 1).
4 Emissions system not functioning properly (not all vehicles, see Chapter 6).
5 Carburetor idle speed and/or mixture not adjusted properly (Chapter 1).
6 Carburetor/fuel injection internal parts excessively worn or damaged (Chapter 4).
7 Low tire pressure or incorrect tire size (Chapter 1).

21 Fuel leakage and/or fuel odor

1   Leak in a fuel feed or vent line (Chapter 4).
2   Tank overfilled. Fill only to automatic shut-off.
3   Emissions system clogged or damaged (Chapter 6).
4   Vapor leaks from system lines (Chapter 4).
5   Carburetor/fuel injection internal parts excessively worn or out of adjustment (Chapter 4).

Cooling system

22 Overheating

1   Insufficient coolant in system (Chapter 1).
2   Water pump drivebelt defective or not adjusted properly (Chapter 1).
3   Radiator core blocked or radiator grille dirty and restricted (Chapter 3).
4   Thermostat faulty (Chapter 3).
5   Fan blades broken or cracked (Chapter 3).
6   Radiator cap not maintaining proper pressure. Have cap pressure tested by gas station or repair shop.
7   Ignition timing incorrect (Chapter 1).
8   On 1986 models with a 2.3 liter engine and dealer installed accessory A/C, overheating may be caused by an incorrect cooling fan.

23 Overcooling

1   Thermostat faulty (Chapter 3).
2   Inaccurate temperature gauge (Chapter 12).

24 External coolant leakage

1   Deteriorated or damaged hoses or loose clamps. Replace hoses and/or tighten clamps at hose connections (Chapter 1).
2   Water pump seals defective. If this is the case, water will drip from the weep hole in the water pump body (Chapter 3).
3   Leakage from radiator core or header tank. This will require the radiator to be professionally repaired (see Chapter 3 for removal procedures).
4   Engine drain plugs or water jacket core plugs leaking (see Chapter 2).

25 Internal coolant leakage

Note: *Internal coolant leaks can usually be detected by examining the oil. Check the dipstick and inside of the rocker arm cover(s) for water deposits and an oil consistency like that of a milkshake.*
1   Leaking cylinder head gasket. Have the cooling system pressure tested.

2   Cracked cylinder bore or cylinder head. Dismantle engine and inspect (Chapter 2).

26 Coolant loss

1   Too much coolant in system (Chapter 1).
2   Coolant boiling away due to overheating (see Section 22).
3   Internal or external leakage (see Sections 24 and 25).
4   Faulty radiator cap. Have the cap pressure tested.

27 Poor coolant circulation

1   Inoperative water pump. A quick test is to pinch the top radiator hose closed with your hand while the engine is idling, then let it loose. You should feel the surge of coolant if the pump is working properly (Chapter 3).
2   Restriction in cooling system. Drain, flush and refill the system (Chapter 1). If necessary, remove the radiator (Chapter 3) and have it reverse flushed.
3   Water pump drivebelt defective or not adjusted properly (Chapter 1).
4   Thermostat sticking (Chapter 3).

Clutch

28 Fails to release (pedal pressed to the floor - shift lever does not move freely in and out of Reverse)

1   Clutch fork off ball stud. Look under the vehicle, on the left side of transmission.
2   Clutch plate warped or damaged (Chapter 8).
3   Clutch hydraulic system low or has air in system and needs to be bled (Chapter 8).

29 Clutch slips (engine speed increases with no increase in vehicle speed)

1   Clutch plate oil soaked or lining worn. Remove clutch (Chapter 8) and inspect.
2   Clutch plate not seated. It may take 30 or 40 normal starts for a new one to seat.
3   Pressure plate worn (Chapter 8).

30 Grabbing (chattering) as clutch is engaged

1   Oil on clutch plate lining. Remove (Chapter 8) and inspect. Correct any leakage source.
2   Worn or loose engine or transmission mounts. These units move slightly when clutch is released. Inspect mounts and bolts.
3   Worn splines on clutch plate hub. Remove clutch components (Chapter 8) and inspect.
4   Warped pressure plate or flywheel. Remove clutch components and inspect.

31 Squeal or rumble with clutch fully engaged (pedal released)

1   Release bearing binding on transmission bearing retainer. Remove clutch components (Chapter 8) and check bearing. Remove any burrs or nicks, clean and relubricate before reinstallation.
2   Weak linkage return spring. Replace the spring.

32 Squeal or rumble with clutch fully disengaged (pedal depressed)

1   Worn, defective or broken release bearing (Chapter 8).
2   Worn or broken pressure plate springs (or diaphragm fingers) (Chapter 8).
3   Air in hydraulic line (Chapter 8).

33 Clutch pedal stays on floor when disengaged

1   Bind in linkage or release bearing. Inspect linkage or remove clutch components as necessary.
2   Clutch hydraulic cylinder faulty or there is air in the system.

Manual transmission

Note: *All the following references are to Chapter 7, unless noted.*

34 Noisy in Neutral with engine running

1   Input shaft bearing worn.
2   Damaged main drive gear bearing.
3   Worn countershaft bearings.
4   Worn or damaged countershaft end play shims.

35 Noisy in all gears

1   Any of the above causes, and/or:
2   Insufficient lubricant (see checking procedures in Chapter 1).

36 Noisy in one particular gear

1   Worn, damaged or chipped gear teeth for that particular gear.
2   Worn or damaged synchronizer for that particular gear.

37 Slips out of high gear

1   Transmission mounting bolts loose.
2   Shift rods not working freely.
3   Damaged mainshaft pilot bushing.
4   Dirt between transmission case and engine or misalignment of transmission.

38 Difficulty in engaging gears

1 Loose, damaged or out-of-adjustment shift linkage. Make a thorough inspection, replacing parts as necessary.
2 Air in hydraulic system (Chapter 8)

39 Oil leakage

1 Excessive amount of lubricant in transmission (see Chapter 1 for correct checking procedures). Drain lubricant as required.
2 Side cover loose or gasket damaged.
3 Rear oil seal or speedometer oil seal in need of replacement.
4 Clutch hydraulic system leaking (Chapter 8).

Automatic transmission

Note: *Due to the complexity of the automatic transmission, it is difficult for the home mechanic to properly diagnose and service this component. For problems other than the following, the vehicle should be taken to a dealer or reputable repair shop.*

40 General shift mechanism problems

1 Chapter 7 deals with checking and adjusting the shift linkage on automatic transmissions. Common problems which may be attributed to poorly adjusted linkage are:
Engine starting in gears other than Park or Neutral
Indicator on shifter pointing to a gear other than the one actually being used
Vehicle moves when in Park
2 Refer to Chapter 7 to adjust the linkage.

41 Transmission will not downshift with accelerator pedal pressed to the floor

Chapter 7 deals with adjusting the TV cable to enable the transmission to downshift properly.

42 Transmission slips, shifts rough, is noisy or has no drive in forward or reverse gears

1 There are many probable causes for the above problems, but the home mechanic should be concerned with only one possibility - fluid level.
2 Before taking the vehicle to a repair shop, check the level and condition of the fluid as described in Chapter 1. Correct fluid level as necessary or change the fluid and filter if needed. If the problem persists, have a professional diagnose the probable cause.
3 A Technical Service Bulletin has been issued for several problems such as burnt fluid (after towing), leaks, engine stalling after

transmission is put in gear, slipping, high shift lever effort, etc.

43 Fluid leakage

1 Automatic transmission fluid is a deep red color. Fluid leaks should not be confused with engine oil, which can easily be blown by air flow to the transmission.
2 To pinpoint a leak, first remove all built-up dirt and grime from around the transmission. Degreasing agents and/or steam cleaning will achieve this. With the underside clean, drive the vehicle at low speeds so air flow will not blow the leak far from its source. Raise the vehicle and determine where the leak is coming from. Common areas of leakage are:
a) *Pan: Tighten mounting bolts and/or replace pan gasket as necessary (see Chapters 1 and 7).*
b) *Filler pipe: Replace the rubber seal where pipe enters transmission case.*
c) *Transmission oil lines: Tighten connectors where lines enter transmission case and/or replace lines.*
d) *Vent pipe: Transmission overfilled and/or water in fluid (see checking procedures, Chapter 1).*
e) *Speedometer connector: Replace the O-ring where speedometer cable enters transmission case (Chapter 7).*
3 On 1986 models, if fluid is leaking from the vent assembly, take the vehicle to a dealer service department.

Driveshaft

44 Oil leak at front of driveshaft

Defective transmission rear oil seal. See Chapter 7 for replacement procedures. While this is done, check the splined yoke for burrs or a rough condition which may be damaging the seal. Burrs can be removed with crocus cloth or a fine whetstone.

45 Knock or clunk when the transmission is under initial load (just after transmission is put into gear)

1 Loose or disconnected rear suspension components. Check all mounting bolts, nuts and bushings (Chapter 10).
2 Loose driveshaft bolts. Inspect all bolts and nuts and tighten them to the specified torque.
3 Worn or damaged universal joint bearings. Check for wear (Chapter 8).

46 Metallic grating sound consistent with vehicle speed

Pronounced wear in the universal joint bearings. Check as described in Chapter 8.

47 Vibration

Note: *Before assuming that the driveshaft is at fault, make sure the tires are perfectly balanced and perform the following test.*
1 Install a tachometer inside the vehicle to monitor engine speed as the vehicle is driven. Drive the vehicle and note the engine speed at which the vibration (roughness) is most pronounced. Now shift the transmission to a different gear and bring the engine speed to the same point.
2 If the vibration occurs at the same engine speed (rpm) regardless of which gear the transmission is in, the driveshaft is NOT at fault since the driveshaft speed varies.
3 If the vibration decreases or is eliminated when the transmission is in a different gear at the same engine speed, refer to the following probable causes.
4 Bent or dented driveshaft. Inspect and replace as necessary (Chapter 8).
5 Undercoating or built-up dirt, etc. on the driveshaft. Clean the shaft thoroughly and recheck.
6 Worn universal joint bearings. Remove and inspect (Chapter 8).
7 Driveshaft and/or companion flange out-of-balance. Check for missing weights on the shaft. Remove the driveshaft (Chapter 8) and reinstall 180° from original position, then retest. Have the driveshaft professionally balanced if the problem persists.

Axles

48 Noise

1 Road noise. No corrective procedures available.
2 Tire noise. Inspect the tires and check tire pressures (Chapter 1).
3 Rear wheel bearings loose, worn or damaged (Chapter 8).
4 On 1986 models, if noise is heard during turns, take the vehicle to a dealer service department.

49 Vibration

See probable causes under Driveshaft. Proceed under the guidelines listed for the driveshaft. If the problem persists, check the rear wheel bearings by raising the rear of the vehicle and spinning the wheels by hand. Listen for evidence of rough (noisy) bearings. Remove and inspect (Chapter 8).

50 Oil leakage

1 Pinion seal damaged (Chapter 8).
2 Axleshaft oil seals damaged (Chapter 8).
3 Differential inspection cover leaking. Tighten the bolts or replace the gasket as required (Chapters 1 and 8).

Brakes

Note: *Before assuming that a brake problem exists, make sure that the tires are in good condition and inflated properly (see Chapter 1), that the front end alignment is correct and that the vehicle is not loaded with weight in an unequal manner.*

51 Vehicle pulls to one side during braking

1   Defective, damaged or oil contaminated brake pads or shoes on one side. Inspect as described in Chapter 9.
2   Excessive wear of brake shoe or pad material or drum/disc on one side. Inspect and correct as necessary.
3   Loose or disconnected front suspension components. Inspect and tighten all bolts to the specified torque (Chapter 10).
4   Defective drum brake or caliper assembly. Remove the drum or caliper and inspect for a stuck piston or other damage (Chapter 9).

52 Noise (high-pitched squeal with the brakes applied)

1   Disc brake pads worn out. The noise comes from the wear sensor rubbing against the disc. Replace the pads with new ones immediately (Chapter 9).
2   On 1986 and 1987 models, if squealing occurs during the first few brake applications when the brakes are cold or in the morning after an overnight rain, the rear brake shoes may require modification.

53 Excessive brake pedal travel

1   Partial brake system failure. Inspect the entire system (Chapter 9) and correct as required.
2   Insufficient fluid in the master cylinder. Check (Chapter 1), add fluid and bleed the system if necessary (Chapter 9).
3   Brakes not adjusting properly. Make a series of starts and stops with the vehicle is in Reverse. If this does not correct the situation, remove the drums and inspect the self-adjusters (Chapter 9).

54 Brake pedal feels spongy when depressed

1   Air in the hydraulic lines. Bleed the brake system (Chapter 9).
2   Faulty flexible hoses. Inspect all system hoses and lines. Replace parts as necessary.
3   Master cylinder mounting bolts/nuts loose.
4   Master cylinder defective (Chapter 9).

55 Excessive effort required to stop vehicle

1   Power brake booster not operating properly (Chapter 9).
2   Excessively worn linings or pads. Inspect and replace if necessary (Chapters 1 and 9).
3   One or more caliper pistons or wheel cylinders seized or sticking. Inspect and rebuild as required (Chapter 9).
4   Brake linings or pads contaminated with oil or grease. Inspect and replace as required (Chapters 1 and 9).
5   New pads or shoes installed and not yet seated. It will take a while for the new material to seat against the drum (or rotor).

56 Pedal travels to the floor with little resistance

Little or no fluid in the master cylinder reservoir caused by leaking wheel cylinder(s), leaking caliper piston(s), loose, damaged or disconnected brake lines. Inspect the entire system and correct as necessary.

57 Brake pedal pulsates during brake application

1   Wheel bearings not adjusted properly or in need of replacement (Chapter 1).
2   Caliper not sliding properly due to improper installation or obstructions. Remove and inspect (Chapter 9).
3   Rotor or drum defective. Remove the rotor or drum (Chapter 9) and check for excessive lateral runout, out-of-round and parallelism. Have the drum or rotor resurfaced or replace it with a new one.

Suspension and steering systems

58 Vehicle pulls to one side

1   Tire pressures uneven (Chapter 1).
2   Defective tire (Chapter 1).
3   Excessive wear in suspension or steering components (Chapter 10).
4   Front end out of alignment.
5   Front brakes dragging. Inspect the brakes as described in Chapter 9.

59 Shimmy, shake or vibration

1   Tire or wheel out-of-balance or out-of-round. Have professionally balanced.
2   Loose, worn or out-of-adjustment wheel bearings (Chapters 1 and 8).
3   Shock absorbers and/or suspension components worn or damaged (Chapter 10).

60 Excessive pitching and/or rolling around corners or during braking

1   Defective shock absorbers. Replace as a set (Chapter 10).
2   Broken or weak springs and/or suspension components. Inspect as described in Chapter 10.

61 Excessively stiff steering

1   Lack of fluid in power steering fluid reservoir (Chapter 1).
2   Incorrect tire pressures (Chapter 1).
3   Lack of lubrication at steering joints (Chapter 1).
4   Front end out of alignment.
5   See Section 63.

62 Excessive play in steering

1   Loose front wheel bearings (Chapter 1).
2   Excessive wear in suspension or steering components (Chapter 10).
3   Steering gearbox out of adjustment (Chapter 10).

63 Lack of power assistance

1   Steering pump drivebelt faulty or not adjusted properly (Chapter 1).
2   Fluid level low (Chapter 1).
3   Hoses or lines restricted. Inspect and replace parts as necessary.
4   Air in power steering system. Bleed the system (Chapter 10).

64 Excessive tire wear (not specific to one area)

1   Incorrect tire pressures (Chapter 1).
2   Tires out-of-balance. Have professionally balanced.
3   Wheels damaged. Inspect and replace as necessary.
4   Suspension or steering components excessively worn (Chapter 10).

65 Excessive tire wear on outside edge

1   Inflation pressures incorrect (Chapter 1).
2   Excessive speed in turns.
3   Front end alignment incorrect (excessive toe-in). Have professionally aligned.
4   Suspension arm bent or twisted (Chapter 10).

66 Excessive tire wear on inside edge

1   Inflation pressures incorrect (Chapter 1).
2   Front end alignment incorrect (toe-out). Have professionally aligned.
3   Loose or damaged steering components (Chapter 10).

67 Tire tread worn in one place

1   Tires out-of-balance.
2   Damaged or buckled wheel. Inspect and replace if necessary.
3   Defective tire (Chapter 1).

Chapter 1
Tune-up and routine maintenance

Contents

Specifications

Recommended lubricants and fluids

Note: *Listed here are manufacturer recommendations at the time this manual was written. Manufacturers occasionally upgrade their fluid and lubricant specifications, so check with your local auto parts store for current recommendations*

Engine oil type	API grade "certified for gasoline engines"
Engine oil viscosity	See accompanying chart
Engine oil capacity	
1987 3.0L V6 engine	4.5 qts
All others	5.0 qts

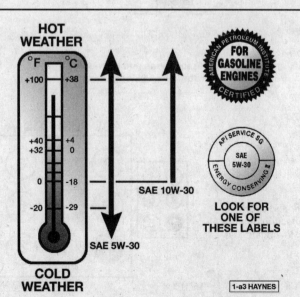

HOT WEATHER

°F °C

+100 +38

+40 +4
+32 0

0 -18

-20 -29

SAE 10W-30

SAE 5W-30

COLD WEATHER

FOR GASOLINE ENGINES — AMERICAN PETROLEUM INSTITUTE CERTIFIED

API SERVICE SG — SAE 5W-30 — ENERGY CONSERVING II

LOOK FOR ONE OF THESE LABELS

ENGINE OIL VISCOSITY RECOMMENDATIONS

1-a3 HAYNES

Recommended lubricants and fluids

Brake fluid type.. DOT 3 heavy brake fluid
Power steering fluid type .... Type F automatic transmission fluid
Transmission fluid type
    Except 1997 automatic transmission .... MERCON automatic transmission fluid
    1997 automatic transmission .... MERCON V automatic transmission fluid
Differential oil .... Hypoid lubricant, 90W
Coolant type .... Aluminum-compatible ethylene glycol-based antifreeze and water
Cooling system capacity
    Four-cylinder engine .... 6.8 qts
    2.8L V6 engine.. 7.6 qts
    3.0L V6 engine.. 11.8 qts
    4.0L V6 engine.. 12.6 qts

* Trak-Lok axles add 4 oz. of friction modifier when oil is changed.

General

Alternator/power steering/air conditioning drivebelt tension
    New belt .... 150 to 190 lbs
    Used belt .... 140 to 160 lbs
Radiator cap pressure
    Standard... 16 psi
    Lower limit (must hold pressure) .... 13 psi
    Upper limit (must relieve pressure)................................... 18 psi

Brakes

Front disc brake pad thickness (minimum)................................. 1/8 in
Rear drum brake shoe lining thickness (minimum)..................... 1/16 in

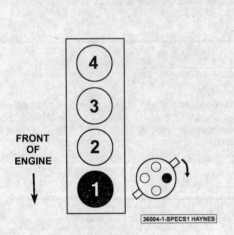

Cylinder locations and distributor rotation - 2.3L engine

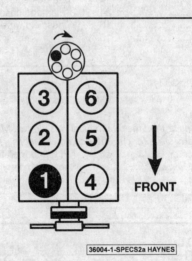

Cylinder locations and distributor rotation - 2.8L engine

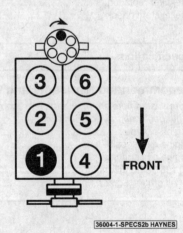

Cylinder locations and distributor rotation - 3.0L engine

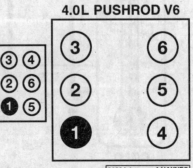

Firing order and spark plug wire connection points - 4.0L V6 engine

4.0L PUSHROD V6

Ignition system

Spark plug type and gap*
    Four-cylinder engine .... Motorcraft AWSF44C or equivalent @ 0.044 inch
    2.8L V6 engine.. Motorcraft AWSF44C or equivalent @ 0.035 inch
    3.0L V6 engine
        1986 and 1987 .... Motorcraft AWSF32C or equivalent @ 0.044 inch
        1988 and later .... Motorcraft AWSF32P or equivalent @ 0.044 inch
    4.0L V6 engine.. Motorcraft AWSF42P or equivalent @ 0.054 inch
Firing order
    Four-cylinder engine 1-3-4-2
    V6 engines (all) 1-4-2-5-3-6
Idle speed .... Refer to *Vehicle Emission Control Information* label

* *Refer to* Vehicle Emission Control *label in the engine compartment; use the information there if it differs from that listed here.*

Torque specifications

Ft-lbs (unless otherwise noted)

Wheel lug nuts .... 80 to 105
Spark plugs.. 72 to 120 in-lbs
Oil pan drain plug .... 15 to 25
Engine block drain plug .... 60 to 96 in-lbs
Carburetor-mounted fuel filter .... 84 to 108 in-lbs
Front hub adjusting nut
    Step 1 .... 17 to 25
    Step 2 .... back off 1/2 turn
    Step 3 .... 10 to 15 in-lbs
Automatic transmission pan bolts .... 96 to 120 in-lbs

1 Introduction

This Chapter is designed to help the home mechanic maintain the Aerostar with the goals of maximum performance, economy, safety and reliability in mind.

Included is a master maintenance schedule, followed by procedures dealing specifically with each item on the schedule. Visual checks, adjustments, component replacement and other helpful items are included. Refer to the accompanying illustrations of the engine compartment and the underside of the vehicle for the locations of various components.

Servicing your vehicle in accordance with the mileage/time maintenance schedule and the step-by-step procedures will result in a planned maintenance program that should produce a long and reliable service life. Keep in mind that it is a comprehensive plan, so maintaining some items but not others at the specified intervals will not produce the same results.

As you service your vehicle, you will discover that many of the procedures can - and should - be grouped together because of the nature of the particular procedure you're performing or because of the proximity of two otherwise unrelated components to one another.

For example, if the vehicle is raised for chassis lubrication, you should inspect the exhaust, suspension, steering and fuel systems while you're under the vehicle. When you're rotating the tires, it makes good sense to check the brakes since the wheels are already removed. Finally, let's suppose you have to borrow or rent a torque wrench. Even if you only need it to tighten the spark plugs,

you might as well check the torque of as many critical fasteners as time allows.

The first step in this maintenance program is to prepare yourself before the actual work begins. Read through all the procedures you're planning to do, then gather up all the parts and tools needed. If it looks like you might run into problems during a particular job, seek advice from a mechanic or an experienced do-it-yourselfer.

2 Tune-up general information

The term *tune-up* is used in this manual to represent a combination of individual operations rather than one specific procedure.

If, from the time the vehicle is new, the routine maintenance schedule is followed closely and frequent checks are made of fluid levels and high wear items, as suggested throughout this manual, the engine will be kept in relatively good running condition and the need for additional work will be minimized.

More likely than not, however, there will be times when the engine is running poorly due to lack of regular maintenance. This is even more likely if a used vehicle, which has not received regular and frequent maintenance checks, is purchased. In such cases, an engine tune-up will be needed outside of the regular routine maintenance intervals.

The first step in any tune-up or diagnostic procedure to help correct a poor running engine is a cylinder compression check. A compression check (see Chapter 2 Part E) will help determine the condition of internal engine components and should be used as a guide for tune-up and repair procedures. If, for instance, a compression check indicates serious internal engine wear, a conventional

tune-up will not improve the performance of the engine and would be a waste of time and money. Because of its importance, the compression check should be done by someone with the right equipment and the knowledge to use it properly.

The following procedures are those most often needed to bring a generally poor running engine back into a proper state of tune.

Minor tune-up

Check all engine related fluids (Section 4)
Check all underhood hoses (Section 9)
Check and adjust the drivebelts (Section 10)
Clean, inspect and test the battery (Section 11)
Check the air filter (Section 13)
Check the PCV valve (Section 14)
Check the cooling system (Section 17)
Replace the spark plugs (Section 26)
Inspect the distributor cap and rotor (Section 27)
Inspect the spark plug and coil wires (Section 27)
Check and adjust the idle speed (Chapter 4)

Major tune-up

All items listed under Minor tune-up plus . . .
Check the EGR system (Chapter 6)
Check the ignition system (Chapter 5)
Check the charging system (Chapter 5)
Replace the air and PCV filters (Sections 13 and 14)
Check the fuel system (Section 15)
Replace the distributor cap and rotor (Section 27)
Replace the spark plug wires (Section 27)

3 Maintenance schedule

The following maintenance intervals are based on the assumption that the vehicle owner will be doing the maintenance or service work, as opposed to having a dealer service department do the work. Although the time/mileage intervals are loosely based on factory recommendations, most have been shortened to ensure, for example, that such items as lubricants and fluids are checked/changed at intervals that promote maximum engine/driveline service life. Also, subject to the preference of the individual owner interested in keeping his or her vehicle in peak condition at all times, and with the vehicle's ultimate resale in mind, many of the maintenance procedures may be performed more often than recommended in the following schedule. We encourage such owner initiative. When the vehicle is new it should be serviced initially by a factory authorized dealer service department to protect the factory warranty. In many cases the initial maintenance check is done at no cost to the owner.

Every 250 miles or weekly, whichever comes first

Check the engine oil level (Section 4)
Check the engine coolant level (Section 4)
Check the windshield washer fluid level (Section 4)
Check the brake and clutch fluid levels (Section 4)
Check the tires and tire pressures (Section 5)

Every 3000 miles or 3 months, whichever comes first

All items listed above plus . . .
Check the power steering fluid level (Section 6)
Check the automatic transmission fluid level (Section 7)
Change the engine oil and oil filter (Section 8)

Every 6000 miles or 6 months, whichever comes first

All items listed above plus . . .
Inspect/replace the underhood hoses (Section 9)
Check/adjust the drivebelts (Section 10)
Check/service the battery (Section 11)
Check/regap the spark plugs (Section 26)

Every 12,000 miles or 12 months, whichever comes first

All items listed above plus . . .
Inspect/replace the windshield wiper blades (Section 12)
Replace the air filter (Section 13)
Check the PCV valve and filter (Section 14)
Check the fuel system (Section 15)
Replace the fuel filter (Section 16)
Inspect the cooling system (Section 17)
Inspect the exhaust system (Section 18)
Rotate the tires (Section 19)
Inspect the steering and suspension components (Section 20)
Inspect the brakes (Section 21)
Lubricate the parking brake cable (Section 21)
Lubricate the automatic transmission control linkage (Section 22)
Check/replenish the manual transmission lubricant (Section 23)
Check the front wheel bearings (Section 24)
Check the rear axle (differential) oil level (Section 25)

Every 24,000 miles or 24 months, whichever comes first

Check/replace the spark plug wires, distributor cap and rotor (Section 27)
Change the automatic transmission fluid and filter (Section 28)
Check the choke and lubricate the linkage (Section 29)
Service the cooling system (drain, flush and refill) (Section 30)
Replace standard-type spark plugs (Section 26)

Every 30,000 miles or 30 months, whichever comes first

Lubricate the chassis components (Section 31)
Change the rear axle (differential) oil (Section 32)

Every 60,000 miles or 60 months, whichever comes first

Replace platinum-type spark plugs (Section 26)

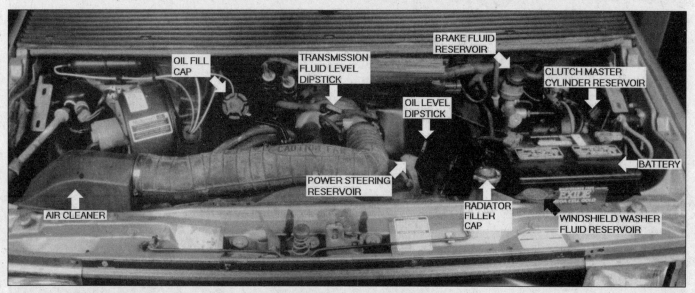

OIL FILL CAP

TRANSMISSION FLUID LEVEL DIPSTICK

BRAKE FLUID RESERVOIR

CLUTCH MASTER CYLINDER RESERVOIR

OIL LEVEL DIPSTICK

POWER STEERING RESERVOIR

RADIATOR FILLER CAP

BATTERY

WINDSHIELD WASHER FLUID RESERVOIR

AIR CLEANER

Typical fluid check and fill locations

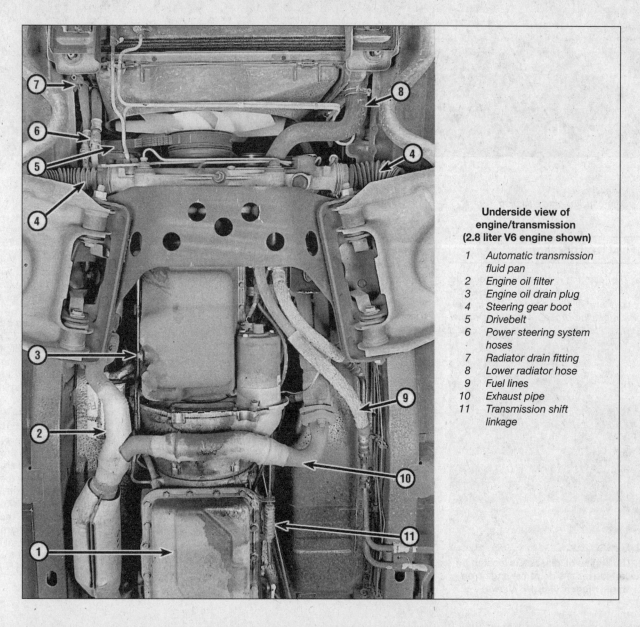

Underside view of engine/transmission (2.8 liter V6 engine shown)

1   *Automatic transmission fluid pan*
2   *Engine oil filter*
3   *Engine oil drain plug*
4   *Steering gear boot*
5   *Drivebelt*
6   *Power steering system hoses*
7   *Radiator drain fitting*
8   *Lower radiator hose*
9   *Fuel lines*
10 *Exhaust pipe*
11 *Transmission shift linkage*

Typical rear underside component layout

1	Exhaust pipe	3	Parking brake cable	5	Differential cover
2	Differential oil check/fill plug	4	Rear shock absorber		

4   Fluid level checks

Note: *The following are fluid level checks to be done on a 250 mile or weekly basis. Additional fluid level checks can be found in specific maintenance procedures which follow.*

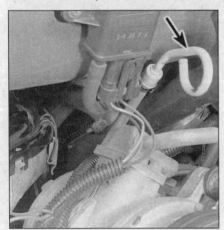

4.2 The engine oil dipstick is located on the left (driver's) side of the engine on most models (2.8L V6 shown)

Regardless of intervals, be alert to fluid leaks under the vehicle which would indicate a fault to be corrected immediately.

1   Fluids are an essential part of the lubrication, cooling, brake, clutch and windshield washer systems. Because the fluids gradually become depleted and/or contaminated during normal operation of the vehicle, they must be periodically replenished. *See Recommended lubricants and fluids at the beginning of this Chapter before adding fluid to any of the following components.* **Note:** *The vehicle must be on level ground when fluid levels are checked.*

Engine oil

Refer to illustrations 4.2, 4.4, 4.6a and 4.6b

2   The engine oil level is checked with a dipstick **(see illustration)** that extends through a tube and into the oil pan at the bottom of the engine.

3   The oil level should be checked before the vehicle has been driven, or about 15 minutes after the engine has been shut off. If the oil is checked immediately after driving the vehicle, some of the oil will remain in the upper engine components, resulting in an inaccurate reading on the dipstick.

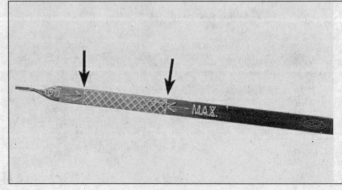

4.4 The oil level should be in the crosshatched area - if it's below the ADD line, add enough oil to bring the level into the crosshatched area (DO NOT add oil if the level is at the MAX line!)

4.6a Oil is added to the engine through the tube near the dipstick (use a funnel to prevent spills)

4.6b The engine oil filler tube cap is clearly marked to avoid any possibility of confusion

4.9 The combined windshield washer/coolant reservoir used on these models features separate tanks for each fluid - DO NOT add engine coolant to the windshield washer reservoir!

4.14 The windshield washer reservoir cap is clearly marked (it's also the larger of the two reservoir caps)

4   Pull the dipstick from the tube and wipe all the oil from the end with a clean rag or paper towel. Insert the clean dipstick all the way back into the tube, then pull it out again. Note the oil at the end of the dipstick. Add oil as necessary to keep the level between the ADD mark and the MAX mark on the dipstick, in the crosshatched area **(see illustration)**.

5   Don't overfill the engine by adding too much oil - it may result in oil fouled spark plugs, oil leaks or oil seal failures.

6   Oil is added to the engine after removing the filler cap **(see illustrations)**. An oil can spout or funnel may help to reduce spills.

7   Checking the oil level is an important preventive maintenance step. A consistently low oil level indicates oil leakage through damaged seals, defective gaskets or past worn rings or valve guides. If the oil looks milky in color or has water droplets in it, the cylinder head gasket(s) may be blown or the head(s) or block may be cracked. The engine should be checked immediately. The condition of the oil should also be checked. Whenever you check the oil level, slide your thumb and index finger up the dipstick before wiping off the oil. If you see small dirt or metal particles clinging to the dipstick, the oil should be changed (Section 8).

Engine coolant

Refer to illustration 4.9

Warning: *Don't allow antifreeze to come in contact with your skin or painted surfaces of the vehicle. Flush contaminated areas immediately with plenty of water. Don't store new coolant or leave old coolant lying around where it's accessible to children or pets - they're attracted by its sweet taste. Ingestion of even a small amount of coolant can be fatal! Wipe up garage floor and drip pan coolant spills immediately. Keep antifreeze containers covered and repair leaks in your cooling system as soon as they're discovered.*

8   All vehicles covered by this manual are equipped with a combined windshield washer/coolant recovery reservoir located in

the left front corner of the engine compartment. The coolant side of the reservoir is connected by a hose to the radiator filler neck. If the engine overheats, coolant escapes through a valve in the radiator cap and travels through the hose into the reservoir. As the engine cools, the coolant is automatically drawn back into the system to maintain the correct level.

9   The coolant level in the reservoir should be checked regularly. **Warning:** *Do not remove the radiator cap to check the coolant level when the engine is warm!* The level in the reservoir varies with the temperature of the engine. When the engine is cold, the coolant level should be at or slightly above the COLD mark on the reservoir. Once the engine has warmed up, the level should be at or near the HOT mark. If it isn't, allow the engine to cool, then remove the small cap from the reservoir and add a 50/50 mixture of ethylene glycol-based antifreeze and water **(see illustration)**. **Caution:** *Do not add coolant to the windshield washer reservoir!*

10   Drive the vehicle and recheck the coolant level. If only a small amount of coolant is required to bring the system up to the proper level, water can be used. However, repeated additions of water will dilute the antifreeze and water solution. In order to maintain the proper ratio of antifreeze and water, always top up the coolant level with the correct mixture. An empty plastic milk jug or bleach bottle makes an excellent container for mixing coolant. Don't use rust inhibitors or additives.

11   If the coolant level drops consistently, there may be a leak in the system. Inspect the radiator, hoses, filler cap, drain plugs and water pump (see Section 17). If no leaks are noted, have the radiator cap pressure tested by a service station.

12   If you have to remove the radiator cap, wait until the engine has cooled, then wrap a thick cloth around the cap and turn it to the first stop. If coolant or steam escapes, let the engine cool down longer, then remove the cap.

13   Check the condition of the coolant as well. It should be relatively clear. If it's brown or rust colored, the system should be drained, flushed and refilled. Even if the coolant appears to be normal, the corrosion inhibitors wear out, so it must be replaced at the specified intervals.

Windshield washer fluid

Refer to illustration 4.14

14   Fluid for the windshield washer system is located in a plastic reservoir in the engine compartment **(see illustration)**.

15   In milder climates, plain water can be used in the reservoir, but it should be kept no more than 2/3 full to allow for expansion if the water freezes. In colder climates, use windshield washer system antifreeze, available at any auto parts store, to lower the freezing point of the fluid. Mix the antifreeze with water in accordance with the manufacturer's directions on the container. **Caution:** *Don't use cooling system antifreeze - it will damage the vehicle's paint!*

16   To help prevent icing in cold weather, warm the windshield with the defroster before using the washer.

4.19 The brake fluid level should be kept just above the rib on the translucent plastic reservoir (arrow)

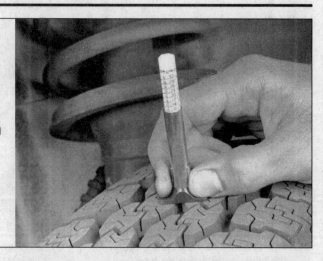

5.2 A tire tread depth indicator should be used to monitor tire wear - they're available at auto parts stores and service stations and cost very little

Battery electrolyte

17   All vehicles with which this manual is concerned are equipped with a battery which is permanently sealed (except for vent holes) and has no filler caps. Water doesn't have to be added to these batteries at any time. If an after-market battery has been installed, the caps on the top of the battery should be removed periodically to check for a low water level. This check is most critical during the warm summer months.

Brake and clutch fluid

Refer to illustration 4.19

18   The brake master cylinder is mounted on the front of the power booster unit in the engine compartment. The clutch fluid reservoir used on manual transmissions is mounted to the left of the brake booster, on the inner fender panel. To check the fluid level, unscrew the cap and remove the rubber insert. The fluid level should be even with the step on the side of the reservoir.

19   The fluid inside the brake reservoir is readily visible. The level should be above the MIN mark on the reservoir **(see illustration)**. If a low level is indicated, be sure to wipe the top of the reservoir with a clean rag to prevent contamination of the brake system before removing the cap.

20   When adding fluid, pour it carefully into the reservoir to avoid spilling it onto surrounding painted surfaces. Be sure the spec-

ified fluid is used, since mixing different types of brake fluid can cause damage to the system. See *Recommended lubricants and fluids* at the front of this Chapter or your owner's manual. **Warning:** *Brake fluid can harm your eyes and damage painted surfaces, so use extreme caution when handling or pouring it. Do not use brake fluid that has been standing open or is more than one year old. Brake fluid absorbs moisture from the air. Excess moisture can cause a dangerous loss of braking effectiveness.*

21   At this time the fluid and master cylinder can be inspected for contamination. The system should be drained and refilled if deposits, dirt particles or water droplets are seen in the fluid.

UNDERINFLATION

CUPPING

OVERINFLATION

Cupping may be caused by:
- Underinflation and/or mechanical irregularities such as out-of-balance condition of wheel and/or tire, and bent or damaged wheel.
- Loose or worn steering tie-rod or steering idler arm.
- Loose, damaged or worn front suspension parts.

INCORRECT TOE-IN OR EXTREME CAMBER

5.3 This chart will help you determine the condition of your tires, the probable cause(s) of abnormal wear and the corrective action necessary

FEATHERING DUE TO MISALIGNMENT

5.4a If a tire loses air on a steady basis, check the valve core first to make sure it's snug (special inexpensive wrenches are commonly available at auto parts stores)

5.4b If the valve core is tight, raise the corner of the vehicle with the low tire and spray a soapy water solution on the tread as the tire is slowly turned - leaks will cause small bubbles to appear

22   After filling the reservoir to the proper level, make sure the cover is on tight to prevent fluid leakage.

23   The brake fluid level in the master cylinder will drop slightly as the pads and the brake shoes at each wheel wear down during normal operation. If the master cylinder requires repeated additions to keep it at the proper level, it's an indication of leakage in the brake system, which should be corrected immediately. Check all brake lines and connections (see Section 21 for more information).

24   If you discover the reservoir empty or nearly empty, the brake system should be bled (Chapter 9).

5   Tire and tire pressure checks

Refer to illustrations 5.2, 5.3, 5.4a, 5.4b and 5.8

1   Periodic inspection of the tires may spare you the inconvenience of being stranded with a flat tire. It can also provide you with vital information regarding possible problems in the steering and suspension systems before major damage occurs.

2   The original tires on this vehicle are equipped with 1/2-inch wide wear bands that will appear when tread depth reaches 1/16-inch, at which point the tires can be considered worn out. Tread wear can be monitored with a simple, inexpensive device known as a tread depth indicator **(see illustration)**.

3   Note any abnormal tread wear **(see illustration)**. Tread pattern irregularities such as cupping, flat spots and more wear on one side than the other are indications of front end alignment and/or balance problems. If any of these conditions are noted, take the vehicle to a tire shop or service station to correct the problem.

4   Look closely for cuts, punctures and embedded nails or tacks. Sometimes a tire will hold air pressure for a short time or leak down very slowly after a nail has embedded itself in the tread. If a slow leak persists,

check the valve stem core to make sure it is tight **(see illustration)**. Examine the tread for an object that may have embedded itself in the tire or for a "plug" that may have begun to leak (radial tire punctures are repaired with a plug that's installed in a puncture). If a puncture is suspected, it can be easily verified by spraying a solution of soapy water on the tire **(see illustration)**. The soapy solution will bubble if there is a leak. Unless the puncture is unusually large, a tire shop or service station can usually repair the tire.

5   Carefully inspect the inner sidewall of each tire for evidence of brake fluid leakage. If you see any, inspect the brakes immediately.

6   Correct air pressure adds miles to the lifespan of the tires, improves mileage and enhances overall ride quality. Tire pressure cannot be accurately estimated by looking at a tire, especially if it's a radial. A tire pressure gauge is essential. Keep an accurate gauge in the glovebox. The pressure gauges attached to the nozzles of air hoses at gas stations are often inaccurate.

7   Always check tire pressure when the tires are cold. Cold, in this case, means the vehicle has not been driven over a mile in the three hours preceding a tire pressure check. A pressure rise of four to eight pounds is not uncommon once the tires are warm.

8   Unscrew the valve cap protruding from the wheel or hubcap and push the gauge firmly onto the valve stem **(see illustration)**. Note the reading on the gauge and compare the figure to the recommended tire pressure shown on the tire placard on the driver's side door. Be sure to reinstall the valve cap to keep dirt and moisture out of the valve stem mechanism. Check all four tires and, if necessary, add enough air to bring them up to the recommended pressure.

9   Don't forget to keep the spare tire inflated to the specified pressure (refer to your owner's manual or the tire sidewall). Note that the pressure recommended for the compact spare is higher than for the tires on the vehicle.

6   Power steering fluid level check

Refer to illustrations 6.5a and 6.5b

1   Check the power steering fluid level periodically to avoid steering system problems, such as damage to the pump. **Caution:** *DO NOT hold the steering wheel against either stop (extreme left or right turn) for more than five seconds. If you do, the power steering pump could be damaged!*

2   The power steering pump, located at the front of the engine, is equipped with a twist-off cap with an integral fluid level dipstick (see the engine compartment component illustrations at the front of this Chapter if necessary).

3   Park the vehicle on level ground and apply the parking brake.

4   Run the engine until it reaches normal operating temperature. With the engine at idle, turn the steering wheel back-and-forth several times to get any air out of the steering system. Shut the engine off, remove the cap by turning it counterclockwise, wipe the dipstick clean and reinstall the cap.

5   Remove the cap again and note the fluid

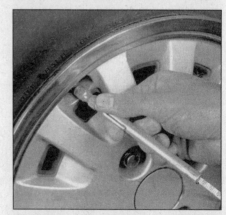

5.8 To extend the life of your tires, check the air pressure at least once a week with an accurate gauge (don't forget the spare!)

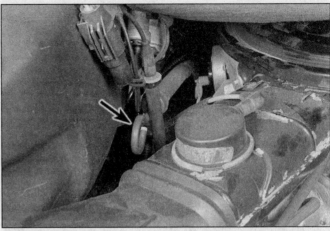

6.5 Once the engine is warmed up and the wheel has been turned back-and-forth a few times to rid the system of air, pull the dipstick out and wipe it off, reinsert it and verify that the fluid level is in the FULL HOT range (be sure to use the proper range, one side is for checking the fluid cold) - if it isn't, add enough fluid to bring the level between the two lines

7.4 The automatic transmission fluid dipstick (arrow) is accessible from under the hood - it's routed through a long tube to the transmission

level. It must be between the two lines designating the Full Hot range **(see illustration)** (be sure to use the proper temperature range on the dipstick when checking the fluid level - the Full Cold lines on the reverse side of the dipstick are only usable when the engine is cold).

6   Add small amounts of fluid until the level is correct. **Caution:** *Do not overfill the pump. If too much fluid is added, remove the excess with a clean syringe or suction pump.*

7   Check the power steering hoses and connections for leaks and wear (see Section 9).

8   Check the condition and tension of the power steering pump drivebelt (see Section 10).

7   Automatic transmission fluid level check

Refer to illustrations 7.4 and 7.6

1   The automatic transmission fluid level should be carefully maintained. Low fluid level can lead to slipping or loss of drive, while overfilling can cause foaming and loss of fluid. Either condition can cause transmission damage.

2   Since transmission fluid expands as it heats up, the fluid level should only be checked when the transmission is warm (at normal operating temperature). If the vehicle has just been driven over 20 miles (32 km), the transmission can be considered warm. **Caution:** *If the vehicle has just been driven for a long time at high speed or in city traffic in hot weather, or if it's been pulling a trailer, an accurate fluid level reading cannot be obtained. Allow the transmission to cool down for about 30 minutes.* You can also check the transmission fluid level when the transmission is cold. If the vehicle hasn't been driven for over five hours and the fluid is about room temperature (70 to 95 degrees F), the transmission is cold. However, the fluid level is normally checked with the transmission warm to ensure accurate results.

3   Immediately after driving the vehicle, park it on a level surface, apply the parking brake and start the engine. While the engine is idling, depress the brake pedal and move the selector lever through all the gear ranges, beginning and ending in Park.

4   Locate the automatic transmission dipstick tube at the front of the engine compartment **(see illustration)**.

5   With the engine still idling, pull the dipstick out of the tube, wipe it off with a clean rag, push it all the way back into the tube and withdraw it again, then note the fluid level.

6   If the transmission is cold, the level should be in the room temperature range on the dipstick (between the two circles); if it's warm, the fluid level should be in the operating temperature range (between the two lines) **(see illustration)**. If the level is low, add the specified automatic transmission fluid through the dipstick tube - use a funnel to prevent spills.

7   Add just enough of the recommended fluid to fill the transmission to the proper level. Add the fluid a little at a time and keep checking the level until it's correct.

8   The condition of the fluid should also be checked along with the level. If the fluid is black or a dark reddish-brown color, or if it smells burned, it should be changed (Section 28). If you're in doubt about its condition, purchase some new fluid and compare the two for color and smell.

8   Engine oil and filter change

Refer to illustrations 8.3, 8.9, 8.14 and 8.18

1   Frequent oil changes are the most important preventive maintenance procedure that can be done by the home mechanic. As engine oil ages, it becomes diluted and contaminated, which leads to premature engine wear.

2   Although some sources recommend oil filter changes every other oil change, the minimal cost of an oil filter and the fact that it's not hard to change dictate that a new filter be used every time the oil is changed.

3   Gather together all necessary tools and materials before beginning this procedure **(see illustration)**.

4   You should have plenty of clean rags and newspapers handy to mop up any spills. Access to the underside of the vehicle is greatly improved if the vehicle can be lifted

7.6 If the automatic transmission fluid is cold, the level should be between the two circles; if it's at operating temperature, the level should be between the two lines

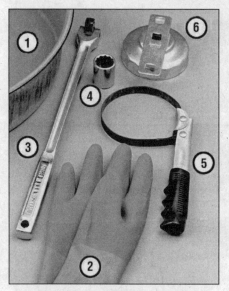

8.3 These tools are required when changing the engine oil and filter

1   **Drain pan** - *It should be fairly shallow in depth, but wide in order to prevent spills*
2   **Rubber gloves** - *When removing the drain plug and filter it is inevitable that you will get oil on your hands (the gloves will prevent burns)*
3   **Breaker bar** - *Sometimes the oil drain plug is pretty tight and a long breaker bar is needed to loosen it*
4   **Socket** - *To be used with the breaker bar or a ratchet (must be the correct size to fit the drain plug)*
5   **Filter wrench** - *This is a metal band-type wrench, which requires clearance around the filter to be effective*
6   **Filter wrench** - *This type fits on the bottom of the filter and can be turned with a ratchet or breaker bar (different size wrenches are available for different types of filters)*

on a hoist, driven onto ramps or supported by jackstands. **Warning:** *Do not work under a vehicle which is supported only by a bumper, hydraulic or scissors-type jack!*
5   If this is your first oil change, get under the vehicle and familiarize yourself with the locations of the oil drain plug and the oil filter. The engine and exhaust components will be warm during the actual work, so note how they are situated to avoid touching them when working under the vehicle.
6   Warm the engine to normal operating temperature. If the new oil or any tools are needed, use this warm-up time to gather everything necessary for the job. The correct type of oil for your application can be found in *Recommended lubricants and fluids* at the beginning of this Chapter.
7   With the engine oil warm (warm engine oil will drain better and more built-up sludge will be removed with it), raise and support the vehicle - make sure it's safely supported!
8   Move all necessary tools, rags and

8.9 Use a socket or box-end wrench to remove the engine oil drain plug - DO NOT use an open-end wrench, as the corners on the plug head can be easily rounded off!

8.18 Lubricate the gasket with clean engine oil before installing the filter on the engine

newspapers under the vehicle. Set the drain pan under the drain plug. Keep in mind that the oil will initially flow from the pan with some force; position the pan accordingly.
9   Being careful not to touch any of the hot exhaust components, use a wrench to remove the drain plug near the bottom of the oil pan **(see illustration)**. Depending on how hot the oil is, you may want to wear gloves while unscrewing the plug the final few turns.
10   Allow the old oil to drain into the pan. It may be necessary to move the pan as the oil flow slows to a trickle.
11   After all the oil has drained, wipe off the drain plug with a clean rag. Small metal particles may cling to the plug and would immediately contaminate the new oil.
12   Clean the area around the drain plug opening and reinstall the plug. Tighten the plug securely with the wrench. If a torque wrench is available, use it to tighten the plug.
13   Move the drain pan into position under the oil filter.
14   Use the filter wrench to loosen the oil filter **(see illustration)**. Chain or metal band filter wrenches may distort the filter canister,

8.14 The oil filter is usually on very tight and will require a special wrench for removal - DO NOT use the wrench to tighten the new filter

but it doesn't matter since the filter will be discarded anyway.
15   Completely unscrew the old filter. Be careful; it's full of oil. Empty the oil inside the filter into the drain pan.
16   Compare the old filter with the new one to make sure they're the same type.
17   Use a clean rag to remove all oil, dirt and sludge from the area where the oil filter mounts to the engine. Check the old filter to make sure the rubber gasket isn't stuck to the engine. If the gasket is stuck to the engine, remove it.
18   Apply a light coat of clean oil to the rubber gasket on the new oil filter **(see illustration)**.
19   Attach the new filter to the engine. Over-tightening the filter will damage the gasket, so don't use a filter wrench. Most filter manufacturers recommend tightening the filter by hand only. Normally they should be tightened 3/4-turn after the gasket contacts the block, but be sure to follow the directions on the filter or container.
20   Remove all tools, rags, etc. from under the vehicle, being careful not to spill the oil in the drain pan, then lower the vehicle.
21   Move to the engine compartment and locate the oil filler cap.
22   If an oil can spout is used, push the spout into the top of the oil can and pour the fresh oil through the filler opening. A funnel may also be used.
23   Pour four quarts of fresh oil into the engine. Wait a few minutes to allow the oil to drain into the pan, then check the level on the oil dipstick (see Section 4 if necessary). If the oil level is above the ADD mark, start the engine and allow the new oil to circulate.
24   Run the engine for only about a minute and then shut it off. Immediately look under the vehicle and check for leaks at the oil pan drain plug and around the oil filter. If either is leaking, tighten with a bit more force.
25   With the new oil circulated and the filter now completely full, recheck the level on the dipstick and add more oil as necessary.

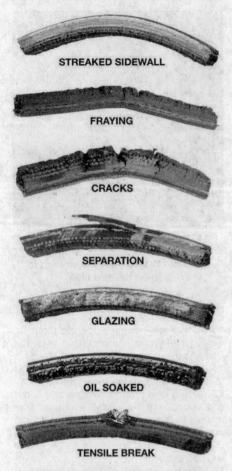

STREAKED SIDEWALL

FRAYING

CRACKS

SEPARATION

GLAZING

OIL SOAKED

TENSILE BREAK

10.3a Here are some of the common problems associated with drivebelts (check the belts very carefully to prevent an untimely breakdown)

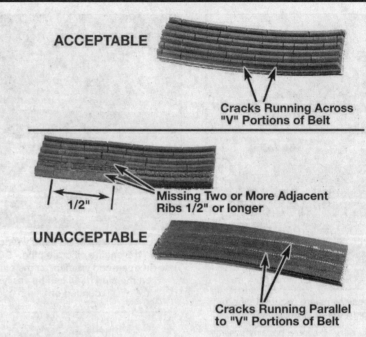

ACCEPTABLE

Cracks Running Across "V" Portions of Belt

1/2"

Missing Two or More Adjacent Ribs 1/2" or longer

UNACCEPTABLE

Cracks Running Parallel to "V" Portions of Belt

10.3b Small cracks in the underside of a V-ribbed belt are acceptable - lengthwise cracks, or missing pieces that cause the belt to make noise, are cause for replacement

26   During the first few trips after an oil change, make it a point to check frequently for leaks and proper oil level.

27   The old oil drained from the engine cannot be reused in its present state and should be disposed of. Oil reclamation centers, auto repair shops and gas stations will normally accept the oil, which can be refined and used again. After the oil has cooled it can be drained into a suitable container (capped plastic jugs, topped bottles, milk cartons, etc.) for transport to one of these disposal sites.

9   Underhood hose check and replacement

Caution: *Replacement of air conditioning hoses must be left to a dealer service department or air conditioning shop that has the equipment to depressurize the system safely. Never remove air conditioning components or hoses until the system has been depressurized.*

General

1   High temperatures in the engine compartment can cause the deterioration of the rubber and plastic hoses used for engine, accessory and emission systems operation. Periodic inspection should be made for cracks, loose clamps, material hardening and leaks.

2   Information specific to the cooling system hoses can be found in Section 9.

3   Some, but not all, hoses are secured to the fittings with clamps. Where clamps are used, check to be sure they haven't lost their tension, allowing the hose to leak. If clamps aren't used, make sure the hose has not expanded and/or hardened where it slips over the fitting, allowing it to leak.

Vacuum hoses

4   It's quite common for vacuum hoses, especially those in the emissions system, to be color coded or identified by colored stripes molded into them. Various systems require hoses with different wall thicknesses, collapse resistance and temperature resistance. When replacing hoses, be sure the new ones are made of the same material.

5   Often the only effective way to check a hose is to remove it completely from the vehicle. If more than one hose is removed, be sure to label the hoses and fittings to ensure correct installation.

6   When checking vacuum hoses, be sure to include any plastic T-fittings in the check. Inspect the fittings for cracks and the hose where it fits over the fitting for distortion, which could cause leakage.

7   A small piece of vacuum hose (1/4-inch inside diameter) can be used as a stethoscope to detect vacuum leaks. Hold one end of the hose to your ear and probe around vacuum hoses and fittings, listening for the -hissing" sound characteristic of a vacuum leak.

Warning: *When probing with the vacuum hose stethoscope, be very careful not to come into contact with moving engine components such as the drivebelts, cooling fan, etc.*

Fuel hose

Warning: *There are certain precautions which must be taken when inspecting or servicing fuel system components. Work in a well ventilated area and do not allow open flames (cigarettes, appliance pilot lights, etc.) or bare light bulbs near the work area. Mop up any spills immediately and do not store fuel soaked rags where they could ignite. On vehicles equipped with fuel injection, the fuel system is under pressure, so if any fuel lines are to be disconnected, the pressure in the system must be relieved first (see Chapter 4 for more information).*

8   Check all rubber fuel lines for deterioration and chafing. Check especially for cracks in areas where the hose bends and just before fittings, such as where a hose attaches to the fuel filter.

9   High quality fuel line, usually identified by the word *Fluroelastomer* printed on the hose, should be used for fuel line replacement. Never, under any circumstances, use unreinforced vacuum line, clear plastic tubing or water hose for fuel lines.

10   Spring-type clamps are commonly used on fuel lines. These clamps often lose their tension over a period of time, and can be "sprung" during removal. Replace all spring-type clamps with screw clamps whenever a hose is replaced.

Metal lines

11   Sections of metal line are often used for

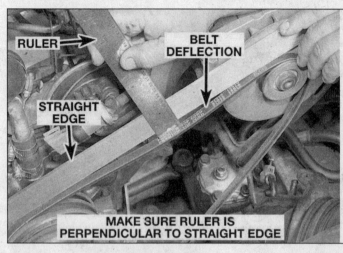

10.5 Measuring drivebelt deflection with a straightedge and ruler

fuel line between the fuel pump and carburetor or fuel injection unit. Check carefully to be sure the line has not been bent or crimped and that cracks have not started in the line.

12   If a section of metal fuel line must be replaced, only seamless steel tubing should be used, since copper and aluminum tubing don't have the strength necessary to withstand normal engine vibration.

13   Check the metal brake lines where they enter the master cylinder and brake proportioning unit (if used) for cracks in the lines or loose fittings. Any sign of brake fluid leakage calls for an immediate thorough inspection of the brake system.

---

10   Drivebelt check, adjustment and replacement

Refer to illustrations 10.3a, 10.3b, 10.5, 10.6a, 10.6b and 10.13

1   The accessory drivebelts, also referred to as V-belts or simply fan belts, are located at the front of the engine. The condition and tension of the drivebelts are critical to the operation of the engine and accessories. Excessive tension causes bearing wear, while insufficient tension produces slippage, noise,

component vibration and belt failure. Because of their composition and the high stresses to which they are subjected, drivebelts stretch and deteriorate as they get older. As a result, they must be periodically checked and adjusted.

Check

2   The number and type of belts used on a particular vehicle depends on the accessories installed. Various types of drivebelts are used on these models. V-ribbed belts are often referred to as serpentine belts because of the winding path they follow between various drive, accessory and idler pulleys.

3   With the engine off, open the hood and locate the drivebelts. Using a flashlight, check each belt for separation of the rubber plies from each side of the core, a severed core, separation of the ribs from the rubber, cracks, torn or worn ribs and cracks in the inner ridges of the ribs. Also check for fraying and glazing, which gives the belt a shiny appearance **(see illustrations)**. Both sides of each belt should be inspected, which means you'll have to twist them to check the undersides. Use your fingers to feel a belt where you can't see it. If any of the above conditions are evident, replace the belt as described below.

4   To check the tension of each belt in accordance with the manufacturer's recommendations, install a drivebelt tension gauge, available at most auto parts stores. Measure the tension in accordance with the tension gauge instructions and compare your measurement to the specified drivebelt tension for either a used or new belt. **Note:** *A "new" belt is defined as any belt which has not been run; a "used" belt is one that has been run for more than ten minutes.* Later four-cylinder and all 4.0L V6 engines are equipped with an automatic tensioner which requires only that the wear indicator remain between the MIN and MAX lines on the tensioner. If the belt is damaged or the indicator is not between the lines, the belt should be replaced. This job should be done by a dealer service department because the tensioner is spring loaded.

5   The special gauge is the most accurate way to check belt tension. However, if you don't have a gauge, and cannot borrow one, the following "rule-of-thumb" method is recommended as an alternative. Lay a straightedge across the longest free span (the distance between two pulleys) of the belt. Push down firmly on the belt at a point half way between the pulleys and see how much the belt moves (deflects). Measure the deflection with a ruler **(see illustration)**. The belt should deflect 1/8 to 1/4-inch if the distance from pulley center-to-pulley center is less than 12-inches; it should deflect from 1/8 to 3/8-inch if the distance from pulley center-to-pulley center is over 12-inches.

Adjustment

6   If adjustment is required to make the drivebelt tighter or looser, it's done by moving an idler pulley or the belt driven accessory on the bracket **(see illustrations)**.

7   For each component, there will be a locking bolt and pivot bolt or nut. Both must be loosened slightly to enable you to move the component.

8   After the bolts have been loosened, move the component away from the engine to tighten the belt or toward the engine to loosen the belt. Hold the accessory in posi-

10.6a Some drivebelts are adjusted by loosening the bolts and moving an idler pulley bracket

10.6b Other drivebelts are tensioned by loosening a lock bolt (A) and turning an adjustment bolt (B)

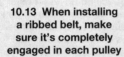

10.13 When installing a ribbed belt, make sure it's completely engaged in each pulley

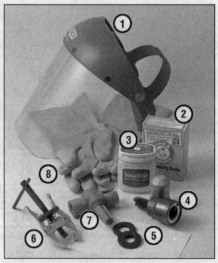

11.1 Tools and materials required for battery maintenance

1   **Face shield/safety goggles -** *When removing corrosion with a brush, the acidic particles can easily fly up into your eyes*
2   **Baking soda -** *A solution of baking soda and water can be used to neutralize corrosion*
3   **Petroleum jelly -** *A layer of this on the battery posts will help prevent corrosion*
4   **Battery post/cable cleaner -** *This wire brush cleaning tool will remove all traces of corrosion from the battery posts and cable clamps*
5   **Treated felt washers -** *Placing one of these on each post, directly under the cable clamps, will help prevent corrosion*
6   **Puller -** *Sometimes the cable clamps are very difficult to pull off the posts, even after the nut/bolt has been completely loosened. This tool pulls the clamp straight up and off the post without damage*
7   **Batty post/cable cleaner -** *Here is another cleaning tool which is a slightly different version of number 4 above, but it does the same thing*
8   **Rubber gloves -** *Another safety item to consider when servicing the battery; remember that's acid inside the battery!*

tion and check the belt tension. If it's correct, tighten the two bolts until snug, then recheck the tension. If it's still correct, tighten the two bolts completely.

9   To adjust the drivebelt on some components, loosen the pivot nut and the locking screw or locknut, then turn the adjusting screw to tension the belt.

10   It will often be necessary to use some sort of pry bar to move the accessory while the belt is adjusted. If this must be done to gain the proper leverage, be very careful not to damage the component being moved, or the part being pried against.

11   Run the engine for about 15 minutes, then recheck the belt tension.

Replacement

12   To replace a belt, follow the above procedures for drivebelt adjustment but slip the belt off the pulleys and remove it. If you're replacing the water pump, air pump or alternator drivebelts, the power steering pump/air conditioning compressor belt must come off first. Since belts tend to wear out more or less at the same time, it's a good idea to replace all of them at the same time. Mark each belt and the corresponding pulley grooves so the replacement belts can be installed properly.

13   Take the old belts with you when purchasing new ones in order to make a direct comparison for length, width and design. When replacing a V-ribbed drivebelt (the wide one used to drive the power steering pump and A/C compressor), make sure that it fits properly into the pulley grooves - it must be completely engaged **(see illustration)**.

14   Adjust the belts as described earlier in this Section.

11   Battery check and maintenance

Refer to illustrations 11.1, 11.8a, 11.8b, 11.8c and 11.8d

Warning: *Certain precautions must be followed when checking and servicing the battery. Hydrogen gas, which is highly flam-* *mable, is always present in the battery cells, so keep lighted tobacco and all other open flames and sparks away from the battery. The electrolyte in the battery cells is actually dilute sulfuric acid, which will cause injury if splashed on your skin or in your eyes. It will also ruin clothes and painted surfaces. When removing the battery cables, always detach the negative cable first and hook it up last!*

1   Battery maintenance is an important procedure which will help ensure that you're not stranded because of a dead battery. Several tools are required for this procedure **(see illustration).**

2   Before servicing the battery, always turn the engine and all accessories off and disconnect the cable from the negative terminal.

3   A sealed (sometimes called maintenance-free) battery is standard equipment on these vehicles. The cell caps cannot be removed, no electrolyte checks are required and water cannot be added to the cells. However, if an after-market battery that requires regular maintenance has been installed, the following procedure can be used.

4   Check the electrolyte level in each of the battery cells. It must be above the plates. There's usually a split-ring indicator in each cell to indicate the correct level. If the level is low, add distilled water only, then install the cell caps. **Caution:** *Overfilling the cells may cause electrolyte to spill over during periods of heavy charging, causing corrosion and damage to nearby components.*

5   If the positive terminal and cable clamp on your vehicle's battery is equipped with a rubber protector, make sure that it's not torn or damaged. It should completely cover the terminal.

6   The external condition of the battery should be checked periodically. Look for damage such as a cracked case.

7   Check the tightness of the battery cable clamps to ensure good electrical connections and inspect the entire length of each cable, looking for cracked or abraded insulation and frayed conductors.

8   If corrosion (visible ·as white, fluffy deposits) is evident, remove the cables from the terminals, clean them with a battery brush and reinstall them **(see illustrations)**. Corro- sion can be kept to a minimum by installing specially treated washers available at auto parts stores or by applying a layer of petroleum jelly or grease to the terminals and cable clamps after they are assembled.

9   Make sure that the battery carrier is in good condition and that the hold-down clamp bolt is tight. If the battery is removed (see Chapter 5 for the removal and installation procedure), make sure that no parts remain in the bottom of the carrier when it's reinstalled. When reinstalling the hold-down clamp, don't overtighten the bolt.

10   Corrosion on the carrier, battery case

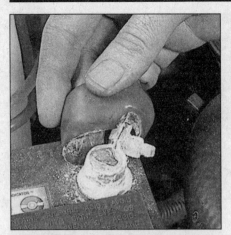

11.8a Battery terminal corrosion usually appears as light, fluffy powder

11.8b Removing a cable from the battery post with a wrench - sometimes a special battery pliers is required for this procedure if corrosion has caused deterioration of the nut hex (always remove the ground cable first and hook it up last!)

11.8c Regardless of the type of tool used to clean the battery posts, a clean, shiny surface should be the result

and surrounding areas can be removed with a solution of water and baking soda. Apply the mixture with a small brush, let it work, then rinse it off with plenty of clean water.

11   Any metal parts of the vehicle damaged by corrosion should be coated with a zinc-based primer, then painted.

12   Additional information on the battery, charging and jump starting can be found in Chapter 5 and at the front of this manual.

12   Windshield wiper blade check and replacement

Refer to illustrations 12.4, 12.6 and 12.8

1   Road film can build up on the wiper blades and affect their efficiency, so they should be washed regularly with a mild detergent solution.

Check

2   The windshield wiper and blade assembly should be inspected periodically. Even if

you do not use your wipers, the sun and elements will dry out the rubber portions, causing them to crack and break apart. If inspection reveals hardened or cracked rubber, replace the wiper blades. If inspection reveals nothing unusual, wet the windshield, turn the wipers on, allow them to cycle several times, then shut them off. An uneven wiper pattern across the glass or streaks over clean glass indicate that the blades should be replaced.

3   The operation of the wiper mechanism can loosen the fasteners, so they should be checked and tightened, as necessary, at the same time the wiper blades are checked (see Chapter 12 for further information regarding the wiper mechanism).

Blade assembly replacement

Note: *The blade assembly has a rectangular hole located directly above the wiper arm*

mounting pin with no apparent provision for release. The hole serves for removal, since the release is internal.

4   Cycle the wiper assembly to a position on the windshield where removal of the blade assembly can be performed without difficulty. Turn the ignition key off at the desired position. With the blade assembly resting on the windshield, insert a small standard screwdriver into the rectangular hole on top of the blade and push down on the coil spring inside the hole. While pressing down with the screwdriver, pull the wiper blade from the wiper arm pin **(see illustration)**.

5   To install the blade assembly, push it onto the pin until it snaps into place. Be sure that the blade assembly is securely attached to the wiper arm.

Blade element replacement

6   At one end of the rubber blade element, insert a standard screwdriver between the

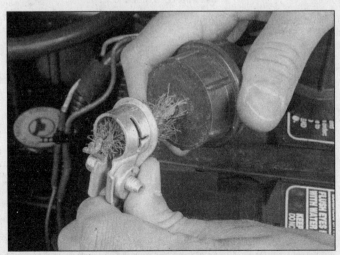

11.8d When cleaning the cable clamps, all corrosion must be removed (the inside of the clamp is tapered to match the taper on the post, so don't remove too much material)

12.4 Push down with a screwdriver blade as shown to release the wiper blade from the arm

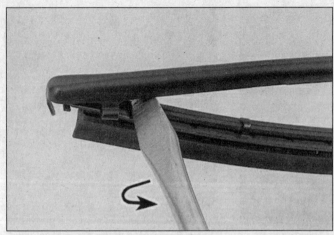

12.6　Insert the screwdriver about 1/8-inch and twist it to disengage the wiper blade

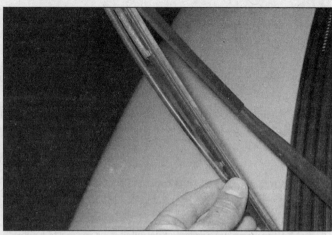

12.8　Install the blade element by sliding the metal backing into the four retaining tabs, then twist it into the fifth and end tab

blade and the metal backing strip **(see illustration)**. Press down and in, then twist the screwdriver clockwise to release the element from the retaining tab.

7　Slide the blade element out of the remaining tabs until the element is completely detached from the frame.

8　To install the element, slide the metal backing strip into four of the retaining tabs, then twist the backing strip into the fifth and end tab **(see illustration)**.

9　Make sure that all the tabs are locked onto the metal backing strip before installing the blade on the wiper arm.

13　Air filter replacement

1　Open the hood.

Carburetor-equipped vehicles

Refer to illustrations 13.2 and 13.3

2　Remove the wing nut from the top of the housing, disconnect the clip and lift the cover off **(see illustration)**.

3　Remove the filter element **(see illustration)**.

4　Wipe down the inside of the air cleaner housing with a clean cloth.

5　On models so equipped, use a screwdriver to pry the PCV filter element out of the housing in the air cleaner. If the PCV filter is oily or dirty, replace it with a new one.

6　Place the new air filter element in the housing. If the element is marked TOP be sure the marked side faces up.

7　Install the cover.

Fuel injected vehicles

8　Loosen the clamp and detach the tube from the air cleaner assembly. You may have to detach or unplug vacuum hoses and wires to gain access to the cover.

9　Remove the screws and lift off the air cleaner assembly cover.

10　Lift the filter element out of the housing.

11　Place the new filter element in the housing. If the element is marked TOP be sure the marked side faces up.

12　Position the cover on the housing and make sure it's seated all the way around, then install and tighten the cover screws.

13　Reconnect the tube and any wires or vacuum hoses that were disconnected.

14　PCV valve check

Note: *To maintain efficient operation of the PCV system, clean the hoses and check the PCV valve at the intervals recommended in the maintenance schedule. For additional information on the PCV system, refer to Chapter 6.*

1　Locate the PCV valve. On V6 engines it's installed in one of the valve covers, secured by a grommet. On four-cylinder engines it's installed in a hose leading to the upper intake manifold.

2　To check the valve, first pull it out of the rocker arm cover or tube and shake it - if it rattles, reinstall it.

3　Start the engine and allow it to idle, then disconnect the PCV hose from the air cleaner housing and feel for vacuum at the hose. If vacuum is felt, the PCV valve/system is working properly.

4　If no vacuum is felt, the oil filler cap, hoses or rocker arm cover gasket may be leaking or the PCV valve may be bad. Check for vacuum leaks at the valve, filler cap, filter assembly (if used) and all hoses.

5　Pull straight up on the valve to remove it. Check the rubber grommet in the rocker

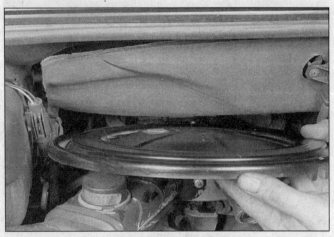

13.2　To replace the air filter, remove the housing cover . . .

13.3　. . . and pull the filter out

arm cover for cracks and distortion. If it's damaged, replace it.

6   If the valve is clogged, the hoses are also probably plugged. Remove the hose between the valve and the intake manifold and the hose between the filter and the air cleaner housing and clean them with solvent.

7   After cleaning the hoses, inspect them for damage, wear and deterioration. Make sure the hoses fit snugly on the fittings.

8   If necessary, install a new PCV valve. **Note:** *The elbow is not part of the PCV valve. A new valve will not include the elbow. The original must be transferred to the new valve. If a new elbow is purchased, it may be necessary to soak it in warm water for up to an hour to slip it onto the new valve. Do not attempt to force the elbow onto the valve or it will break.*

9   Install the clean PCV system hoses.

---

15   Fuel system check

Warning: *Gasoline is extremely flammable, so take extra precautions when you work on any part of the fuel system. Don't smoke or allow open flames or bare light bulbs near the work area, and don't work in a garage where a gas-type appliance (such as a water heater or clothes dryer) is present. Since gasoline is carcinogenic, wear latex gloves when there's a possibility of being exposed to fuel, and, if you spill any fuel on your skin, rinse it off immediately with soap and water. Mop up any spills immediately and do not store fuel-soaked rags where they could ignite. The fuel system on fuel-injected models is under constant pressure, so, if any fuel lines are to be disconnected, the fuel pressure in the system must be relieved first (see Chapter 4 for more information). When you perform any kind of work on the fuel system, wear safety glasses and have a Class B type fire extinguisher on hand.*

1   If you smell gasoline while driving or after the vehicle has been sitting in the sun, inspect the fuel system immediately.

2   Remove the gas filler cap and inspect it for damage and corrosion. The gasket should have an unbroken sealing imprint. If the gasket is damaged or corroded, install a new cap.

3   Inspect the fuel feed and return lines for cracks. Make sure that the connections between the fuel lines and the carburetor or fuel injection system and between the fuel lines and the in-line fuel filter are tight. **Warning:** *If your vehicle is fuel injected, you must relieve the fuel system pressure before servicing fuel system components. The fuel system pressure relief procedure is outlined in Chapter 4.*

4   Since some components of the fuel system - the fuel tank and part of the fuel feed and return lines, for example - are underneath the vehicle, they can be inspected more easily with the vehicle raised on a hoist. If that's not possible, raise the vehicle and support it on jackstands.

5   With the vehicle raised and safely supported, inspect the gas tank and filler neck for punctures, cracks and other damage. The connection between the filler neck and the

tank is particularly critical. Sometimes a rubber filler neck will leak because of loose clamps or deteriorated rubber. Inspect all fuel tank mounting brackets and straps to be sure that the tank is securely attached to the vehicle. **Warning:** *Do not, under any circumstances, try to repair a fuel tank (except rubber components). A welding torch or any open flame can easily cause fuel vapors inside the tank to explode!*

6   Carefully check all rubber hoses and metal lines leading away from the fuel tank. Check for loose connections, deteriorated hoses, crimped lines and other damage. Repair or replace damaged sections as necessary (Chapter 4).

---

16   Fuel filter replacement

Warning: *Gasoline is extremely flammable, so take extra precautions when you work on any part of the fuel system. Don't smoke or allow open flames or bare light bulbs near the work area, and don't work in a garage where a gas-type appliance (such as a water heater or clothes dryer) is present. Since gasoline is carcinogenic, wear latex gloves when there's a possibility of being exposed to fuel, and, if you spill any fuel on your skin, rinse it off immediately with soap and water. Mop up any spills immediately and do not store fuel-soaked rags where they could ignite. The fuel system on fuel-injected models is under constant pressure, so, if any fuel lines are to be disconnected, the fuel pressure in the system must be relieved first (see Chapter 4 for more information). When you perform any kind of work on the fuel system, wear safety glasses and have a Class B type fire extinguisher on hand.*

Fuel injected vehicles

Warning: *Before removing the fuel filter, the fuel system pressure must be relieved. See Chapter 4.*

1   Locate the fuel filter on the left (driver's side) frame rail, near the rear of the engine. Inspect the fittings at both ends of the filter to see if they're clean. If more than a light coating of dust is present, clean the fittings before proceeding.

2   Removal of the hairpin clip from each fitting is a two-stage procedure. First, spread the two clip legs apart about 1/8-inch to disengage them, then push in on them. Pull on the other end of the clip to detach it from the fitting. **Caution:** *Do not use any tools or you may damage the plastic clips or fittings.*

3   Once both hairpin clips are released, grasp the fuel hoses, one at a time, and pull them straight off the filter.

4   After the hoses have been detached, check the clips for damage and distortion. If they were damaged in any way during removal, new ones must be used when the hoses are reattached to the new filter (if new clips are packaged with the filter, be sure to use them in place of the originals).

5   Remove the screws and detach the filter

and retainer from the bracket. **Note:** *Later models do not use a retainer or bracket.*

6   Remove the rubber insulator ring. On later models, simply loosen the filter retaining clamp enough to allow the filter to pass through.

7   Note that the fuel flow arrow on the filter points toward the open end of the retainer - the new filter must be installed in the same way.

8   Install the new filter in the retainer or filter clamp.

9   Install the rubber insulator ring. If the filter moves freely after installing the retainer, replace the insulators with new ones.

10   Install the filter retainer in the bracket and tighten the screws securely. On filter clamp types, tighten the clamp.

11   Carefully push each hose onto the filter until it's seated against the collar on the fitting, then install the hairpin clips. The triangular shaped side of each clip must point away from the filter. Make sure the clips are securely attached to the hose fittings - if they come off, the hoses could back off the filter and a fire could result!

12   Start the engine and check for fuel leaks.

Carburetor-equipped vehicles

13   Remove the air cleaner housing from the top of the carburetor to gain access to the filter.

14   Place a rag under the filter, then remove the fuel line from the filter while holding the filter hex with a back-up wrench. A flare nut wrench should be used if available - it will prevent rounding off the fuel line fitting hex.

15   Unscrew the filter from the carburetor. Before installation, apply one drop of thread sealant to the 2nd and 3rd threads (from the end) of the new filter.

16   Position the new filter and thread it into the carburetor - be very careful not to cross thread the fitting. Attach the fuel line to the filter and tighten the fitting securely.

17   Start the engine and check for fuel leaks at the fittings while the engine idles for two minutes. If fuel leaks are noted, tighten the appropriate fitting(s) slightly.

18   Reinstall the air cleaner housing.

---

17   Cooling system check

Refer to illustration 17.4

1   Many major engine failures can be attributed to a faulty cooling system. If the vehicle is equipped with an automatic transmission, the cooling system also cools the transmission fluid and thus plays an important role in prolonging transmission life.

2   The cooling system should be checked with the engine cold. Do this before the vehicle is driven for the day or after the engine has been shut off for at least three hours.

3   Remove the radiator cap by turning it to the left until it reaches a stop. If you hear a hissing sound (indicating there is still pressure in the system), wait until it stops. Now press down on the cap with the palm of your hand and continue turning to the left until the cap

Check for a chafed area that could fail prematurely.

Check for a soft area indicating the hose has deteriorated inside.

Overtightening the clamp on a hardened hose will damage the hose and cause a leak.

Check each hose for swelling and oil-soaked ends. Cracks and breaks can be located by squeezing the hose.

17.4 Hoses, like drivebelts, have a habit of failing at the worst possible time - to prevent the inconvenience of a blown radiator or heater hose, inspect them carefully as shown here

can be removed. Thoroughly clean the cap, inside and out, with clean water. Also clean the filler neck on the radiator. All traces of corrosion should be removed. The coolant inside the radiator should be relatively transparent. If it's rust colored, the system should be drained and refilled (Section 30). If the coolant level isn't up to the top, add additional antifreeze/coolant mixture (see Section 4).

4   Carefully check the large upper and lower radiator hoses along with the smaller diameter heater hoses which run from the engine to the firewall. Inspect each hose along its entire length, replacing any hose which is cracked, swollen or shows signs of deterioration. Cracks may become more apparent if the hose is squeezed **(see illustration)**. Regardless of condition, it's a good idea to replace hoses with new ones every two years.

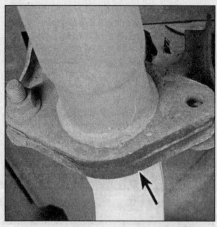

18.2 Inspect the exhaust system for conditions such as this - the stud broke off, creating an exhaust leak, evidenced by the carbon deposits

5   Make sure that all hose connections are tight. A leak in the cooling system will usually show up as white or rust colored deposits on the areas adjoining the leak. If wire-type clamps are used at the ends of the hoses, it may be a good idea to replace them with more secure screw-type clamps.
6   Use compressed air or a soft brush to remove bugs, leaves, etc. from the front of the radiator or air conditioning condenser. Be careful not to damage the delicate cooling fins or cut yourself on them.
7   Every other inspection, or at the first indication of cooling system problems, have the cap and system pressure tested. If you don't have a pressure tester, most gas stations and repair shops will do this for a minimal charge.

18 Exhaust system check

Refer to illustration 18.2
1   With the engine cold (at least three hours after the vehicle has been driven), check the complete exhaust system from the engine to the end of the tailpipe. Ideally, the inspection should be done with the vehicle on a hoist to permit unrestricted access. If a hoist isn't available, raise the vehicle and support it securely on jackstands.
2   Check the exhaust pipes and connections for evidence of leaks, severe corrosion and damage. Make sure that all brackets and hangers are in good condition and tight **(see illustration)**.
3   At the same time, inspect the underside of the body for holes, corrosion, open seams, etc. which may allow exhaust gases to enter the passenger compartment. Seal all body openings with silicone or body putty.
4   Rattles and other noises can often be traced to the exhaust system, especially the mounts and hangers. Try to move the pipes, muffler and catalytic converter. If the components can come in contact with the body or

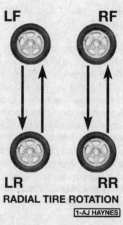

LF   RF

LR   RR

RADIAL TIRE ROTATION
1-AJ HAYNES

19.2 Recommended tire rotation pattern

suspension parts, secure the exhaust system with new mounts.
5   Check the running condition of the engine by inspecting inside the end of the tailpipe. The exhaust deposits here are an indication of engine state-of-tune. If the pipe is black and sooty or coated with white deposits, the engine is in need of a tune-up, including a thorough fuel system inspection and adjustment.

19 Tire rotation

Refer to illustration 19.2
1   The tires should be rotated at the specified intervals and whenever uneven wear is noticed. Since the vehicle will be raised and the tires removed anyway, check the brakes also (Section 21).
2   Radial tires must be rotated in a specific pattern **(see illustration)**.
3   Refer to the information in Jacking and towing at the front of this manual for the proper procedure to follow when raising the vehicle and changing a tire. If the brakes are to be checked, don't apply the parking brake.
4   The vehicle must be raised on a hoist or supported on jackstands to get all four wheels off the ground. Make sure the vehicle is safely supported!
5   After the rotation procedure is finished, check and adjust the tire pressures as necessary and be sure to check the lug nut tightness.

20 Steering and suspension check

Refer to illustrations 20.11 and 20.12
Note: *The steering linkage and suspension components should be checked periodically. Worn or damaged suspension and steering linkage components can result in excessive and abnormal tire wear, poor ride quality and vehicle handling and reduced fuel economy. For detailed illustrations of the steering and suspension components, refer to Chapter 10.*

20.11 To check the steering gear mounts and tie-rod connections for play, grasp each front tire like this and try to move it back-and-forth - if play is noted, check the steering gear mounts and make sure that they're tight; if either tie-rod is worn or bent, replace it

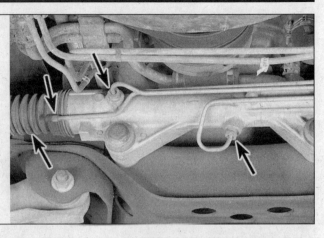

20.12 Check the steering gear hydraulic line fittings and boots for leaks and damage (arrows) and make sure the mounting bolts are tight

Shock absorber check

1   Park the vehicle on level ground, turn the engine off and set the parking brake. Check the tire pressures.

2   Push down at one corner of the vehicle, then release it while noting the movement of the body. It should stop moving and come to rest in a level position within one or two bounces.

3   If the vehicle continues to move up-and-down or if it fails to return to its original position, a worn or weak shock absorber is probably the reason.

4   Repeat the above check at each of the three remaining corners of the vehicle.

5   Raise the vehicle and support it on jackstands. The front jackstands should be positioned under the suspension (control) arms.

6   Check the shock absorbers for evidence of fluid leakage. A light film of fluid is no cause for concern. Make sure that any fluid noted is from the shocks and not from some other source. If leakage is noted, replace the shocks as a set.

7   Check the shock absorbers to be sure they're securely mounted and undamaged. Check the upper mounts for damage and wear. If damage or wear is noted, replace the shocks as a set.

8   If shock absorbers must be replaced, refer to Chapter 10 for the procedure.

Steering and suspension check

9   Visually inspect the steering system components for damage and distortion. Look for leaks and damaged seals, boots and fittings.

10   Have an assistant grasp the lower edge of the tire and move the wheel in-and-out while you look for movement at the balljoints. If there's any movement, the suspension balljoints must be replaced.

11   Grasp each front tire at the front and

rear edges, push in at the rear, pull out at the front and feel for play in the steering system components **(see illustration)**. If any free play is noted, check the steering gear mounts and the tie-rod balljoints for looseness. If the steering gear mounts are loose, tighten them. If the tie-rods are loose, the balljoints may be worn (check to make sure the nuts are tight). Additional steering and suspension system information and illustrations can be found in Chapter 10.

12   Check the steering gear for leaks at the fittings, damaged boots and loose boot clamps **(see illustration)**.

21 Brake check

Refer to illustrations 21.11, 21.15a and 21.15b

Note: *In addition to the specified intervals, the brake system should be inspected each time the wheels are removed or a malfunction is indicated. Because of the obvious safety implications, the following brake system checks are some of the most important maintenance procedures you can perform on your vehicle.*

Symptoms of brake system problems

1   Some disc brake pads have built-in wear indicators which should make a high pitched squealing or scraping noise when they're worn to the replacement point. When you hear this noise, replace the pads immediately or expensive damage to the discs could result.

2   Any of the following symptoms could indicate a potential brake system defect. The vehicle pulls to one side when the brake pedal is depressed, the brakes make squealing or dragging noises when applied, brake travel is excessive, the pedal pulsates and brake fluid leaks are noted (usually on the inner side of the tire or wheel). If any of these conditions are noted, inspect the brake system immediately.

Brake lines and hoses

Note: *Steel tubing is used throughout the brake system, with the exception of flexible, reinforced hoses at the front wheels and as connectors at the rear axle. Periodic inspec-*

tion of these lines is very important.

3   Park the vehicle on level ground and turn the engine off.

4   Remove the wheel covers. Loosen, but do not remove, the lug nuts on all four wheels.

5   Raise the vehicle and support it securely on jackstands.

6   Remove the wheels (see *Jacking and towing* at the front of this manual, or refer to your owner's manual, if necessary).

7   Check all brake hoses and lines for cracks, chafing of the outer cover, leaks, blisters and distortion. Check all threaded fittings for leaks and make sure the brake hose mounting bolts and clips are secure.

8   If leaks or damage are discovered, they must be fixed immediately. Refer to Chapter 9 for detailed repair procedures.

Front disc brakes

9   If it hasn't already been done, raise the front of the vehicle and support it securely on jackstands. Apply the parking brake and remove the front wheels.

10   The disc brake calipers, which contain the pads, are now visible. Each caliper has an outer and an inner pad - all pads should be checked.

11   Note the pad thickness by looking through the inspection hole in the caliper **(see illustration)**. If the lining material is 1/8-

21.11 The front disc brake pad lining thickness (arrows) can be checked through the caliper inspection hole

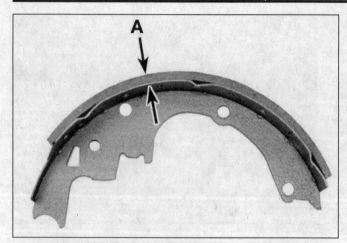

21.15a The rear brake shoe lining thickness (A) is measured from the outer surface of the lining to the metal shoe

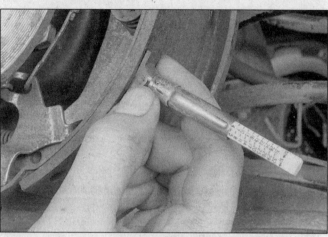

21.15b The lining thickness measured to the rivet head can be checked with a tire tread depth gauge

inch thick or less, or if it is tapered from end-to-end, the pads should be replaced (see Chapter 9). Keep in mind that the lining material is riveted or bonded to a metal plate or shoe - the metal portion is not included in the measurement.

12   Check the condition of the brake disc. Look for score marks, deep scratches and overheated areas (they will appear blue or discolored). If damage or wear is noted, the disc can be removed and resurfaced by an automotive machine shop or replaced with a new one. Refer to Chapter 9 for more detailed inspection and repair procedures.

Rear drum brakes

13   Refer to Chapter 9 and remove the rear brake drums.

14   **Warning:** *Brake dust produced by lining wear and deposited on brake components may contain asbestos, which is hazardous to your health. DO NOT blow it out with compressed air and DO NOT inhale it! DO NOT use gasoline or solvents to remove the dust. Brake system cleaner should be used to flush the dust into a drain pan. After the brake components are wiped clean with a damp rag, dispose of the contaminated rag(s) and solvent in a covered and labeled container. Try to use non-asbestos replacement parts whenever possible.*

15   Note the thickness of the lining material on the rear brake shoes **(see illustrations)** and look for signs of contamination by brake fluid and grease. If the lining material is within 1/16-inch of the recessed rivets or metal shoes, replace the brake shoes with new ones. The shoes should also be replaced if they are cracked, glazed (shiny lining surfaces) or contaminated with brake fluid or grease. See Chapter 9 for the replacement procedure.

16   Check the shoe return and hold-down springs and the adjusting mechanism to make sure they're installed correctly and in good condition. Deteriorated or distorted springs, if not replaced, could allow the linings to drag and wear prematurely.

17   Check the wheel cylinders for leakage by carefully peeling back the rubber boots. If brake fluid is noted behind the boots, the wheel cylinders must be replaced (see Chapter 9).

18   Check the drums for cracks, score marks, deep scratches and hard spots, which will appear as small discolored areas. If imperfections cannot be removed with emery cloth, the drums must be resurfaced by an automotive machine shop (see Chapter 9 for more detailed information).

19   Refer to Chapter 9 and install the brake drums.

20   Install the wheels, but don't lower the vehicle yet.

Parking brake

Note: *The parking brake cable and linkage should be periodically checked and lubricated. This maintenance procedure helps prevent the parking brake cable adjuster or the linkage from binding and adversely affecting the operation or adjustment of the parking brake.*

Lubrication

21   Set the parking brake.

22   Apply multi-purpose grease to the parking brake linkage, adjuster assembly, connectors and the areas of the parking brake cable that come in contact with the other parts of the vehicle.

23   Release the parking brake and repeat the lubrication procedure.

24   Remove the jackstands and lower the vehicle.

25   Tighten the wheel lug nuts to the specified torque and install the wheel covers.

Check

26   The easiest, and perhaps most obvious, method of checking the parking brake is to park the vehicle on a steep hill with the parking brake set and the transmission in Neutral. If the parking brake cannot prevent the vehicle from rolling, refer to Chapter 9 and adjust it as directed.

22   Automatic transmission control linkage lubrication

1   Raise the vehicle, support it securely on jackstands and locate the shift cable on the left side of the transmission.

2   Clean the linkage and pivot points at the upper end of the cable.

3   Lubricate the shift linkage and pivot points with multi-purpose grease.

23   Manual transmission lubricant level check

Note: *The transmission lubricant level and quality shouldn't deteriorate under normal driving conditions. However, it's recommended that you check the level occasionally. The most convenient time would be when the vehicle is raised for another reason, such as an engine oil change.*

1   The transmission has a check/fill plug which must be removed to check the lubricant level. If the vehicle is raised to gain access to the plug, be sure to support it safely on jackstands - DO NOT crawl under a vehicle which is supported only by a jack!

2   Remove the plug from the transmission and use your little finger to reach inside the housing and feel the lubricant level. It should be at or very near the bottom of the plug hole.

3   If it isn't, add the recommended lubricant through the plug hole with a syringe or squeeze bottle.

4   Install and tighten the plug securely and check for leaks after the first few miles of driving.

24   Front wheel bearing check, repack and adjustment

Refer to illustrations 24.1, 24.3, 24.15 and 24.16

1   In most cases the front wheel bearings

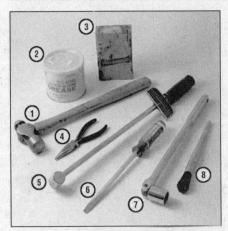

24.1 Tools and materials needed for front wheel bearing maintenance

1   *Hammer - A common hammer will do just fine*
2   *Grease - High-temperature grease which is formulated specially for front wheel bearings should be used*
3   *Wood block - If you have a scrap piece of 2x4, it can be used to drive the new seal into the hub*
4   *Needle-nose pliers - Used to straighten and remove the cotter pin in the spindle*
5   *Torque wrench - This is very important in this procedure; if the bearing is too tight, the wheel won't turn freely - if it is too loose, the wheel will 'wobble' on the spindle. Either way, it could mean extensive damage*
6   *Screwdriver - Used to remove the seal from the hub (a long screwdriver would be preferred)*
7   *Socket/breaker bar - Needed to loosen the nut on the spindle if it is extremely tight*
8   *Brush - Together with some clean solvent, this will be used to remove old grease from the hub and spindle*

will not need servicing until the brake pads are changed. However, the bearings should be checked whenever the front of the vehicle is raised for any reason. Several items, including a torque wrench and special grease, are required for this procedure **(see illustration)**.
2   With the vehicle securely supported on jackstands, spin each wheel and check for noise, rolling resistance and free play.
3   Move the wheel in-and-out on the spindle **(see illustration)**. If there's any noticeable movement, the bearings should be checked and then repacked with grease or replaced if necessary.
4   Remove the wheel.
5   Remove the brake caliper (Chapter 9) and hang it out of the way on a piece of wire. **Warning:** *DO NOT allow the brake caliper to hang by the rubber hose!*
6   Pry the grease cap out of the hub with a screwdriver or hammer and chisel.

24.3 To check the wheel bearings, try to move the tire in and out - if play is noted, or if the bearings feel rough or sound noisy when the tire is rotated, replace them

7   Straighten the bent ends of the cotter pin, then pull the cotter pin out of the retainer and spindle. Discard the cotter pin and use a new one during reassembly.
8   Remove the adjusting nut and washer from the end of the spindle.
9   Pull the hub assembly out slightly, then push it back into its original position. This should force the outer bearing off the spindle enough so it can be removed.
10   Pull the hub off the spindle.
11   Use a screwdriver to pry the grease seal out of the rear of the hub. As this is done, note how the seal is installed.
12   Remove the inner wheel bearing from the hub.
13   Use solvent to remove all traces of old grease from the bearings, hub and spindle. A small brush may prove helpful; however make sure no bristles from the brush embed themselves inside the bearing rollers. Allow the parts to air dry.
14   Carefully inspect the bearings for cracks, heat discoloration, worn rollers, etc. Check the bearing races inside the hub for wear and damage. If the bearing races are defective, the hubs should be taken to a machine shop with the facilities to remove the old races and press new ones in. Note that the bearings and races come as matched sets and old bearings should never be installed on new races.
15   Use high-temperature front wheel bearing grease to pack the bearings. Work the grease completely into the bearings, forcing it between the rollers, cone and cage from the back side **(see illustration)**.
16   Apply a thin coat of grease to the spindle at the outer bearing seat, inner bearing seat, shoulder and seal seat **(see illustration)**.
17   Put a small quantity of grease inboard of each bearing race inside the hub. Using your finger, form a dam at these points to provide extra grease availability and to keep thinned grease from flowing out of the bearing.

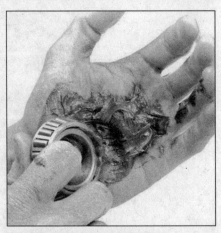

24.15 Pack the wheel bearing by working the grease into the rollers from the back side

18   Place the grease-packed inner bearing into the rear of the hub and put a little more grease outboard of the bearing.
19   Place a new seal over the inner bearing and tap the seal evenly into place with a hammer and block of wood until it's flush with the hub.
20   Carefully place the hub assembly onto the spindle and push the grease-packed outer bearing into position.
21   Install the washer and adjusting nut. Tighten the nut only slightly.
22   Spin the hub in a forward direction to seat the bearings and remove any grease or burrs which could cause excessive bearing play later.
23   While spinning the wheel, tighten the adjusting nut to the specified torque (step 1 in the Specifications).
24   Loosen the nut 1/2 turn, no more.
25   Tighten the nut to the specified torque (step 3 in the Specifications). Install a new cotter pin through the hole in the spindle and retainer. If the nut slots don't line up, loosen the nut slightly until they do. From the hand-

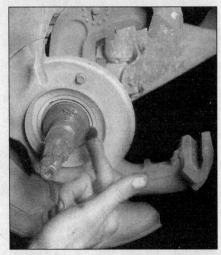

24.16 Apply a thin coat of grease to the spindle

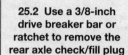
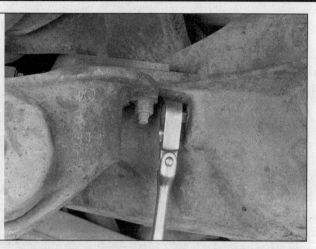

25.2 Use a 3/8-inch drive breaker bar or ratchet to remove the rear axle check/fill plug

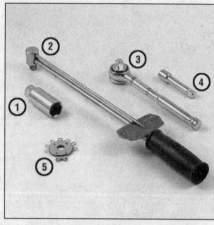

26.2 Tools required for changing spark plugs

1   *Spark plug socket - This will have special padding inside to protect the spark plug's porcelain insulator*
2   *Torque wrench - Although not mandatory, using this tool is the best way to ensure the plugs are tightened properly*
3   *Ratchet - Standard hand tool to fit the spark plug socket*
4   *Extension - Depending on model and accessories, you may need special extensions and universal joints to reach one or more of the plugs*
5   *Spark plug gap gauge - This gauge for checking the gap comes in a variety of styles. Make sure the gap for your engine is included.*

tight position, the nut should not be loosened more than one-half flat to install the cotter pin.
26   Bend the ends of the cotter pin until they're flat against the nut. Cut off any extra length which could interfere with the grease cap.
27   Install the grease cap, tapping it into place with a hammer.
28   Install the caliper (Chapter 9).
29   Install the tire/wheel assembly on the hub and tighten the lug nuts.
30   Check the bearings in the manner described earlier in this Section.
31   Lower the vehicle.

25   Rear axle (differential) oil level check

Refer to illustration 25.2

1   The differential has a check/fill plug which must be removed to check the oil level. If the vehicle is raised to gain access to the plug, be sure to support it safely on jackstands - DO NOT crawl under the vehicle when it's supported only by the jack.
2   Remove the oil check/fill plug from the side of the differential **(see illustration)**.
3   Use your little finger as a dipstick to make sure the oil level is even with the bottom of the plug hole. If not, use a syringe to add the recommended lubricant until it just starts to run out of the opening. On some models a tag is located in the area of the plug which gives information regarding lubricant type, particularly on models equipped with a limited slip differential.
4   Install the plug and tighten it securely.

26   Spark plug replacement

Refer to illustrations 26.2, 26.5a, 26.5b, 26.6 and 26.10

Note: *Access to the right (passenger) side spark plugs on models equipped with V6 engines is very limited. The two forward plugs (cylinders one and two) can be reached through the fenderwell. The rear spark plug*

(cylinder three) can be reached from the passenger compartment, once the engine cover has been removed. The use of a flexible-head ratchet would be helpful when changing these plugs.
1   The spark plugs are located on the side(s) of the engine.
2   In most cases, the tools necessary for spark plug replacement include a spark plug socket which fits onto a ratchet (spark plug sockets are padded inside to prevent damage to the porcelain insulators on the new plugs), various extensions and a gap gauge to check and adjust the gaps on the new plugs **(see illustration)**. A special plug wire removal tool is available for separating the wire boots from the spark plugs, but it isn't absolutely necessary. A torque wrench should be used to tighten the new plugs.
3   The best approach when replacing the spark plugs is to purchase the new ones in advance, adjust them to the proper gap and replace the plugs one at a time. When buying the new spark plugs, be sure to obtain the correct plug type for your particular engine. This information can be found on the Emission Control Information label located under the hood and in the factory owner's manual. If differences exist between the plug specified on the emissions label and in the owner's manual, assume that the emissions label is correct.
4   Allow the engine to cool completely before attempting to remove any of the plugs. While you're waiting for the engine to cool, check the new plugs for defects and adjust the gaps.
5   The gap is checked by inserting a wire type gauge between the electrodes at the tip of the plug **(see illustration)**. The gap between the electrodes should be the same as the one specified on the Emissions Control Information label. The wire should just slide between the electrodes with a slight amount of drag. If the gap is incorrect, use the adjuster on the gauge body to bend the curved side electrode slightly until the proper gap is obtained **(see illustration)**. If the side electrode is not exactly over the center electrode, bend it with the adjuster until it is. Check for cracks in the porcelain insulator (if

any are found, the plug should not be used).
6   With the engine cool, remove the spark plug wire from one spark plug. Pull only on the boot at the end of the wire - do not pull on the wire. A plug wire removal tool should be used if available **(see illustration)**.
7   If compressed air is available, use it to blow any dirt or foreign material away from the spark plug hole. The idea here is to eliminate the possibility of debris falling into the cylinder as the spark plug is removed.
8   Place the spark plug socket over the plug and remove it from the engine by turning it in a counterclockwise direction.
9   Compare the spark plug to those shown in the photos on the inside back cover to get an indication of the general running condition of the engine.
10   Thread one of the new plugs into the hole until you can no longer turn it with your fingers, then tighten it with a torque wrench (if available) or the ratchet. It might be a good idea to slip a short length of rubber hose over the end of the plug to use as a tool to thread it into place **(see illustration)**. The hose will grip the plug well enough to turn it, but will start to slip if the plug begins to cross-thread in the hole - this will prevent damaged threads and the accompanying repair costs.
11   Before pushing the spark plug wire onto the end of the plug, inspect it following the

26.5a Spark plug manufacturers recommend using a wire type gauge when checking the gap - if the wire doesn't slide between the electrodes with a slight drag, adjustment is required

26.5b To change the gap, bend the side electrode only, as indicated by the arrows, and be very careful not to crack or chip the porcelain insulator surrounding the center electrode

procedures outlined in Section 27.

12   Attach the plug wire to the new spark plug, again using a twisting motion on the boot until it's seated on the spark plug.

13   Repeat the procedure for the remaining spark plugs, replacing them one at a time to prevent mixing up the spark plug wires.

27   Spark plug wire, distributor cap and rotor check and replacement

Refer to illustrations 27.11 and 27.12

Spark plug wires

Note: *Every time a wire is detached from a spark plug, the distributor cap or the coil, silicone dielectric compound (a white grease available at auto parts stores) must be applied to the inside of the boot before reconnection. Use a small standard screwdriver to coat the entire inside surface of each boot with a thin layer of the compound.*

1   The spark plug wires should be checked and, if necessary, replaced at the same time new spark plugs are installed.

2   The easiest way to identify bad wires is to make a visual check while the engine is running. In a dark, well-ventilated garage, start the engine and look at each plug wire. Be careful not to come into contact with any moving engine parts. If there is a break in the wire, you will see arcing or a small spark at the damaged area. If arcing is noticed, make a note to obtain new wires.

3   The spark plug wires should be inspected one at a time, beginning with the spark plug for the number one cylinder to prevent confusion. On the four-cylinder engine, the number one cylinder is at the front of the engine. On V6 engines, it's at the front of the right (passenger side) bank. Clearly label each original plug wire with a piece of tape marked with the correct num-

ber. The plug wires must be reinstalled in the correct order to ensure proper engine operation.

4   Disconnect the plug wire from the first spark plug. A removal tool can be used **(see illustration 26.6)**, or you can grab the wire boot, twist it slightly and pull the wire free. *Don't pull on the wire itself, only on the rubber boot.*

5   Push the wire and boot back onto the end of the spark plug. It should fit snugly. If it doesn't, detach the wire and boot once more and use a pair of pliers to carefully crimp the metal connector inside the wire boot until it does.

6   Using a clean rag, wipe the entire length of the wire to remove built-up dirt and grease.

7   Once the wire is clean, check for burns, cracks and other damage. Don't bend the wire sharply or you might break the conductor.

8   Disconnect the wire from the distributor or coil pack. Again, pull only on the rubber boot. Check for corrosion and a tight fit. Replace the wire in the distributor/coil pack.

9   Inspect the remaining spark plug wires, making sure that each one is securely fastened at the distributor/coil pack and spark plug when the check is complete.

10   If new spark plug wires are required, purchase a set for your specific engine model. Pre-cut wire sets with the boots already installed are available. Remove and replace the wires one at a time to avoid mix-ups in the firing order.

Distributor cap and rotor (except 4.0L)

Note: *It's common practice to install a new distributor cap and rotor each time new spark plug wires are installed. If you're planning to install new wires, install a new cap and rotor also. But if you are planning to reuse the existing wires, be sure to inspect the cap and rotor to make sure they're in good condition.*

11   Remove the mounting screws and detach the cap from the distributor. Check it

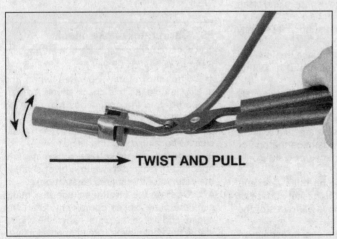

26.6 When removing the spark plug wires, grasp the boots only and use a twisting/pulling motion

TWIST AND PULL

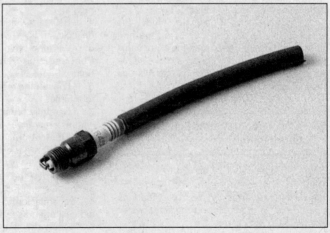

26.10 A length of snug-fitting ID rubber hose will save time and prevent damaged threads when installing the spark plugs

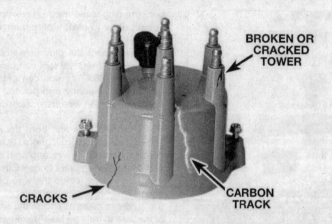

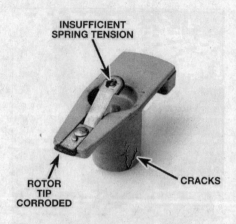

27.12 The ignition rotor should be checked for wear and corrosion as indicated here (if in doubt about its condition, buy a new one)

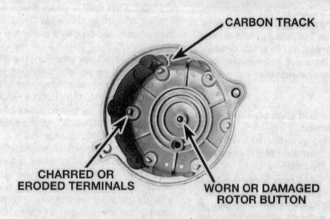

27.11 Shown here are some of the common defects to look for when inspecting the distributor cap (if in doubt about its condition, install a new one)

for cracks, carbon tracks and worn, burned or loose terminals **(see illustration)**

12   Check the rotor for cracks and carbon tracks. Make sure the center terminal spring tension is adequate and look for corrosion and wear on the rotor tip **(see illustration).**

13   Replace the cap and rotor if damage or defects are found. Note that on some models the rotor is held on the shaft by two screws and is indexed so it can be installed only one way. Before installing the cap, apply silicone dielectric compound to the rotor tip.

14   When installing a new cap, remove the wires from the old cap one at a time and attach them to the new cap in the exact same location- don't simultaneously remove all the wires from the old cap or firing order mix-ups may occur.

28   Automatic transmission fluid and filter change

Refer to illustrations 28.7 and 28.10
Caution: *The use of transmission fluid other than the type listed in the Specifications could result in transmission malfunctions or failure.*

1   At the specified intervals, the transmission fluid should be drained and replaced.

Since the fluid will remain hot long after driving, perform this procedure only after the engine has cooled down completely.

2   Before beginning work, purchase the specified transmission fluid (see *Recommended lubricants and fluids* at the front of this Chapter), a new filter and gaskets.

3   Other tools necessary for this job include jackstands to support the vehicle in a raised position, a drain pan capable of holding at least eight pints, newspapers and clean rags.

4   Raise the vehicle and support it securely on jackstands.

5   With the drain pan in place, remove the front and side transmission pan mounting bolts.

6   Loosen the rear pan bolts approximately four turns.

7   Carefully pry the transmission pan loose with a screwdriver, allowing the fluid to drain **(see illustration).** Don't damage the pan or transmission gasket surfaces or leaks could develop.

8   Remove the remaining bolts, pan and gasket. Carefully clean the gasket surface of the transmission to remove all traces of the old gasket and sealant.

9   Drain the fluid from the transmission pan, clean it with solvent and dry it with compressed air.

10   Remove the filter from the mount inside the transmission **(see illustration).**

11   Install a new filter and gasket. Tighten the mounting bolts securely.

12   Make sure the gasket surface on the transmission pan is clean, then install a new gasket. Put the pan in place against the transmission and install the bolts. Working around the pan, tighten each bolt a little at a time until the final torque figure is reached. Don't overtighten the bolts!

13   Lower the vehicle and add about three quarts of automatic transmission fluid through the filler tube (Section 7).

14   With the transmission in Park and the parking brake set, run the engine at a fast idle, but don't race it.

15   Move the gear selector through each range and back to Park. Check the fluid level. If the level is low, add enough fluid to bring the level to the specified range (see Section 7). Add fluid a little at a time to prevent overfilling.

16   Check the fluid level again when the transmission is hot (see Section 7).

17   Check under the vehicle for leaks during the first few trips.

29   Carburetor choke check

Refer to illustration 29.3

1   The choke only operates when the engine is cold, so this check should be performed before the engine has been started for the day.

2   Open the hood and remove the air cleaner housing cover and filter.

3   Locate the choke plate (the flat plate attached by small screws to a pivot shaft) in the carburetor throat **(see illustration).**

4   Operate the throttle linkage and make sure the plate closes completely. Start the engine and watch the plate - when the engine starts, the choke plate should open slightly.

5   Allow the engine to continue running at idle speed. As the engine warms up to oper-

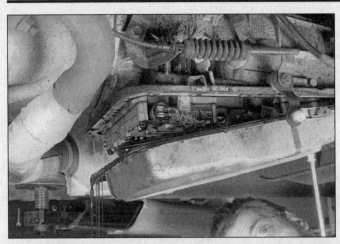

28.7 With the rear bolts loosened and holding it in place, lower the front of the transmission pan and allow the fluid to drain

28.10 The transmission filter is held in place with small bolts

ating temperature, the plate should slowly open.

6   After a few minutes, the choke plate should be fully open to the vertical position.

7   Note that the engine speed corresponds to the plate opening angle. With the plate closed, the engine should run at a fast idle speed. As the plate opens, the engine speed will decrease. The fast idle speed is controlled by the fast idle cam, and even though the choke plate is open completely, the idle speed will remain high until the throttle is opened, releasing the fast idle cam. Check the drop in idle speed as the choke plate opens by occasionally "blipping" the throttle.

8   If the choke doesn't work as described, shut off the engine and check the shaft and linkage for deposits which could cause binding. Use a spray-on choke cleaning solvent to remove the deposits as you operate the linkage. This should loosen up the linkage and the shaft and allow the choke to work properly. If the choke still fails to function correctly, the choke bimetal assembly is malfunctioning and the carburetor may have to

be overhauled. Refer to Chapter 4 for further information.

9   At regular intervals, clean and lubricate the choke shaft, the fast idle cam and linkage and the vacuum diaphragm pulldown rod to ensure good choke performance.

30   Cooling system servicing (draining, flushing and refilling)

Refer to illustration 30.4
Warning: *Antifreeze is a poisonous solution, so be careful not to spill any of the coolant mixture on the vehicle's paint or your skin. If this happens, rinse immediately with plenty of clean water. Consult local authorities regarding proper disposal procedures for antifreeze before draining the cooling system. In many areas, reclamation centers have been established to collect used oil and coolant mixtures.*

1   Periodically, the cooling system should be drained, flushed and refilled to replenish

the antifreeze mixture and prevent formation of rust and corrosion, which can impair the performance of the cooling system and cause engine damage. When the cooling system is serviced, all hoses and the radiator cap should be checked and replaced if necessary.

Draining

2   Apply the parking brake and block the wheels. If the vehicle has just been driven, wait several hours to allow the engine to cool down before beginning this procedure.

3   Once the engine is completely cool, remove the radiator cap.

4   Move a large container under the radiator drain to catch the coolant **(see illustration)**. Attach a 3/8-inch diameter hose to the drain fitting to direct the coolant into the container, then open the drain fitting (a pair of pliers may be required to turn it).

5   After the coolant stops flowing out of the radiator, move the container under the engine block drain plug. Remove the plug and allow the coolant in the block to drain.

6   While the coolant is draining, check the

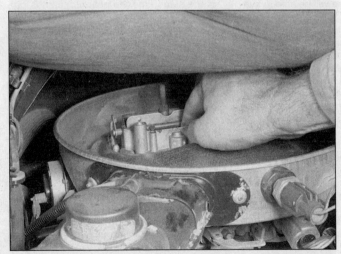

29.3 With the air cleaner cover removed, the choke plate can be checked for proper operation

30.4 The radiator drain fitting (arrow) is located at the right rear corner of the radiator - before opening the valve, push a short section of 3/8-inch diameter rubber hose onto the plastic fitting to prevent the coolant from splashing as it drains

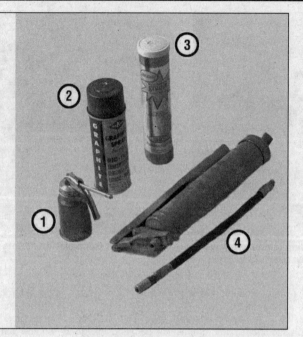

31.1 Materials required for chassis and body lubrication

1   *Engine oil - Light engine oil in a can like this can be used for door and hood hinges*
2   *Graphite spray - Used to lubricate lock cylinders*
3   *Grease - Grease, in a variety of types and weights, is available for use in a grease gun. Check the Specifications for your requirements.*
4   *Grease gun - A common grease gun, shown here with a detachable hose and nozzle, is needed for chassis lubrication. After use, clean it thoroughly!*

condition of the radiator hoses, heater hoses and clamps (refer to Section 17 if necessary).
7   Replace any damaged clamps or hoses (refer to Chapter 3 for detailed replacement procedures).

Flushing

8   Once the system is completely drained, flush the radiator with fresh water from a garden hose until water runs clear at the drain. The flushing action of the water will remove sediments from the radiator but will not remove rust and scale from the engine and cooling tube surfaces.
9   These deposits can be removed by the chemical action of a cooling system cleaner. Follow the procedure outlined in the manufacturer's instructions. If the radiator is severely corroded, damaged or leaking, it should be removed (Chapter 3) and taken to a radiator repair shop.
10   Remove the overflow hose from the coolant recovery reservoir. Drain the reservoir and flush it with clean water, then reconnect the hose.

Refilling

11   Close and tighten the radiator drain. Install and tighten the block drain plug.
12   Place the heater temperature control in the maximum heat position.
13   Slowly add new coolant (a 50/50 mixture of water and antifreeze) to the radiator until it's full. Add coolant to the reservoir up to the lower mark.
14   Leave the radiator cap off and run the engine in a well-ventilated area until the thermostat opens (coolant will begin flowing through the radiator and the upper radiator hose will become hot).
15   Turn the engine off and let it cool. Add more coolant mixture to bring the level back up to the lip on the radiator filler neck.

16   Squeeze the upper radiator hose to expel air, then add more coolant mixture if necessary. Replace the radiator cap.
17   Start the engine, allow it to reach normal operating temperature and check for leaks.

31   Chassis lubrication

Refer to illustrations 31.1 and 31.6
1   Refer to Recommended lubricants and fluids at the front of this Chapter to obtain the necessary grease, etc. You'll also need a grease gun **(see illustration)**. Occasionally plugs will be installed rather than grease fittings. If so, grease fittings will have to be purchased and installed.
2   Look under the vehicle and see if grease fittings or plugs are installed. If there are plugs, remove them and buy grease fittings, which will thread into the component. A dealer or auto parts store will be able to supply the correct fittings. Straight, as well as angled, fittings are available.
3   For easier access under the vehicle, raise it with a jack and place jackstands under the frame. Make sure it's safely supported by the stands. If the wheels are to be removed at this interval for tire rotation or brake inspection, loosen the lug nuts slightly while the vehicle is still on the ground.
4   Before beginning, force a little grease out of the nozzle to remove any dirt from the end of the gun. Wipe the nozzle clean with a rag.
5   With the grease gun and plenty of clean rags, crawl under the vehicle and begin lubricating the components.
6   Wipe the suspension balljoint grease fitting nipple clean and push the nozzle firmly over it **(see illustration)**. Squeeze the trigger on the grease gun to force grease into the component. The upper balljoints should be lubricated until the rubber seal is firm to the

touch. Don't pump too much grease into the fitting as it could rupture the seal. For all other suspension and steering components, continue pumping grease into the fitting until it oozes out of the joint between the two components. If it escapes around the grease gun nozzle, the nipple is clogged or the nozzle is not completely seated on the fitting. Resecure the gun nozzle to the fitting and try again. If necessary, replace the fitting with a new one.
7   Wipe the excess grease from the components and the grease fitting. Repeat the procedure for the remaining fittings.
8   While you're under the vehicle, clean and lubricate the parking brake cable, along with the cable guides and levers. This can be done by smearing some of the chassis grease onto the cable and its related parts with your fingers.
9   Open the hood and smear a little chassis grease on the hood latch mechanism. Have an assistant pull the hood release lever from inside the vehicle as you lubricate the cable at the latch.
10   Lubricate all the hinges (door, hood, etc.) with engine oil to keep them in proper working order.
11   The key lock cylinders can be lubricated with spray graphite or silicone lubricant, which is available at auto parts stores.
12   Lubricate the door weatherstripping with silicone spray. This will reduce chafing and retard wear.

32   Rear axle (differential) oil change

Refer to illustration 32.6
1   Some differentials can be drained by removing the drain plug, while on others it's necessary to remove the cover plate on the differential housing. As an alternative, a hand suction pump can be used to remove the differential lubricant through the filler hole. If there is no drain plug and a suction pump isn't available, be sure to obtain a new gasket at the same time the gear lubricant is purchased.
2   Raise the vehicle and support it securely on jackstands. Move a drain pan, rags, newspapers and wrenches under the vehicle.
3   Remove the check/fill plug from the differential **(see illustration 25.2)**.
4   If equipped with a drain plug, remove the plug and allow the differential oil to drain completely. After the oil has drained, install the plug and tighten it securely.
5   If a suction pump is being used, insert the flexible hose. Work the hose down to the bottom of the differential housing and pump the oil out.
6   If the differential is being drained by removing the cover plate, remove the bolts on the lower half of the plate. Loosen the bolts on the upper half and use them to keep the cover loosely attached **(see illustration)**. Allow the oil to drain into the pan, then completely remove the cover.
7   Using a lint-free rag, clean the inside of

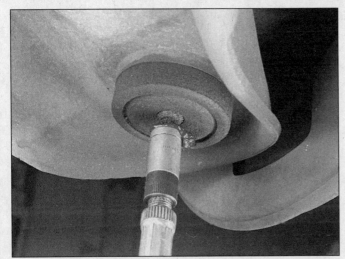

31.6 After wiping the grease fitting clean, push the nozzle firmly into place and pump the grease into the component - usually about two pumps of the gun will be sufficient

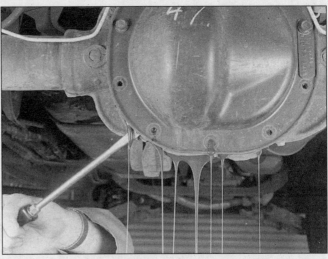

32.6 With the differential cover retained by two loosened bolts at the top, pry it loose and allow the oil to drain - don't damage the gasket sealing surfaces in the process!

the cover and the accessible areas of the differential housing. As this is done, check for chipped gears and metal particles in the lubricant, indicating that the differential should be more thoroughly inspected and/or repaired.

8   Thoroughly clean the gasket mating surfaces of the differential housing and the cover plate. Use a gasket scraper or putty knife to remove all traces of the old gasket.

9   Apply a thin layer of RTV sealant to the cover flange, then press a new gasket into position on the cover. Make sure the bolt holes align properly.

10   Place the cover on the differential housing and install the bolts. Tighten the bolts securely.

11   On all models, use a hand pump, syringe or funnel to fill the differential housing with the specified lubricant until it's level with the bottom of the plug hole.

12   Install the check/fill plug and tighten it securely.

Notes

Chapter 2 Part A
Four-cylinder engine

Contents

Specifications

General

Cylinder numbers (front-to-rear)	1-2-3-4
Firing order	1-3-4-2

Camshaft

Bearing journal diameter	1.7713 to 1.7720 in
Bearing clearance	0.001 to 0.003 in
End play	0.001 to 0.007 in

Torque specifications

Ft-lbs (unless otherwise indicated)

Camshaft sprocket bolt	50 to 71
Crankshaft pulley/sprocket bolt	103 to 133
Rear camshaft retaining plate bolt	72 to 108 in-lbs
Auxiliary shaft sprocket bolt	28 to 40
Auxiliary shaft retaining plate screws	72 to 108 in-lbs
Timing belt outer cover bolts	72 to 108 in-lbs
Timing belt tensioner adjustment bolt	14 to 21
Timing belt tensioner pivot bolt	28 to 40
Camshaft cover bolts	72 to 96 in-lbs
Cylinder head bolts	
Step 1	50 to 60
Step 2	80 to 90
Flywheel/driveplate bolts	56 to 64
Intake manifold bolts	
Step 1	60 to 84 in-lbs
Step 2	14 to 21
Exhaust manifold bolts	
Step 1	60 to 84 in-lbs
Step 2	16 to 23
Oil pan-to-engine block bolts	120 in-lbs
Oil pan-to-transmission bolts	29 to 40
Oil pump bolts	14 to 21
Oil pump pick-up tube bolts	14 to 21
Engine mount-to-block bolts	45 to 60
Engine mount-to-frame bolts	
Left side	71 to 94
Right side	52 to 67

FRONT OF ENGINE

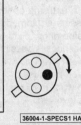

36004-1-SPECS1 HAYNES

Cylinder locations and distributor rotation

1   General information

This Part of Chapter 2 is devoted to in-vehicle repair procedures for the 2.3L OHC four-cylinder engine.

Information concerning engine removal and installation, as well as engine block and cylinder head overhaul, is in Part E of this Chapter.

The following repair procedures are based on the assumption that the engine is installed in the vehicle. If the engine has been removed from the vehicle and mounted on a stand, many of the steps included in this Part of Chapter 2 will not apply.

The specifications included in this Part of Chapter 2 apply only to the engine and procedures in this Part. The specifications necessary for rebuilding the block and cylinder head are found in Part E.

2   Repair operations possible with the engine in the vehicle

Many major repair operations can be accomplished without removing the engine from the vehicle.

Clean the engine compartment and the exterior of the engine with some type of pressure washer before any work is done. A clean engine will make the job easier and will help keep dirt out of the internal areas of the engine.

Depending on the components involved, it may be a good idea to remove the hood and engine cover to improve access to the engine as repairs are performed (refer to Chapter 11 if necessary).

If vacuum, exhaust, oil or coolant leaks develop, indicating a need for gasket or seal replacement, the repairs can generally be made with the engine in the vehicle. The intake and exhaust manifold gaskets, oil pan gasket and cylinder head gasket are all accessible with the engine in place.

Engine components such as the intake and exhaust manifolds, the oil pan, the oil pump, the auxiliary shaft, the water pump, the starter motor, the alternator, the distributor and the fuel system components can be removed for repair with the engine in place.

Since the cylinder head can be removed without pulling the engine, camshaft and valve component servicing can also be accomplished with the engine in the vehicle.

In extreme cases caused by a lack of necessary equipment, repair or replacement of piston rings, pistons, connecting rods and rod bearings is possible with the engine in the vehicle. However, this practice is not recommended because of the cleaning and preparation work that must be done to the components involved.

3   Camshaft cover - removal and installation

1   Disconnect the negative cable from the battery.

2   Remove the air cleaner (Chapter 4).

3   Remove the engine cover from inside the vehicle (Chapter 11).

4   Identify and tag the various wires and hoses which cross over the camshaft cover, including the spark plug wires. Do this carefully as they all must be repositioned correctly during installation.

5   Remove the throttle linkage, the throttle body assembly and the upper intake manifold (Chapter 4).

6   Remove the bolts and one stud and separate the cover from the engine. It may be necessary to break the gasket seal by tapping the cover with a soft-face hammer. If it's really stuck, use a knife, gasket scraper or chisel to remove it, but be very careful not to damage the gasket sealing surfaces of the cover or head.

7   Place clean rags in the camshaft gallery to keep foreign material out of the engine.

8   Remove all traces of gasket material from the cover and head. Be careful not to nick or gouge the surfaces. Clean the mating surfaces with lacquer thinner or acetone.

9   Reinstall the cover with a new gasket - no sealant is required. Install the bolts and tighten them to the specified torque in a crisscross pattern.

10   Reinstall the intake manifold (Chapter 4) and the remaining components previously removed or disconnected.

4   Valve train components - removal and installation

Note: *Broken valve springs and defective valve stem seals can be replaced without removing the cylinder head. Two special tools and a compressed air source are normally required to perform this operation, so read through this Section carefully and rent or buy the tools before beginning the job. If compressed air isn't available, a length of nylon rope can be used to keep the valves from falling into the cylinder during this procedure.*

1   Refer to Section 3 and remove the camshaft cover.

2   Remove the spark plug from the cylinder which has the defective component. If all of the valve stem seals are being replaced, all of the spark plugs should be removed (Chapter 1).

3   Turn the crankshaft until the piston in the affected cylinder is at top dead center on the compression stroke (refer to Section 16 for instructions). If you're replacing all of the valve stem seals, begin with cylinder number one and work on the valves for one cylinder at a time. Move from cylinder-to-cylinder following the firing order sequence (1-3-4-2).

4   Thread an adapter into the spark plug hole and connect an air hose from a compressed air source to it. Most auto parts stores can supply the air hose adapter. **Note:** *Many cylinder compression gauges utilize a screw-in fitting that may work with your air hose quick-disconnect fitting.*

5   Apply compressed air to the cylinder. The valves for that cylinder should be held in place by the air pressure. If the valve faces or seats are in poor condition, leaks may prevent the air pressure from retaining the valves - refer to the alternative procedure below.

6   If you don't have access to compressed air, an alternative method can be used. Position the piston at a point approximately 45-degrees before TDC on the compression stroke, then feed a long piece of nylon rope through the spark plug hole until it fills the combustion chamber. Be sure to leave the end of the rope hanging out of the engine so it can be removed easily. Use a large breaker bar and socket to rotate the crankshaft in the normal direction of rotation until **slight** resistance is felt.

7   Stuff shop rags into the cylinder head holes above and below the valves to prevent parts and tools from falling into the engine, then use a valve spring compressor to compress the spring.

8   With the spring compressed, slide the cam follower over the lash adjuster to remove it. Also remove the lash adjuster.

9   Still compressing the spring, remove the keepers. Release the valve spring tool and remove the spring retainer, damper assembly and valve spring.

10   Remove and discard the valve stem seal. **Note:** *If air pressure fails to hold the valve in the closed position during this operation, the valve face or seat is probably damaged. If so, the cylinder head will have to be removed for additional repair operations.*

11   Wrap a rubber band or tape around the top of the valve stem so the valve will not fall into the cylinder, then release the air pressure.

12   Inspect the valve stem for damage. Rotate the valve in the guide and check the end for eccentric movement, which would indicate that the valve is bent.

13   Move the valve up-and-down in the guide and make sure it doesn't bind. If the valve stem binds, either the valve is bent or the guide is damaged. In either case, the head will have to be removed for repair.

14   Reapply air pressure to the cylinder to retain the valve in the closed position, then remove the tape or rubber band from the valve stem. Lubricate the valve stem with engine oil and install a new umbrella type guide seal, if used.

15   Place the plastic installation cap over the end of the valve stem.

16   Start the valve stem seal carefully over the cap and push the seal down until the jacket touches the top of the guide.

17   Remove the plastic cap and use the installation tool or two screwdrivers to bottom the seal on the valve guide.

18   Install the valve spring, damper and retainer, compress the spring and install the keepers.

19   Install the lash adjuster and cam follower.

20   Disconnect the air hose and remove the adapter from the spark plug hole. If a rope

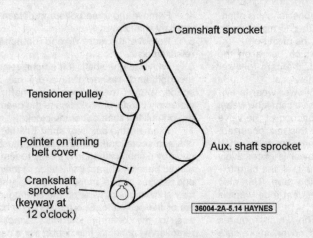

5.14 Timing mark details

- Camshaft sprocket
- Tensioner pulley
- Pointer on timing belt cover
- Aux. shaft sprocket
- Crankshaft sprocket (keyway at 12 o'clock)

36004-2A-5.14 HAYNES

5.23 Before starting the engine, remove the plug from the belt cover and recheck the timing marks

was used in place of air pressure, pull it out of the cylinder. **Caution:** *Do not turn the crankshaft backwards to release the rope. Turning the crankshaft backwards will cause the timing belt to jump time.*

21　Reinstall the various components previously removed.

22　Start and run the engine, then check for oil leaks and unusual sounds coming from the camshaft cover area.

5　Timing belt - removal, installation and adjustment

Refer to illustrations 5.14 and 5.23

1　Disconnect the negative cable from the battery.

2　Remove the fan shroud (Chapter 3).

3　Remove the fan and the water pump pulley (Chapter 3).

4　Remove the drivebelts.

5　Drain the cooling system (Chapter 1) and remove the upper radiator hose (Chapter 3).

6　Remove the thermostat housing and gasket (Chapter 3).

7　Position the number one piston at top dead center on the compression stroke (Section 16). **Caution:** *Always turn the crankshaft in the direction of normal rotation (clockwise, viewed from the front). Backward rotation may cause the timing belt to jump teeth. DO NOT turn the crankshaft during this procedure, after the number one piston is at TDC, until after the timing belt is reinstalled!*

8　Make sure the distributor rotor is pointing at the number one plug wire terminal.

9　Remove the crankshaft drivebelt pulley and belt guide.

10　Remove the four bolts and one screw and detach the outer timing belt cover.

11　Loosen the belt tensioner adjustment bolt, then position a belt tension releasing tool or a breaker bar on the tensioner roll pin and retract the belt tensioner. Tighten the adjustment bolt to hold the tensioner in the retracted position.

12　Remove the timing belt and inspect it for wear and damage. If it's worn or damaged, replace the belt. **Caution:** *If in doubt about the belt condition, install a new one.*

13　Check the sprockets for wear, cracks and burrs on the teeth. If the sprockets are worn or damaged, they can be detached after removing the mounting bolts. Make sure the new sprockets are installed correctly. Tighten the mounting bolts to the specified torque. **Note:** *Install a new camshaft sprocket bolt or use Teflon sealing tape on the old bolt.*

14　Make sure the valve timing mark on the camshaft sprocket is aligned with the pointer **(see illustration)**. Install the timing belt over the crankshaft sprocket, then over the auxiliary shaft sprocket.

15　Pull the belt up (pulling slack from between the auxiliary and camshaft sprockets) and install it on the camshaft sprocket in a counterclockwise direction so no slack will be between the two sprockets. Align the belt fore and aft on the sprockets.

16　Loosen the tensioner adjustment bolt to allow the tensioner to move against the belt. If the spring does not have enough tension to move the roller against the belt (the belt hangs loose), it may be necessary to insert a breaker bar between the tensioner and roll pin and push the roller against the belt. Tighten the bolt to the specified torque. **Note:** *The spring cannot be used to set belt tension. A wrench must be used on the tensioner assembly.*

17　Remove the spark plugs. Slowly rotate the crankshaft two complete turns in the direction of normal rotation to make sure the belt is seated and to remove any slack. While turning the crankshaft, feel and listen for valve-to-piston contact. Don't force the crankshaft - if binding is felt, recheck all work or seek professional advice!

18　Tighten the tensioner adjustment and pivot bolts to the specified torque. Recheck the alignment of the valve timing marks.

19　Install the timing belt cover and tighten the mounting bolts to the specified torque.

20　The remaining installation steps are the

reverse of removal. Be sure to tighten the crankshaft pulley/sprocket bolt to the specified torque.

21　Bring the number one piston to top dead center on the compression stroke (Section 16).

22　Remove the distributor cap and make sure the distributor rotor is pointing at the number one spark plug wire terminal.

23　Remove the access plug from the timing belt cover **(see illustration)**.

24　Look through the access hole in the belt cover to be sure that the timing mark on the camshaft sprocket is lined up with the pointer.

25　Make sure the timing mark on the crankshaft pulley aligns with the TDC mark on the belt cover.

26　Reinstall the belt cover access plug.

6　Camshaft - removal, inspection and installation

Removal

1　Disconnect the negative cable at the battery and remove the engine cover (Chapter 11).

2　Remove the camshaft cover, the timing belt cover and the timing belt (Sections 3 and 5).

3　Remove the spring clip from the hydraulic valve lash adjuster end of each cam follower.

4　Using a valve spring compressor, compress the valve spring and slide out the cam follower (see Section 4 if necessary). Keep the cam followers in order so they can be reinstalled in their original locations.

5　Lift out the hydraulic lash adjusters, keeping each one with its respective cam follower.

6　Remove the camshaft sprocket mounting bolt and washer. A long screwdriver placed through one of the holes in the sprocket will prevent the camshaft from turning.

7   Draw off the sprocket with a puller, then remove the belt guide.

8   Remove the sprocket locating pin from the end of the camshaft.

9   Remove the camshaft retaining plate from the rear bearing pedestal.

10   Raise the vehicle and support it securely on jackstands.

11   Position a floor jack under the engine. Place a block of wood on the jack pad. Remove the left and right engine mount bolts and nuts. Raise the engine as high as it will go. Place blocks of wood between the engine mounts and frame brackets and remove the jack. **Warning:** *DO NOT place your hands under the engine where they would be crushed if the jack failed!*

12   Using a hammer and a brass or aluminum drift, drive the camshaft out toward the front of the engine, taking the front seal with it. Be very careful not to damage the camshaft bearings and journals as it's pushed out.

Inspection

Camshaft and bearings

13   After the camshaft has been removed from the engine, cleaned with solvent and dried, inspect the bearing journals for uneven wear, pits and galling. If the journals are damaged, the bearing inserts in the head are probably damaged as well. Both the camshaft and bearings will have to be replaced with new ones. Measure the inside diameter of each camshaft bearing and record the results (take two measurements, 90 degrees apart, at each bearing).

14   Measure the camshaft bearing journals with a micrometer to determine if they're excessively worn or out-of-round. If they're more than 0.005-inch out-of-round, the camshaft should be replaced with a new one. Subtract the bearing journal diameters from the corresponding bearing inside diameter measurements to obtain the oil clearance. If it's excessive, new bearings must be installed. **Note:** *Camshaft bearing replacement requires special tools and expertise that place it outside the scope of the do-it-yourselfer. Take the head to an automotive machine shop to ensure that the job is done correctly.*

15   Check the camshaft lobes for heat discoloration, score marks, chipped areas, pitting and uneven wear. If the lobes are in good condition the camshaft can be reused.

16   Make sure the camshaft oil passages are clear and clean.

17   To check the thrust plate for wear, install the camshaft in the cylinder head and position the thrust plate at the rear. Using a dial indicator, check the total end play by tapping the camshaft carefully back and-forth. If the end play is outside the specified limit, replace the thrust plate with a new one.

Cam followers

18   Check the faces of the cam followers (which bear on the cam lobes) for signs of pitting, score marks and other forms of wear. They should fit snugly on the pivot bolt.

19   Inspect the face which bears on the valve stem. If it's pitted, the cam follower must be replaced with a new one.

20   If excessive cam follower wear is evident (and possibly excessive cam lobe wear), it may be due to a malfunction of the valve drive lubrication tube. If this has occurred, replace the tube and the cam follower. If more than one cam follower is excessively worn, replace the camshaft, all the cam followers and the lubrication tube. This also applies where excessive cam lobe wear is found.

21   During any operation which requires removal of the camshaft cover, make sure that oil is being discharged from the lubrication tube nozzles by cranking the engine with the starter motor.

Installation

22   Liberally coat the camshaft journals and bearings with engine assembly lube or moly-base grease, then carefully install the camshaft in the cylinder head.

23   Install the retaining plate and screws.

24   Lubricate the new camshaft oil seal with engine oil and carefully tap it into place at the front of the cylinder head with a large socket and a hammer. Make sure it enters the bore squarely and seats completely.

25   Install the belt guide and pin at the front end of the camshaft, then carefully tap on the sprocket.

26   Install a new sprocket bolt or use Teflon sealing tape on the threads of the old one.

27   Coat the hydraulic lash adjusters with engine assembly lube or moly-base grease, then install them in their original locations.

28   Coat the camshaft lobes and cam followers with engine assembly lube or moly-base grease.

29   Install the timing belt (Section 5).

30   Compress each valve spring and position each cam follower on its respective valve end and adjuster. Install the retaining spring clips.

31   Install the remaining components previously removed.

---

7   Auxiliary shaft - removal, inspection and installation

---

1   Detach the negative battery cable from the battery. Remove the large bolt from the front of the crankshaft and pull the drivebelt pulley off. Remove the bolts/screws and detach the outer timing belt cover.

2   Loosen the auxiliary shaft sprocket bolt. If the shaft turns, immobilize the sprocket by inserting a large screwdriver or 3/8-inch drive extension through one of the sprocket holes. Detach the timing belt (Section 5).

3   Pull off the sprocket (a puller may be required), and remove the pin from the shaft.

4   Remove the three bolts and detach the auxiliary shaft cover.

5   Remove the screws and detach the retaining plate.

6   Withdraw the shaft. If it's tight, reinstall the bolt and washer. Use a pry bar and spacer block to pry out the shaft. Be extremely careful not to damage the bearings as you pull the shaft out of the block.

7   Examine the auxiliary shaft bearing for pits and score marks. Replacement must be done by a shop (although it's easy to remove the old bearing, correct installation of the new one requires special tools). The auxiliary shaft may show signs of wear on the bearing journal or the eccentric. Score marks and damage to the bearing journals *cannot* be removed by grinding. If in doubt, ask a dealer service department to check the auxiliary shaft and give advice on replacement. Examine the gear teeth for wear and damage. If either is evident, a replacement shaft must be obtained.

8   Dip the auxiliary shaft in engine oil before installing it in the block. Tap it in gently with a soft-face hammer to ensure that it's seated. Install the retaining plate and the auxiliary shaft cover.

9   The remainder of the procedure is the reverse of removal. Make sure that the auxiliary shaft pin is in place before installing the sprocket. Tighten the sprocket mounting bolt to the specified torque.

---

8   Front oil seals - replacement

Note: *The camshaft, crankshaft and auxiliary shaft oil seals are all replaced the same way, using the same tools, after the appropriate sprocket has been removed.*

1   Disconnect the negative battery cable from the battery.

2   To remove the camshaft or auxiliary shaft sprockets, refer to Section 6 or 7. A special puller designed for this purpose is available at most auto parts stores.

3   To remove the crankshaft sprocket, use the special tool or a gear puller.

4   A special tool, available at most auto parts stores, or its equivalent may be used to remove all the seals. When using the tool, be sure the jaws are gripping the thin edge of the seal very tightly before operating the screw portion of the tool. If the tool isn't available, a hammer and chisel may be used to remove the seal(s) if care is exercised.

5   Clean the seal bore and shaft surface prior to installation of the new seal. Apply a thin layer of grease to the outer edge of the new seal(s).

6   Install the seal(s) with a special tool, available at most auto parts stores. If the special tool isn't available, you may be able to use a large deep socket and a hammer or a short piece of pipe, the sprocket bolt and a large flat washer.

7   Install the components removed to gain access to the seals.

9 Intake manifold - removal and installation

1 Relieve the fuel system pressure (Chapter 4).
2 Disconnect the negative battery cable from the battery.
3 Drain the cooling system (Chapter 1).
4 Remove the engine cover (Chapter 11).
5 Remove the air cleaner and ducts (Chapter 4).
6 Label and disconnect the vacuum hoses from the throttle body and emissions devices attached to the intake manifold.
7 Remove the throttle body, fuel injection wiring harness, fuel supply manifold, throttle cable, speed control cable (if equipped) and upper intake manifold (Chapter 4).
8 Disconnect the coolant bypass hose from the lower intake manifold.
9 Remove the engine oil dipstick bracket retaining bolt.
10 Remove the four bottom mounting bolts from the intake manifold.
11 Remove the four upper mounting bolts from the manifold. Note that the front two bolts also secure the engine lifting "eye."
12 Detach the manifold from the engine.
13 Clean the manifold and head mating surfaces and check the manifold for cracks. **Note:** *The mating surfaces of the head and manifold must be perfectly clean when the manifold is installed. Gasket removal solvents in aerosol cans are available at most auto parts stores and may be helpful when removing old gasket material that's stuck to the head and manifold. Be careful not to scrape or gouge the sealing surfaces.*
14 Clean and oil the manifold mounting bolts.
15 Position the new intake manifold gasket on the head, using a light coat of gasket sealant to hold it in place.
16 Position the intake manifold, along with the lifting "eye," on the head, install the fasteners and tighten them, working from the center outwards, reaching the specified torque in two steps.
17 The remainder of installation is the reverse of removal.
18 Reconnect the battery and refill the cooling system.
19 Start the engine and let it run while checking for coolant, fuel and vacuum leaks.
20 Reinstall the engine cover.

10 Exhaust manifold - removal and installation

Removal

1 Disconnect the negative battery cable from the battery.
2 Remove the air cleaner and duct assembly if needed to gain working room.
3 Remove the engine cover (Chapter 11).
4 Remove the EGR line bolts at the exhaust manifold and loosen at the EGR tube (Chapter 6).
5 Disconnect the oxygen sensor wire.
6 Remove the bolt attaching the heater hoses to the camshaft cover.
7 Raise the vehicle and support it securely on jackstands. Remove the two bolts attaching the exhaust pipe to the manifold (Chapter 4).
8 Remove the eight exhaust manifold mounting bolts. A special Allen head socket will be needed to remove the bolts.
9 Remove the exhaust manifold out through the bottom of the vehicle.
Installation
10 Using a new gasket, install the manifold and the eight bolts attaching it to the cylinder head. Tighten the bolts (in two or three steps) to the specified torque , working from the center outwards.
11 Install the bolts attaching the exhaust pipe to the exhaust manifold.
12 Position the heater hoses on the camshaft cover.
13 Reconnect the oxygen sensor wire.
14 Install the EGR line at the exhaust manifold.
15 Install the air cleaner and duct assembly.
16 Reconnect the battery cable.
17 Reinstall the engine cover.

11 Cylinder head - removal and installation

Refer to illustrations 11.31 and 11.33

Removal

1 Begin the procedure by positioning the number one piston at top dead center (TDC) on the compression stroke (see Section 16). Remove the engine cover (Chapter 11).
2 Drain the cooling system (Chapter 1). Remove the air cleaner assembly and the air duct.
3 Remove the bolt retaining the heater hose to the camshaft cover.
4 Detach the distributor cap from the distributor.
5 Remove the spark plug wires from the spark plugs and remove the distributor cap and wires as an assembly. Remove the spark plugs.
6 Label and disconnect all vacuum hoses attached to the components on the head.
7 Remove the dipstick tube.
8 Remove the camshaft cover (Section 3).
9 Remove the intake manifold bolts (Section 9).
10 Loosen the alternator mounting bolts, remove the belt from the pulley and remove the bracket bolts.
11 Disconnect the upper radiator hose at both ends and remove it from the engine.
12 Remove the timing belt front cover.
13 Make sure the valve timing is correct (Section 5, Paragraphs 21 through 25), then loosen the timing belt tensioner (Section 5).
14 Remove the timing belt from the camshaft and auxiliary shaft sprockets.
15 Remove the heat stove from the exhaust manifold.
16 Remove the exhaust manifold mounting bolts (Section 10). The manifold itself can remain in place.
17 Remove the timing belt tensioner bolts.
18 Remove the timing belt tensioner spring stop from the cylinder head.
19 Disconnect the oil pressure sending unit lead wire.
20 Refer to Section 9 and remove the intake manifold.
21 Using a new head gasket, outline the cylinders and bolt pattern on a piece of cardboard. Be sure to indicate the front of the engine for reference. Punch holes at the bolt locations.
22 Loosen the cylinder head mounting bolts in 1/4-turn increments until they can be removed by hand. Store the bolts in the cardboard holder as they're removed - this will ensure that they're reinstalled in their original locations.
23 Lift the head off the engine. If it's stuck, don't pry between the head and block. Instead, rap the head with a soft-face hammer or a block of wood and a hammer to break the gasket seal.
24 Place the head on a block of wood to prevent damage to the gasket surface. Refer to Part D for cylinder head disassembly and valve service procedures.

Installation

25 If a new cylinder head is being installed, transfer all external parts from the old cylinder head to the new one.
26 The mating surfaces of the cylinder head and block must be perfectly clean when the head is installed.
27 Use a gasket scraper to remove all traces of carbon and old gasket material, then clean the mating surfaces with lacquer thinner or acetone. If there's oil on the mating surfaces when the head is installed, the gasket may not seal correctly and leaks may develop. Use a vacuum cleaner to remove any debris that falls into the cylinders.
28 Check the block and head mating surfaces for nicks, deep scratches and other damage. If damage is slight, it can be removed with a file; if it's excessive, machining may be the only alternative.
29 Use a tap of the correct size to chase the threads in the head bolt holes. Mount each bolt in a vise and run a die down the threads to remove corrosion and restore the threads. Dirt, corrosion, sealant and damaged threads will affect critical head bolt torque readings.
30 Position the new gasket over the dowel pins in the block.
31 Make a mark on the cylinder head indicating the position of the pin in the end of the camshaft. Turn the camshaft until the pin is in the five o'clock position - this must be done

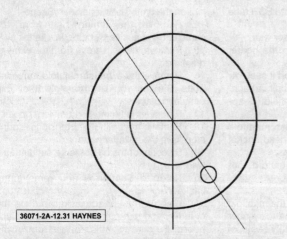

36071-2A-12.31 HAYNES

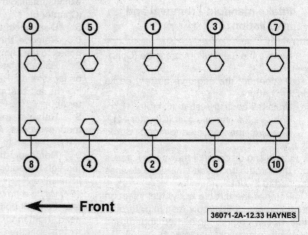

← **Front**

36071-2A-12.33 HAYNES

11.31 Be sure to turn the camshaft until the pin is in the 5 o'clock position as shown here before installing the head

11.33 Cylinder head bolt TIGHTENING sequence

to avoid damage to the valves when the head is installed **(see illustration)**.

32　Carefully position the head on the block without disturbing the gasket.

33　Install the cylinder head bolts and tighten them in sequence **(see illustration)** to the specified torque.

34　Return the camshaft to its original position (with the pin aligned with the mark you made on the head).

35　The remainder of installation is the reverse of the removal procedure.

36　Change the engine oil and filter (Chapter 1).

12　Oil pan - removal and installation

Note: *Vehicles equipped with an automatic transmission have an oil pan which can be removed out the front of the engine compartment. Those with a manual transmission must be removed out the rear.*

1　Disconnect the negative battery cable from the battery.

2　Remove the air cleaner and hoses as an assembly.

3　Remove the oil dipstick.

4　Remove the engine mount nuts.

5　Remove the oil cooler lines at the radiator, if so equipped.

6　Remove the two bolts retaining the fan shroud to the radiator and remove the shroud.

7　Remove the radiator mounting bolts (automatic transmission only).

8　Drain the radiator and remove the hoses. Move the radiator up and wire it to the hood to provide working clearance (automatic transmission only).

9　Raise and support the vehicle on jackstands.

10　Drain the oil and remove the oil filter.

11　Remove the starter motor (Chapter 5).

12　Disconnect the exhaust manifold tube at the inlet pipe bracket (Chapter 4).

13　Remove the transmission mount-to-crossmember nuts.

14　Remove the bellcrank from the converter housing (automatic transmission only).

15　Remove the oil cooler lines from the retainer at the engine block (automatic transmission only).

16　Remove the front crossmember (automatic transmission only).

17　Disconnect the right front lower shock absorber mount (manual transmission only).

18　Position a jack under the engine, raise the engine and block it with a piece of wood approximately 2-1/2 inches thick between the motor mount and the frame rail. **Warning:** *Do not place any part of your body under the engine where it could be crushed if the jack fails.*

19　Remove the jack.

20　Position the jack under the transmission and raise it slightly (automatic transmission only).

21　Remove the oil pan mounting bolts and lower the pan.

22　Remove the oil pump and pick-up tube assembly and drop them into the oil pan. Note how the intermediate shaft is installed!

23　Remove the oil pan (out the front on automatics - out the rear on manual).

24　Clean the oil pan and inspect for damage. Remove spacers, if any, that are attached to the oil pan transmission mounting pad.

25　Clean the oil pan gasket surface on the cylinder block and the oil pan flange. These surfaces must be perfectly clean to prevent oil leaks.

26　Clean the oil pump exterior and oil pump pick-up tube screen.

27　To install, follow the instructions which will come with the new gasket to apply RTV sealant. Position the gasket inside the channel of the pan.

28　Place the oil pump and pick-up tube assembly inside the oil pan.

29　Position the oil pan under the engine, reach inside and install the oil pump and

pick-up tube assembly, tightening the bolts to the specified torque (see Section 13).

30　Install the oil pan mounting bolts and hand tighten them only at this time.

31　Tighten the two oil pan-to-transmission bolts to the specified torque, then loosen them 1/2 turn.

32　Tighten the oil pan-to-engine block bolts to the specified torque, working around the pan, tightening each bolt a little at a time.

33　Retighten the remaining two bolts.

Note: *If the oil pan is installed while the transmission is removed, either reinstall the transmission or use a special fixture to insure the oil pan is installed flush with the rear of the block.*

34　The remainder of the installation is the reverse of the removal procedure.

35　Be sure to refill the engine with the proper quantity, grade and viscosity oil.

13　Oil pump - removal and installation

1　Remove the oil pan as described in Section 12. The pump must be removed as a part of the oil pan removal procedure.

2　If there is any possibility that the pump is faulty or if an engine overhaul is being performed, replace the pump. A faulty pump can ruin an otherwise good engine.

3　Prime the pump before installation. Hold it with the pick-up tube up and pour a few ounces of clean oil into the inlet screen. Turn the pump driveshaft by hand until oil comes out the outlet.

4　Install the pump intermediate shaft with the collar end in the engine block. Be sure the shaft seats in the distributor. Hold the shaft in place and install the pump assembly as the oil pan is installed (see Section 12). Be sure to tighten the bolts to the specified torque.

5　Add oil to the engine (Chapter 1), then start it and be sure the oil pressure comes up. If it doesn't come up within 10 or 15 seconds, shut off the engine immediately and find the cause of the problem. **Caution:** *Continued running of the engine without oil pressure will severely damage the moving parts!*

6   Check for oil leaks.

14   Flywheel/driveplate - removal and installation

1   Remove the transmission (Chapter 7).

2   Remove the pressure plate and the clutch disc from the rear of the flywheel as described in Chapter 8 (manual transmission models).

3   Lock the flywheel/driveplate using a large screwdriver in mesh with the starter ring gear and remove the six bolts securing the flywheel to the crankshaft. The bolts should be removed a quarter turn at a time in a diagonal and progressive manner.

4   Mark the mating position of the flywheel/driveplate and crankshaft flange and then remove the flywheel.

5   Remove the engine rear cover plate bolts and ease the rear cover plate over the two dowel pins. Remove the cover plate.

6   If the flywheel ring gear is badly worn or has teeth missing, it should be replaced by an automotive machine shop.

7   Check for cracks and other damage.

8   Installation of the flywheel/driveplate and clutch is the reverse of the removal procedure. Apply oil-resistant sealer sparingly to the bolt threads. Be sure to tighten all bolts to the specified torque. For further information on clutch servicing and pilot bearing replacement, see Chapter 8.

15   Rear crankshaft oil seal - replacement

1   Remove the transmission and all other components necessary, such as the clutch and pressure plate, to get at the oil seal in the back of the engine (Chapters 7 and 8).

2   Remove the flywheel or driveplate (Section 14).

3   Use a small punch to make two holes on opposite sides of the seal and install small sheet metal screws in the seal. Pry on the screws with two large screwdrivers until the seal is removed from the engine. It may be necessary to place small blocks of wood against the block to provide a fulcrum point for prying. **Caution:** *Be careful not to scratch or otherwise damage the crankshaft oil seal surface.*

4   Apply a thin film of engine oil to the mating edges of the seal and the seal bore in the block, as well as the seal lips.

5   Position the seal on a seal installer tool, available at most auto parts stores, or its equivalent, then position the tool and seal at the rear of engine. Install the seal with the spring side toward the engine. Alternate tightening of the bolts to seat the seal properly.

6   If the special seal installer is not available, gently tap the seal into place with a soft-face hammer.

7   The remainder of the installation is the reverse of the removal procedure.

16   Top Dead Center (TDC) for number one piston - locating

1   Top Dead Center (TDC) is the highest point in the cylinder that each piston reaches as it travels up-and-down when the crankshaft turns. Each piston reaches TDC on the compression stroke and again on the exhaust stroke, but TDC generally refers to piston position on the compression stroke. The timing marks are referenced to the number one piston at TDC on the compression stroke.

2   Positioning the piston(s) at TDC is an essential part of many procedures such as in-vehicle valve train service, timing belt replacement and distributor removal.

3   In order to bring any piston to TDC, the crankshaft must be turned using one of the methods outlined below. When looking at the front of the engine, normal crankshaft rotation is clockwise. **Caution:** *Turning the crankshaft backwards may cause the timing belt to jump time.*

a)   *The preferred method is to turn the crankshaft with a large socket and breaker bar attached to the large bolt that's threaded into the front of the crankshaft.*

b)   *A remote starter switch, which may save some time, can also be used. Attach the switch leads to the switch and battery terminals on the solenoid. Once the piston is close to TDC, use a socket and breaker bar as described in the previous paragraph.*

c)   *If an assistant is available to turn the ignition switch to the Start position in short bursts, you can get the piston close to TDC without a remote starter switch. Use a socket and breaker bar as described in Paragraph a) to complete the procedure.*

4   Locate the number one spark plug wire terminal in the distributor cap, then mark the distributor base directly under the terminal.

5   Remove the distributor cap as described in Chapter 1.

6   Turn the crankshaft (see Paragraph 3 above) until the ignition timing mark for TDC on the crankshaft pulley is aligned with the timing mark on the belt cover..

7   The rotor should now be pointing directly at the mark you made earlier. If it isn't, the piston is at TDC on the exhaust stroke.

8   To get the piston to TDC on the compression stroke, turn the crankshaft one complete turn (360 degrees) clockwise. The rotor should now be pointing at the mark. When the rotor is pointing at the number one spark plug wire terminal in the distributor cap (which is indicated by the mark on the distributor) and the ignition timing marks are aligned, the number one piston is at TDC on the compression stroke.

9   After the number one piston has been positioned at TDC on the compression stroke, TDC for any of the remaining cylinders can be located by turning the crankshaft and following the firing order (refer to the Specifications).

17   Engine mounts - removal and installation

1   Remove the fan shroud mounting screws.

2   Using a floor jack and a block of wood between the jack head and oil pan, support the engine.

3   Remove the nuts and washers securing the insulator assemblies (engine mounts) to the frame brackets.

4   Lift the engine sufficiently to disengage the front insulator studs from the crossmember bracket.

5   Remove the insulator assembly-to-block bolts and remove the insulators and brackets.

6   To install, mount the insulators on the block, lower the engine until the insulator stud and locking pin engages and bottoms in the slots in the crossmember and install the nuts and washers.

7   Remove the floor jack and reinstall the fan shroud mounting screws.

Notes

Chapter 2 Part B
2.8L V6 engine

Contents

Specifications

General
Cylinder numbers
    Left (driver's) side 4-5-6
    Right side 1-2-3
Firing order 1-4-2-5-3-6

Camshaft
Camshaft end play.. 0.0008 to 0.004 in
Valve clearance
    Intake.. 0.014 in
    Exhaust... 0.016 in

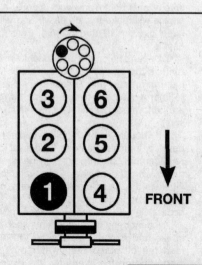

Cylinder locations and distributor rotation - 2.8L engine

FRONT

36004-1-SPECS2a HAYNES

Torque specifications

	Ft-lbs (unless otherwise noted)
Camshaft gear bolt	30 to 36
Camshaft thrust plate bolts	13 to 16
Crankshaft damper/pulley bolt	85 to 96
Cylinder head bolts	
Step 1	29 to 40
Step 2	40 to 51
Step 3	70 to 85
Exhaust manifold bolts	20 to 30
Flywheel-to-crankshaft bolts	47 to 52
Front cover bolts	13 to 16
Intake manifold bolt/nut	
Step 1	36 to 72 in-lbs
Step 2	72 to 132 in-lbs
Step 3	11 to 15
Step 4	15 to 18
Intake manifold stud	10 to 12
Oil pump pick-up tube-to-pump	72 to 120 in-lbs
Oil pump pick-up tube to main bearing cap	12 to 15
Oil pan bolts	60 to 96 in-lbs
Rocker arm cover bolts	36 to 60 in-lbs
Rocker arm shaft support bolts	43 to 50

1   General information

This Part of Chapter 2 is devoted to in-vehicle repair procedures for the 2.8 liter V6 engine. Information concerning engine removal and installation, as well as engine block and cylinder head overhaul, is in Part E of this Chapter.

The following repair procedures are based on the assumption that the engine is installed in the vehicle. If the engine has been removed from the vehicle and mounted on a stand, many of the steps included in this Part of Chapter 2 will not apply.

The Specifications included in this Part of Chapter 2 apply only to the engine and procedures in this Part. The Specifications necessary for rebuilding the block and cylinder head are found in Part E.

2   Repair operations possible with the engine in the vehicle

Many major repair operations can be accomplished without removing the engine from the vehicle. However, due to limited access, extra time and patience will be required!

Clean the engine compartment and the exterior of the engine with some type of pressure washer before any work is done. A clean engine will make the job easier and will help keep dirt out of the internal areas of the engine.

Depending on the components involved, remove the engine cover, the lower instrument trim panel and, if necessary, the hood to improve access to the engine as repairs are performed (refer to Chapter 11 if necessary).

If vacuum, exhaust, oil or coolant leaks develop, indicating a need for gasket or seal replacement, the repairs can generally be made with the engine in the vehicle. The intake and exhaust manifold gaskets, oil pan gasket and cylinder head gaskets are all accessible with the engine in place.

Exterior engine components such as the intake and exhaust manifolds, the oil pan (and the oil pump), the water pump, the starter motor, the alternator, the distributor and the fuel system components can be removed for repair with the engine in place.

Since the cylinder heads can be removed without pulling the engine, valve component servicing can also be accomplished with the engine in the vehicle.

In extreme cases caused by a lack of necessary equipment, repair or replacement of piston rings, pistons, connecting rods and rod bearings is possible with the engine in the vehicle. However, this practice is not recommended because of the cleaning and preparation work that must be done to the components involved.

3   Rocker arm covers - removal and installation

Refer to illustration 3.11

Removal

1   Disconnect the battery, negative cable first, then positive.
2   Remove the engine cover (Chapter 11).
3   Remove the air cleaner and air duct.
4   Remove the spark plug wires after labeling them for reinstallation.

Left side view of engine (removed from vehicle for clarity)

Front view of engine (removed from vehicle for clarity)

5   Remove the PCV valve and hose from the right rocker arm cover.
6   Remove the carburetor choke air deflector plate.
7   Remove the A/C compressor (Chapter 3) and brackets (if so equipped). Set the compressor aside without disconnecting the hoses. **Warning:** *The air conditioning system contains refrigerant under high pressure. Air conditioning hoses should never be loosened or disconnected unless the system has been depressurized by a dealer service department or air conditioning shop.*
8   Remove the rocker arm cover bolts and load distribution washers. Lay out the washers so they can be reinstalled in their original positions.
9   Disconnect the kickdown linkage and transmission dipstick tube (automatic transmission only).
10   Position the thermactor air hose and wiring harness away from the right rocker arm cover.
11   Remove the engine oil filler tube and

bracket (see illustration).
12   Disconnect the vacuum line at the canister purge solenoid (Chapter 6) and disconnect the line routed from the canister to the purge solenoid.
13   Disconnect the power brake booster hose from the firewall mounted booster, if so equipped.
14   Tap the cover with a light plastic hammer to break the seal or gently pry the cover up with a screwdriver.
15   Remove the rocker arm cover.

Installation

16   Clean all old gasket material and sealer from the cylinder head and rocker arm cover gasket surfaces.
17   Apply a thin layer of gasket sealer to the rocker arm cover mating surface and lay the gasket in the cover. Put the rocker arm cover in place and install the mounting bolts and load distribution washers.
18   Connect the kickdown linkage and

transmission dipstick tube (automatic transmission only).
19   After ensuring that all rocker arm cover load distribution washers are installed in their original positions, tighten the rocker arm cover bolts to the specified torque.
20   Install the spark plug wires.
21   Install the PCV valve and hose.
22   Install the carburetor choke air deflector plate.
23   Reposition the thermactor air hose and wiring harness in their original positions.
24   Install the engine oil filler tube.
25   Connect the vacuum line at canister purge solenoid and connect the line routed from the canister to the purge solenoid.
26   Connect the power brake hose, if so equipped.
27   Install the air cleaner.
28   Reconnect the battery cables, positive cable first, then negative.
29   Start the engine and check for oil leaks.
30   Reinstall the engine cover.

Right side view of engine (removed from vehicle for clarity)

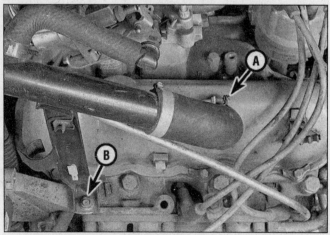

3.11 Loosening the hose clamp (A) and removing the bracket nut (B) will allow the oil filler tube to be removed from the rocker arm cover

4.2 Remove the three rocker shaft support bolts (arrows) by loosening them two turns at a time

4.4 The pushrods should be stored in order of removal so they can be returned to their original locations

4   Rocker arm assembly and pushrods - removal, installation and adjustment

Refer to illustrations 4.2, 4.4 and 4.12

Removal

1   Remove the rocker arm cover (Section 3).
2   Remove the rocker arm shaft support bolts, loosening the bolts two turns at a time until free **(see illustration)**.
3   Lift off the rocker arm shaft assembly and oil baffle.
4   Pull the pushrods out and store them in order **(see illustration)**.

Installation

5   Loosen the valve lash adjusting bolts several turns.
6   Thoroughly lubricate all components with engine oil.
7   Install the pushrods, making sure they are properly seated in the lifters.
8   Install the oil baffle and rocker arm shaft assembly onto the head, lining up the pushrods and adjusting screws.
9   Starting in the middle and working out, tighten the rocker arm shaft support bolts two turns at a time to the specified torque.

Adjustment

10   Place your finger on the intake valve rocker arm for the number five cylinder.
11   Turn the engine over until the pushrod just starts to lift.
12   Using a feeler gauge to check the clearance, turn the self-locking adjusting bolt until the clearance on the number one intake valve is as specified **(see illustration)**.
13   Repeat the procedure on the number one exhaust valve. Note that the clearance (valve lash) is *NOT* the same for the intake and exhaust valves.
14   Turn the engine over until the number three cylinder intake valve pushrod just starts

to lift, then adjust the valves for the number *four* cylinder.
15   Turn the engine over until the number six cylinder intake valve pushrod just starts to lift, then adjust the valves for the number *two* cylinder.
16   Turn the engine over until the number one cylinder intake valve pushrod just starts to lift, then adjust the valves for the number *five* cylinder.
17   Turn the engine over until the number four cylinder intake valve pushrod just starts to lift, then adjust the valves for the number *three* cylinder.
18   Turn the engine over until the number two cylinder intake valve pushrod just starts to lift, then adjust the valves for the number *six* cylinder.

5   Valve spring, retainer and seals - replacement

Refer to illustrations 5.6, 5.8, 5.15, 5.17a, 5.17b and 5.17c
Note: *Broken valve springs and defective valve stem seals can be replaced without removing the cylinder head. Two special tools and a compressed air source are normally*

required to perform this operation, so read through this Section carefully and rent or buy the tools before beginning the job. If compressed air is not available, a length of nylon rope can be used to keep the valves from falling into the cylinder during this procedure.
1   Refer to Section 3 and remove the rocker arm cover from the affected cylinder head. If all of the valve stem seals are being replaced, remove both rocker arm covers.
2   Remove the spark plug from the cylinder which has the defective component. If all of the valve stem seals are being replaced, all of the spark plugs should be removed.
3   Turn the crankshaft until the piston in the affected cylinder is at top dead center on the compression stroke (refer to Section 11 for instructions). If you are replacing all of the valve stem seals, begin with cylinder number one and work on the valves for one cylinder at a time. Move from cylinder-to-cylinder following the firing order sequence (1-4-2-5-3-6).
4   Thread an adapter into the spark plug hole and connect an air hose from a compressed air source to it. Most auto parts stores can supply the air hose adapter. **Note:** *Many cylinder compression gauges utilize a screw-in fitting that may work with your air hose quick-disconnect fitting.*
5   Remove the rocker arm assembly for the

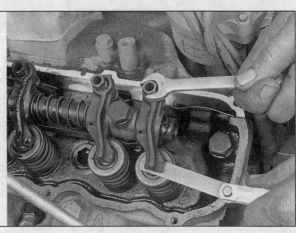

4.12 Turn the adjuster clockwise to decrease clearance between the rocker arm and valve stem

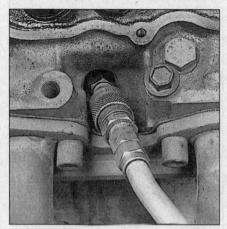

5.6 An adapter is used to attach the compressed air source to the cylinder

5.8 Use needle-nose pliers (shown) or a small magnet to remove the valve spring keepers. Be careful you don't drop them down into the engine!

5.15 Note that the closely wound coils (arrow) go on the bottom (against the cylinder head)

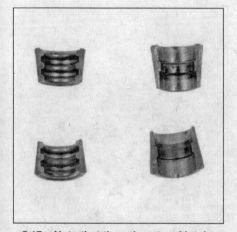

5.17a Note that the exhaust and intake valves have different style keepers

5.17b The grooves in the intake valves (A) and exhaust valves (B) require that the correct keepers be used

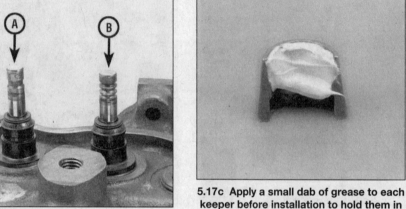

5.17c Apply a small dab of grease to each keeper before installation to hold them in place on the valve stem until the spring is released

head with the defective part and pull out the pushrod. If all of the valve stem seals are being replaced, all of the pushrods should be removed (refer to Section 4).

6   Apply compressed air to the cylinder **(see illustration)**. The valves should be held in place by the air pressure. If the valve faces or seats are in poor condition, leaks may prevent the air pressure from retaining the valves - refer to the alternative procedure below.

7   If you do not have access to compressed air, an alternative method can be used. Position the piston at a point approximately 45-degrees before TDC on the compression stroke, then feed a long piece of nylon rope through the spark plug hole until it fills the combustion chamber. Be sure to leave the end of the rope hanging out of the engine so it can be removed easily. Use a large breaker bar and socket to rotate the crankshaft in the normal direction of rotation until slight resistance is felt.

8   Stuff shop rags into the cylinder head holes above and below the valves to prevent parts and tools from falling into the engine, then use a valve spring compressor to compress the spring/damper assembly. Remove

the keepers with small needle-nose pliers or a magnet **(see illustration)**. **Note:** *A couple of different types of tools are available for compressing the valve springs with the head in place. One type grips the lower spring coils and presses on the retainer as the knob is turned, as shown here.*

9   Remove the spring retainer valve spring assembly, then remove the umbrella-type guide seal. **Note:** *If air pressure fails to hold the valve in the closed position during this operation, the valve face or seat is probably damaged. If so, the cylinder head will have to be removed for additional repair operations.*

10   Wrap a rubber band or tape around the top of the valve stem so the valve will not fall into the cylinder, then release the air pressure. **Note:** *If a rope was used instead of air pressure, turn the crankshaft slightly in the direction opposite normal rotation.*

11   Inspect the valve stem for damage. Rotate the valve in the guide and check the end for eccentric movement, which would indicate that the valve is bent.

12   Move the valve up-and-down in the

guide and make sure it doesn't bind. If the valve stem binds, either the valve is bent or the guide is damaged. In either case, the head will have to be removed for repair.

13   Reapply air pressure to the cylinder to retain the valve in the closed position, then remove the tape or rubber band from the valve stem. If a rope was used instead of air pressure, rotate the crankshaft in the normal direction of rotation until slight resistance is felt.

14   Lubricate the valve stem with engine oil and install a new umbrella-type guide seal.

15   Install the spring/damper assembly and shield in position over the valve. Note that the close wound coils must be against the head **(see illustration)**.

16   Install the valve spring retainer. Compress the valve spring assembly.

17   Position the keepers in the upper groove. Note the difference between intake and exhaust **(see illustrations)**. Apply a small dab of grease to the inside of each keeper to hold it in place if necessary **(see illustration)**. Remove the pressure from the spring tool and make sure the keepers are seated.

6.5 Disconnect the bypass hose (arrow) at the rear of the water pump housing

6.12 A cardboard box can be used to keep the manifold bolts in the proper sequence

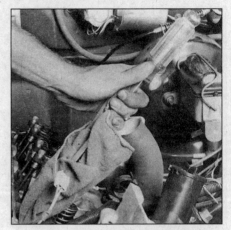

6.13 Use a screwdriver wrapped with a rag to carefully pry on the manifold and break the gasket seal

6.14a Scrape all old gasket material off the mating surfaces

6.14b A rag will help keep pieces of gasket material out of the engine during the cleaning process

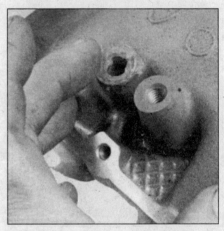

6.16a Apply sealing compound to the bolt bosses prior to installation

18   Disconnect the air hose and remove the adapter from the spark plug hole. If a rope was used in place of air pressure, pull it out of the cylinder.

19   Refer to Section 4 and install the rocker arm assembly and pushrod(s).

20   Install the spark plug(s) and hook up the wire(s).

21   Refer to Section 3 and install the rocker arm cover(s).

22   Start and run the engine, then check for oil leaks and unusual sounds coming from the rocker arm cover area.

6   Intake manifold - removal and installation

Refer to illustrations 6.5, 6.12, 6.13, 6.14a, 6.14b, 6.16a and 6.16b

1   Disconnect the negative cable from the battery.

2   Remove the air cleaner and air ducts.

3   Remove the engine cover under the dash (Chapter 11).

4   Disconnect the throttle cable (Chapter 4).

5   Drain the coolant (Chapter 1). Disconnect and remove the hose from water outlet to radiator (bottom hose). Disconnect the bypass hose from the intake manifold to the water pump housing rear cover **(see illustration)**.

6   Label the spark plug wires (if not already marked by the factory) and remove the distributor cap and plug wires as an assembly. Disconnect the distributor wiring harness.

7   Remove the distributor (Chapter 5).

8   Remove the rocker arm covers (Section 3).

9   Disconnect the fuel line from the fuel filter (Chapter 4).

10   Disconnect and label all wires and hoses that will interfere with manifold removal.

11   Remove the carburetor and EGR spacer (Chapters 4 and 6).

12   Remove the intake manifold bolts and nuts by backing off on the bolts/nuts two turns apiece until loose. **Note:** *During removal of the bolts note the length and location of the bolts so they can be installed in their original location* **(see illustration)**.

13   Tap the manifold lightly with a plastic hammer or wrap a rag around the tip of a screwdriver and very gently pry up on the manifold to break the seal **(see illustration)**.

14   Remove all old gasket material and sealing compound from the manifold and mating surfaces on the heads and engine block. Place a rag in the cavity to prevent debris from falling into the engine **(see illustrations)**.

Installation

15   Apply sealing compound to the joining surfaces and place the intake manifold gasket into position, making sure the tab on the right cylinder bank fits into the cutout of the manifold gasket.

16   Apply sealing compound to the attaching bolt bosses on the intake manifold **(see illustration)** and position the intake manifold. The length of the bolts, using torque sequence numbers is: 1 and 2 - 3-1/4 inches; 3 and 4 - nut; 5, 6 and 7 - 4-1/2 inches; 8 - 4-3/4 inches. Follow the tightening sequence **(see illustration)** and tighten the bolts to the specified torque.

17   The remainder of the installation is the reverse of the removal procedures.

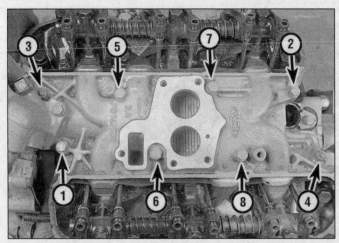

6.16b Intake manifold bolt tightening sequence

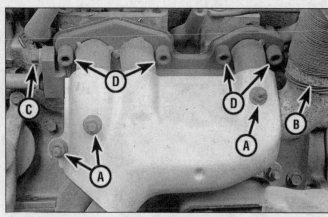

7.2 Right exhaust manifold

A   Heat shield mounting nuts
B   Hot air tube
C   Thermactor tube bolt
D   Manifold mounting bolts

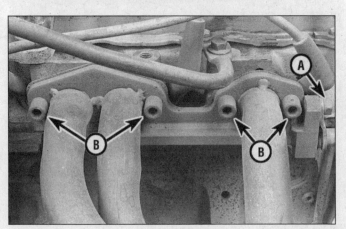

7.3 Left exhaust manifold

A   Thermactor tube bolt
B   Manifold mounting bolts

7.6 An 8 mm Allen wrench is required for removal and installation of the exhaust manifold bolts

18   Refill the cooling system and reconnect the battery.
19   Run the engine and check for leaks.
20   Reinstall the engine cover.

7   Exhaust manifolds - removal and installation

Refer to illustrations 7.2, 7.3 and 7.6

1   Remove the air cleaner and air duct (Chapter 4).
2   Remove the attaching nuts from the exhaust manifold shroud on the right side (see illustration).
3   Disconnect the attaching nuts from the exhaust pipe at the manifold, then remove the thermactor by unbolting the attaching bolt and pulling the tube out of the way (see illustration).
4   Unplug the electrical connector from the choke or disconnect choke (Chapter 4).
5   Disconnect the EGO sensor wire on the left side exhaust manifold (Chapter 4).
6   Remove the manifold attaching bolts (see illustration).
7   Remove the manifold from the cylinder head.

8   Position the manifold in place with new gaskets and install and tighten the attaching bolts to the specified torque. Tighten the bolts two turns at a time, drawing the manifold to the head evenly.
9   Install a new exhaust pipe gasket, then install the exhaust pipe to the exhaust manifold.
10   Position the exhaust manifold shroud on the manifold (right side) and install and tighten the nuts securely.
11   Install the thermactor tube.
12   Install the air cleaner and choke electrical connector.
13   Connect the EGO sensor wire.

8   Cylinder heads - removal and installation

Refer to illustrations 8.5, 8.7, 8.8, 8.9a, 8.9b, 8.10 and 8.12

Removal

1   Disconnect the negative cable at the battery.
2   Drain the coolant (Chapter 1).

8.5 Keep the pushrods in their original order

3   Refer to Section 6 and remove the intake manifold.
4   Remove the rocker arm covers, rocker arm shafts and oil baffles (Sections 3 and 4).
5   Remove the pushrods and keep them in sequence for proper assembly (see illustration).

8.7 A rag-wrapped screwdriver can be used to break the gasket seal - be careful that the head does not fall as this is done

8.8 A back-up nut and two wrenches can be used to remove the head studs

8.9a Scrape all old gasket material from the head mounting surfaces

8.9b Be sure to remove all foreign matter (including fluid) from the head bolt holes (arrows)

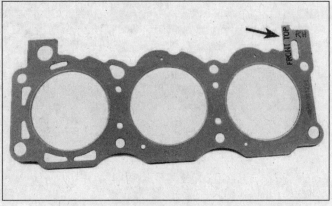

8.10 Be sure the correct head gasket is used for the side you are working on and that the marks face up (arrow)

6   Remove the exhaust manifold(s) (Section 7).

7   Remove the cylinder head attaching bolts and lift off the cylinder head. Discard the head gasket. It may be necessary to wrap a rag around a screwdriver and carefully pry up on the cylinder head to break the seal **(see illustration)**.

8   **Note:** *It may be necessary to remove the cylinder head studs in order to lift up and back on the cylinder head to clear the front cover retainer. To remove the stud, run a nut down to the end of the threads and back it up with another nut, then use the first nut to unscrew the stud* **(see illustration)**.

Installation

9   Clean the cylinder heads, intake manifold, rocker arm cover and cylinder block gasket surfaces **(see illustrations)**.

.10   Place new cylinder head gaskets in position on cylinder block. Gaskets are marked with the words *front* and *top* for correct positioning. Left and right head gaskets are not interchangeable **(see illustration)**.

11   A new head gasket set comes with alignment dowels to make the installation of the head easier. Install alignment dowels in cylinder block and install cylinder head

assemblies on the engine block.

12   Remove the alignment dowels and install cylinder head attaching bolts and studs and tighten in sequence **(see illustration)** to specified torque.

13   Install the intake manifold.

14   Install the exhaust manifolds.

15   Apply engine oil to both ends of the pushrods and install them.

16   Install the oil baffles and rocker arm assemblies.

17   The remainder of the installation is the reverse of the removal procedure.

18   Adjust the valves (Section 4).

19   Refill the cooling system (Chapter 1).

20   Change the oil and filter (Chapter 1).

21   Run the engine and check for leaks.

9   Front crankshaft oil seal - replacement

Refer to illustrations 9.2 and 9.4

1   To replace the front oil seal, drain the coolant and remove the radiator (Chapter 3). Remove the crankshaft hub, pulley and drivebelts. **Note:** *It is not necessary to remove the front cover.*

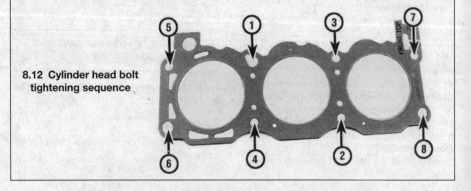

8.12 Cylinder head bolt tightening sequence

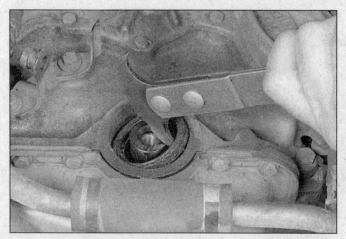

9.2 Removing the front oil seal with a seal puller

9.4 If you lack the special seal installer, you can use a socket and hammer to tap it into place

2   Remove the front cover seal by using a special seal remover, or its equivalent, and an impact slide hammer, or its equivalent **(see illustration)**. Both special tools are available at most auto parts stores.

3   To replace, coat the new seal with engine oil.

4   Slide the new seal onto the crankshaft and, using a front seal installer, available at most auto parts stores, or a socket the same diameter as the seal, drive the seal in until the tool butts against the front cover **(see illustration)**.

5   Replace crankshaft hub and bolt and tighten to the specified torque.

6   Install the drivebelt pulley.

7   Install the radiator and drivebelts.

8   Fill the cooling system (Chapter 1).

9   Operate the engine at a fast idle and check for coolant and oil leaks.

10   Timing gear cover - removal and installation

Refer to illustration 10.8

Note: *The manufacturer recommends that the oil pan be removed before beginning this procedure. However, we found that the cover could be removed for gasket replacement or timing gear inspection without removing the oil pan, as described in the steps which follow. A new pan gasket will be required, however. If removing the cover to replace the timing gears it will be necessary to remove the oil pan.*

Removal

1   Drain the coolant and remove the radiator (Chapter 3).

2   Unbolt the air conditioner compressor (Chapter 3) and power steering pump and bracket, if so equipped (refer to Chapter 10). Tie the compressor out of the way. Do not loosen or disconnect the refrigerant hoses!

3   Loosen the alternator and thermactor pump adjusting bolts and remove the drivebelt (Chapter 1). Remove the alternator (Chapter 5).

4   Remove the fan (Chapter 3).

5   Remove the coolant hoses connected to the water pump, then remove the water pump (Chapter 3).

6   Remove the crankshaft hub from the crankshaft (Section 9).

7   Raise the vehicle and support it securely on jackstands.

8   From under the vehicle remove the five front oil pan bolts **(see illustration)**.

9   Remove the seven front cover bolts and carefully remove the cover by prying the top of the cover loose with a screwdriver. Carefully pry the front cover away from the oil pan. If necessary, remove the guide sleeves from cylinder block and then the spacer plate.

Installation

10   Clean the mating surfaces of all gasket material and apply sealing compound to the gasket surfaces on the block and back of the cover. Be sure to clean the front section of the oil pan.

11   Cut a section of the new pan gasket to match the removed section, coat with sealant and install.

12   If removed, install the guide sleeves with new seal rings lubricated with engine oil, the chamfered end towards the front cover.

13   With sealing compound applied, place the new gasket in position on the front cover.

14   Place the cover on the engine and start all the retaining bolts. The cover may be centered by inserting a special front cover aligner tool, available at most auto parts stores, or equivalent, in the oil seal. If the tool is not available, use the crankshaft hub to align the cover.

15   Tighten the attaching bolts to the specified torque.

16   Remove the aligner, install the hub and tighten the attaching bolt to the specified torque.

17   If removed, install the oil pan as described in Section 16.

18   The remainder of the installation is the reverse of the removal procedure.

11   Top Dead Center (TDC) for number one piston - locating

Refer to illustrations 11.4 and 11.6

1   Top Dead Center (TDC) is the highest point in the cylinder that each piston reaches as it travels up-and-down when the crankshaft turns. Each piston reaches TDC on the compression stroke and again on the exhaust stroke, but TDC generally refers to piston position on the compression stroke. The timing marks on the vibration damper

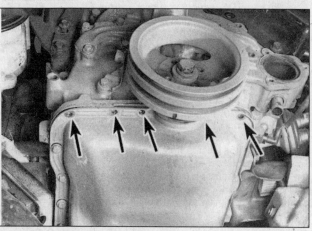

10.8 When removing the timing gear cover without taking off the oil pan, you must remove the five front oil pan mounting bolts (arrows)

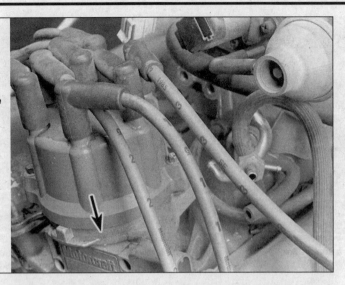

11.4 If your engine still has original-type spark plug wires, they are helpful for locating the number one terminal on the distributor - arrow indicates mark for number one TDC

installed on the front of the crankshaft are referenced to the number one piston at TDC on the compression stroke.

2 Positioning the piston(s) at TDC is an essential part of many procedures such as rocker arm removal, valve adjustment, timing gear replacement and distributor removal.

3 In order to bring any piston to TDC, the crankshaft must be turned using one of the methods outlined below. When looking at the front of the engine, normal crankshaft rotation is clockwise. **Warning:** *Before beginning this procedure, be sure to place the transmission in Neutral and disconnect the coil wire at the distributor cap and ground it to disable the ignition system.*

a) *The preferred method is to turn the crankshaft with a large socket and breaker bar attached to the crankshaft hub bolt that is threaded into the front of the crankshaft.*

b) *A remote starter switch, which may save some time, can also be used. Attach the switch leads to the S (switch) and battery terminals on the starter solenoid. Once the piston is close to TDC, use a socket and breaker bar as described in the previous paragraph.*

c) *If an assistant is available to turn the ignition switch to the Start position in short bursts, you can get the piston close to TDC without a remote starter switch. Use a socket and breaker bar as described in Paragraph a) to complete the procedure.*

4 Make a mark on the distributor housing directly below the number one spark plug wire terminal on the distributor. If the numbers are missing from the spark plug wires **(see illustration)**, trace the wire from the Number One spark plug wire to the distributor cap.

5 Remove the distributor cap as described in Chapter 1.

6 Turn the crankshaft (see Step 3 above) until the TDC mark on the crankshaft hub is aligned with the timing pointer **(see illustration)**.

7 The rotor should now be pointing directly at the mark on the distributor housing. If it is 180° off, the piston is at TDC on the exhaust stroke.

8 To get the piston to TDC on the compression stroke, turn the crankshaft one complete turn (360°) clockwise. The rotor should now be pointing at the mark. When the rotor is pointing at the number one spark plug wire terminal in the distributor cap (which is indicated by the mark on the housing) and the timing marks are aligned, the number one piston is at TDC on the compression stroke.

9 After the number one piston has been positioned at TDC on the compression stroke, TDC for any of the remaining cylinders can be located by turning the crankshaft 120° at a time and following the firing order (refer to the Specifications).

12 Timing gears - removal and installation

Refer to illustrations 12.6a and 12.6b
Note: *Any time a component is to be removed which could effect the timing of the engine, begin the procedure by putting the number one piston at TDC (top dead center) on the*

compression stroke. Check the alignment of all the timing marks. Remove the distributor cap and see that the rotor is pointing towards the number one spark plug wire. Do not rotate engine while the timing gears are off.

1 Remove the oil pan (Section 16).
2 Remove the timing gear cover (Section 10).
3 Remove the radiator (Chapter 3).
4 Remove the bolt retaining the camshaft gear to the camshaft and remove the camshaft gear.
5 Using a special gear puller and shaft protector, available at most auto parts stores, or a two jaw gear puller, remove the crankshaft gear.
6 To install the crankshaft gear, align the keyway in the gear with the Woodruff key, then slide the gear onto the shaft. **Note:** *To install the crankshaft gear we fabricated an installing tool by using a piece of 1-7/8 inch diameter pipe, 3-3/4 inches long and inserted a bolt and washer through the pipe* **(see illustrations)**.
7 With the Woodruff key in the camshaft aligned with the gear keyway and the timing marks (dot to dot) on both gears aligned, install the cam gear.
8 Install the cam gear retaining bolt and washer and tighten to the specified torque.
9 Check camshaft end play (Section 14). If not within specifications replace the thrust plate.
10 Install the timing gear cover, oil pan and radiator.
11 Fill the cooling system with coolant and the crankcase with oil.
12 Operate the engine at a fast idle and check all hose connections and gaskets for leaks.

13 Valve lifters - removal, inspection and installation

Refer to illustrations 13.2a and 13.2b

Removal and installation

1 Remove the cylinder head and related parts (Section 8).

11.6 The timing pointer (arrow) should be aligned with the Top Dead Center (TDC) mark on the crankshaft hub

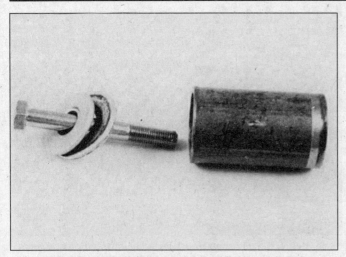

12.6a We fabricated this tool to install the crankshaft timing gear

12.6b As you tighten the bolt the gear slides into place

2   Remove the lifters with a magnet and place them in a rack in sequence so they can be replaced in their original order (see illustrations).

3   Clean the lifters and apply engine oil to both the lifters and bores before installing. Be sure to install the lifters in their original bores.

4   Install the cylinder head and related parts.

Lifter inspection

5   The faces of the lifters which bear on the camshaft should show no signs of pitting, scoring or other forms of wear. They should also not be a loose fit in the lifter bore. Wear is only normally encountered at very high mileages or in cases of neglected engine lubrication.

6   If the lifters show signs of wear they must be replaced as a complete set and the camshaft must be replaced as well. See Chapter 2 Part D for further information.

14 Camshaft end play - check

Caution: *Prying against the camshaft gear, with the valve train load on the camshaft, can break or damage the gear and camshaft.*

1   Remove the rocker arm cover and timing gear cover (Sections 3 and 10).

2   Back off on rocker arm adjusting bolts or loosen the rocker arm shaft stands sufficiently to remove the valve spring loading on the camshaft (Section 4).

3   Push the camshaft toward the rear of the engine.

4   Install a dial indicator so that the indicator is on the camshaft gear attaching bolt washer. Zero the dial indicator.

5   Position a large screwdriver between the camshaft gear and the block.

6   Pull the camshaft forward and release it.

7   Compare the dial indicator reading with specifications. If end play is excessive, replace the thrust plate (see illustration 15.14).

8   Remove the dial indicator.

9   After checking the camshaft end play, adjust the valve clearance (Section 4).

10   Install the rocker arm covers and timing gear cover.

15 Camshaft - removal and installation

Refer to illustrations 15.9, 15.14 and 15.15

Removal

1   Disconnect the negative cable at the battery.

2   Remove the rocker arm covers (Section 3) and the rocker arm assembly (Section 4).

3   Remove the timing gear cover (Section 10). Stuff a rag in the oil pan opening.

4   Label and disconnect the spark plug wires from the spark plugs.

5   Remove the distributor cap and spark plug wires as an assembly.

13.2a Work the lifters out of the block with a magnet

13.2b Keep the lifters in order and replace them in their original locations

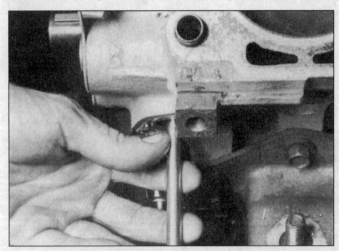

15.9 Remove the fuel pump rod from the block

15.14 Unbolt the camshaft thrust plate (the spacer ring and thrust plate are available in two thicknesses for end play adjustment)

6   Disconnect the distributor wiring harness connector and remove the distributor (Chapter 5).

7   Remove the alternator (Chapter 5).

8   Remove the thermactor pump (Chapter 6).

9   Disconnect the fuel line at the carburetor and remove the carburetor (Chapter 4). Remove the two bolts retaining the fuel pump and set the pump out of the way, then remove the fuel pump rod **(see illustration)**.

10   Remove the intake manifold (Section 6).

11   Lift out push rods and lifters and place them in a marked rack so they can be installed in the same location (Section 13).

12   Remove the camshaft gear attaching bolt and washer.

13   Slide the gear off the camshaft and remove the Woodruff key from the camshaft.

14   Remove the camshaft thrust plate **(see illustration)**.

15   Carefully remove the camshaft from the block, avoiding damage to the camshaft bearings **(see illustration)**.

Installation

16   Oil the camshaft journals and cam lobes with engine oil. Install the spacer ring with the chamfered side toward the camshaft and insert the camshaft Woodruff key.

17   Insert the camshaft into the block, care-fully avoiding damage to the bearing surfaces.

18   Install the thrust plate so that it covers the main oil gallery and tighten the attaching bolts to the specified torque.

19   Install the camshaft gear and tighten the attaching bolt to the specified torque.

20   Check the camshaft end play (Section 14). The spacer ring and thrust plate are available in two thicknesses to permit adjusting the end play.

21   The remainder of the installation is the reverse of the removal procedures.

16 Oil pan - removal and installation

1   Disconnect negative cable at the battery.

2   Remove the air cleaner assembly (Chapter 4).

3   Remove the fan shroud and position it over the fan.

4   Remove the engine cover (Chapter 11).

5   Raise the vehicle and support it securely on jackstands.

6   Drain the engine oil (Chapter 1).

7   Remove the oil filter (Chapter 1)

8   Disconnect the exhaust pipes from the exhaust manifolds.

9   Disconnect the oil cooler bracket and

lower it out of way (if so equipped).

10   Remove the starter cable from the starter.

11   Remove the two attaching bolts and remove the starter motor (Chapter 5)

12   Disconnect and position out of way the transmission oil cooler lines (if so equipped).

13   Disconnect the engine mounts by taking off each of the top nuts (Section 19).

14   Position a jack under the engine oil pan. Place a block of wood between the jack and oil pan so as not to damage the oil pan. Raise engine and install wooden blocks between the front insulator mounts and the cross-member. **Warning:** *Do not place any part of your body under the engine.*

15   Lower the engine onto the blocks and remove the jack.

16   Remove the oil pan bolts and lower the pan out the rear.

17   Clean the gasket surfaces of the engine and oil pan.

18   Apply sealer to the gasket mating surfaces and install the oil pan gaskets. Be sure the wedge seals are in place.

19   Install the oil pan, tightening the bolts to the specified torque.

20   The remainder of installation is the reverse of removal.

17 Oil pump - removal and installation

Refer to illustration 17.2

1   Remove the oil pan as described in Section 16.

2   Remove the oil pickup-to-main bearing cap mounting nuts **(see illustration)**.

3   Remove the oil pump mounting bolts. This requires a TORX bit, available in many auto parts and tool stores.

4   Lift out the oil pump assembly and withdraw the oil pump driveshaft.

5   Prior to installation prime the oil pump by pouring motor oil into the oil pickup and turning the pump shaft by hand.

6   If you are replacing the oil pickup, use a

15.15 Carefully support the camshaft as you remove it from the block

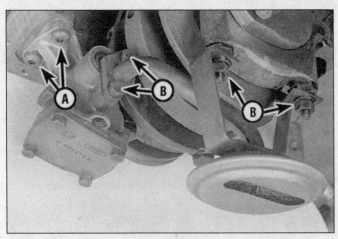

17.2 The oil pump Torx-head attaching bolts (A) and oil pickup mounting nuts (B)

18.6 Pry the seal out carefully with a screwdriver

new gasket and tighten the bolts to the specified torque.
7   Insert the oil pump driveshaft into the block with the pointed end facing inward (the pointed end is closest to the pressed on flange).
8   Install the oil pump with a new gasket and tighten the bolts to the specified torque.
9   Install the oil pickup nuts and tighten them to the specified torque.
10   Install the oil pan (Section 16).
11   Change the oil filter and add oil.
12   Run the engine and check for leaks.

18 Rear crankshaft oil seal - replacement

Refer to illustrations 18.6 and 18.9
1   Remove the transmission by following the appropriate procedures in Chapter 7.
2   Remove the clutch and pressure plate, if so equipped (Chapter 8).
3   Remove the flywheel/driveplate.
4   Remove the bolts and spring washers securing the flywheel housing and rear plate where fitted.
5   Clean the area surrounding the seal thoroughly.
6   Carefully pry the seal out with a screwdriver by working around it a little at a time **(see illustration)**. Caution: *During this operation avoid scratching or otherwise damaging the crankshaft and oil seal bore surface.*
7   Coat the oil seal to cylinder block surface of the oil seal with sealer.

8   Coat the seal contact surface of the oil seal and crankshaft with engine oil.
9   Start the seal in the recess and install it with a special tool, available at most auto parts stores, or an equivalent such as a hammer and socket. Drive the seal into position until it is firmly seated **(see illustration)**.
10   The remainder of the installation is the reverse of the removal procedures.

19 Engine mounts - removal and installation

1   Disconnect the negative cable at the battery.
2   Remove the fan shroud attaching bolts.
3   Support the engine using a wood block and a jack placed under the oil pan.
4   Remove the engine cover (Chapter 11).
5   Raise the vehicle and support it securely on jackstands.
6   Remove the nuts and washers attaching the insulators to the crossmember brackets and lift the engine enough to disengage the insulator stud from the crossmember. **Warning:** *Do not place any part of your body under the engine.*
7   Remove the bolt attaching the fuel pump shield to the left engine bracket, if equipped.
8   Remove the engine insulator assembly to cylinder block attaching bolts and lockwashers and remove the engine mount assembly.
9   To install, position the engine mount

18.9 The installed seal should look like this

assembly to the cylinder block and install the bracket bolts.
10   Install the bolt attaching the fuel pump shield to the left bracket, if equipped.
11   Lower the engine until the insulator stud engages and bottoms in the slot on the crossmember pedestal. Tighten the nuts securely.
12   Remove the block of wood and jack from the engine oil pan.
13   Install the fan shroud attaching bolts.
14   Connect the battery and install the engine cover.

Notes

Chapter 2 Part C
3.0L V6 engine

Contents

Specifications

General

Cylinder numbers (front-to-rear)	
Left (driver's) side	4-5-6
Right side	1-2-3
Firing order	1-4-2-5-3-6

Valves

Valve stem-to-rocker arm clearance	0.088 to 0.189 in (2.23 to 4.77 mm)

Torque specifications

	Ft-lbs
Camshaft sprocket bolt	41 to 51
Camshaft thrust plate bolts	72 to 96
Timing chain cover bolts	15 to 22
Cylinder head bolts (in sequence)	
Through 1994	
Step 1	37
Step 2	68
1995 through 1997	
Step 1	33 to 41
Step 2	63 to 73
Vibration damper bolt	
Through 1994	107
1995 to 1997	92 to 122
Intake manifold bolts (in sequence)	
Step 1	132 in-lbs
Step 2	19
Exhaust manifold bolts	
Step 1	16
Step 2	22
Oil pan bolts	84 to 108 in-lbs
Rocker arm cover bolts	84 to 108 in-lbs
Rocker arm fulcrum bolts	
Through 1994	
Step 1	96 in-lbs
Step 2	24
1995 to 1997	19 to 28
Oil pump-to-block bolts	30 to 40
Oil pump pickup tube-to-oil pump	15 to 22
Engine mount-to-block bolts	33 to 44
Engine mount-to-crossmember nuts	52 to 54

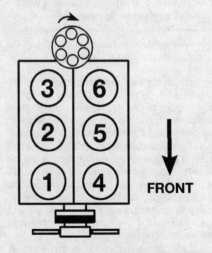

36004-1-SPECS2b HAYNES

Cylinder location and distributor rotation

4.3a Loosen the bolt (arrow) and pivot the rocker arm to the side to remove the pushrod

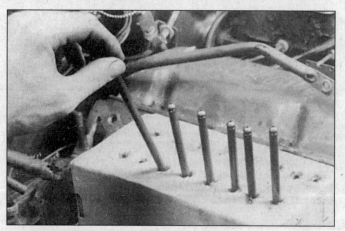

4.3b A perforated cardboard box can be used to store the pushrods to ensure installation in their original locations

1   General information

This Part of Chapter 2 is devoted to in-vehicle repair procedures. Part E covers the removal and installation procedures for the 3.0 liter V6 engines. All information concerning engine block and cylinder head servicing can also be found in Part E of this Chapter.

The repair procedures included in this Part are based on the assumption that the engine is still installed in the vehicle. Therefore, if this information is being used during a complete engine overhaul - with the engine already out of the vehicle and on a stand - many of the steps included here will not apply.

The specifications included in this Part of Chapter 2 apply only to the engine and procedures found here. For specifications regarding engines other than the 3.0 liter V6, see Part A or B, whichever applies. Part E of Chapter 2 contains the specifications necessary for engine block and cylinder head rebuilding procedures.

2   Repair operations possible with the engine in the vehicle

Many major repair operations can be accomplished without removing the engine from the vehicle.

Clean the engine compartment and the exterior of the engine with some type of pressure washer before any work is done. A clean engine will make the job easier and will help keep dirt out of the internal areas of the engine.

If oil or coolant leaks develop, indicating a need for gasket or seal replacement, the repairs can generally be made with the engine in the vehicle. The oil pan gasket, the cylinder head gaskets, intake and exhaust manifold gaskets, timing cover gaskets and the crankshaft oil seals are accessible with the engine in place.

Exterior engine components, such as the water pump, the starter motor, the alternator, the distributor and the EFI components, as well as the intake and exhaust manifolds, can be removed for repair with the engine in place.

Since the cylinder heads can be removed without pulling the engine, valve component servicing can also be accomplished with the engine in the vehicle.

Replacement of, repairs to or inspection of the timing chain and sprockets and the oil pump are all possible with the engine in place.

In extreme cases caused by a lack of necessary equipment, repair or replacement of piston rings, pistons, connecting rods and rod bearings is possible with the engine in the vehicle. However, this practice is not recommended because of the cleaning and preparation work that must be done to the components involved.

3   Rocker arm covers - removal and installation

1   Disconnect the negative cable at the battery.
2   Remove the engine cover under the dash (Chapter 11).
3   Disconnect the spark plug wires on the side(s) you are taking off, leaving them attached to the wire separators.
4   Remove the spark plug wire separators from the rocker arm cover studs.
5   If the right rocker arm cover is being removed, remove the PCV valve (Chapter 6), EGR tube and heater hoses from the valve cover.
6   If the left cover is being removed, take off the oil filler tube and PCV tube.
7   Remove the rocker cover attaching screws and remove the cover. If you must pry the cover off, be careful not to bend the sealing flange.
8   Clean all gasket sealing surfaces thoroughly and lightly oil all fastener threads.
9   Apply a bead of RTV sealant at the cylinder head to intake manifold rail step, two places per rail. Put a new gasket in place.
10   Place the cover on the head and install the attaching bolts and studs, noting the location of the spark plug wire separators.

Tighten all fasteners to the specified torque.
11   The remainder of the installation is the reverse of removal.
12   Reconnect the battery, start the engine and check for leaks.
13   Reinstall the engine cover.

4   Rocker arms and pushrods - removal, inspection, installation and adjustment

Refer to illustrations 4.3a and 4.3b

Removal

1   Remove the rocker arm covers as described in Section 3.
2   Remove the rocker arm fulcrum bolts.
3   Remove the rocker arms. If only the pushrods are being removed, loosen the fulcrum bolts and rotate the rocker arms out of the way of the pushrods **(see illustration)**. Keep the pushrods organized to ensure that they are reinstalled in their original locations **(see illustration)**.

Inspection

4   Check the rocker arms for excessive wear, cracks and other damage, especially where the pushrods and valve stems contact the rocker arm faces.
5   Make sure the hole at the pushrod end of the rocker arm is open.
6   Check the rocker arm fulcrum contact area for wear and galling. If the rocker arms are worn or damaged, replace them with new ones and use new fulcrum seats as well.
7   Inspect the pushrods for cracks and excessive wear at the ends. Roll each pushrod across a flat surface to see if they are bent.

Installation

8   Apply engine oil or assembly lube to the top of the valve stem and the pushrod guide in the cylinder head.
9   Apply engine oil to the rocker arm fulcrum seat and the fulcrum seat socket in the rocker arm.

5.8 Apply air pressure to the cylinder, compress the valve spring and remove the keepers

5.14 Carefully place the seal in position, then tap it into place with a deep socket and hammer

10   Install the pushrods.

11   Install the rocker arms, fulcrums, and fulcrum bolts.

12   Turn the crankshaft until the lifter is all the way down.

13   Tighten the fulcrum bolts to the specified torque in two stages.

14   Replace the rocker arm covers and gaskets as described in Section 3.

15   Start the engine and check for roughness and/or noise.

16   Reinstall the engine cover.

Adjustment

Note: *Adjustment is normally only needed when valves and/or seats have been ground a considerable amount.*

17   Set the engine at Top Dead Center (TDC) on number 1 cylinder. This is position 1.

18   Using a lifter bleed-down tool (or equivalent), available at most auto parts stores, press on the rocker arm until the lifter leaks down.

19   Check the following valves:

Intake - 1, 3 and 6
Exhaust - 1, 2 and 4

20   Check the clearance between the valve stem and rocker arm with a feeler gauge. Compare this to the Specifications and write it down. **Note:** *The arrangement of intake and exhaust valves is as follows:*

Left side - E-I-E-I-E-I
Right side - I-E-I-E-I-E

21   Rotate the crankshaft 360 degrees to position 2 and check the following valves:

Intake - 2, 4 and 5
Exhaust - 3, 5 and 6

5   Valve springs, retainers and seals - replacement

Refer to illustrations 5.8 and 5.14

Note: *Broken valve springs and defective valve stem seals can be replaced without removing the cylinder head. Two special tools and a compressed air source are normally required to perform this operation, so read through this Section carefully and rent or buy*

the tools before beginning the job. If compressed air is not available, a length of nylon rope can be used to keep the valves from falling into the cylinder during this procedure.

1   Refer to Section 3 and remove the rocker arm cover from the affected cylinder head. If all of the valve stem seals are being replaced, remove both rocker arm covers.

2   Remove the spark plug from the cylinder which has the defective component. If all of the valve stem seals are being replaced, all of the spark plugs should be removed.

3   Turn the crankshaft until the piston in the affected cylinder is at top dead center on the compression stroke (refer to Section 11 for instructions). If you are replacing all of the valve stem seals, begin with cylinder number one and work on the valves for one cylinder at a time. Move from cylinder-to-cylinder following the firing order sequence (1-4-2-5-3-6).

4   Thread an adapter into the spark plug hole and connect an air hose from a compressed air source to it. Most auto parts stores can supply the air hose adapter. **Note:** *Many cylinder compression gauges utilize a screw-in fitting that may work with your air hose quick-disconnect fitting.*

5   Remove the bolt, pivot ball and rocker arm for the valve with the defective part and pull out the pushrod. If all of the valve stem seals are being replaced, all of the rocker arms and pushrods should be removed (refer to Section 4).

6   Apply compressed air to the cylinder. The valves should be held in place by the air pressure. If the valve faces or seats are in poor condition, leaks may prevent the air pressure from retaining the valves - refer to the alternative procedure below.

7   If you do not have access to compressed air, an alternative method can be used. Position the piston at a point approximately 45-degrees before TDC on the compression stroke, then feed a long piece of nylon rope through the spark plug hole until it fills the combustion chamber. Be sure to leave the end of the rope hanging out of the engine so it can be removed easily. Use a large breaker bar and socket to rotate the crankshaft in the normal direction of rotation

until slight resistance is felt.

8   Stuff shop rags into the cylinder head holes above and below the valves to prevent parts and tools from falling into the engine, then use a valve spring compressor to compress the spring/damper assembly **(see illustration).** Remove the keepers with small needle-nose pliers or a magnet. **Note:** *A couple of different types of tools are available for compressing the valve springs with the head in place. One type grips the lower spring coils and presses on the retainer as the knob is turned, while the other type utilizes the rocker arm bolt for leverage. Both types work very well, although the lever type is usually less expensive.*

9   Remove the spring retainer and valve spring assembly, then remove the umbrella-type guide seal. **Note:** *If air pressure fails to hold the valve in the closed position during this operation, the valve face or seat is probably damaged. If so, the cylinder head will have to be removed for additional repair operations.*

10   Wrap a rubber band or tape around the top of the valve stem so the valve will not fall into the combustion chamber, then release the air pressure. **Note:** *If a rope was used instead of air pressure, turn the crankshaft slightly in the direction opposite normal rotation.*

11   Inspect the valve stem for damage. Rotate the valve in the guide and check the end for eccentric movement, which would indicate that the valve is bent.

12   Move the valve up-and-down in the guide and make sure it doesn't bind. If the valve stem binds, either the valve is bent or the guide is damaged. In either case, the head will have to be removed for repair.

13   Reapply air pressure to the cylinder to retain the valve in the closed position, then remove the tape or rubber band from the valve stem. If a rope was used instead of air pressure, rotate the crankshaft in the normal direction of rotation until slight resistance is felt.

14   Lubricate the valve stem with engine oil and install a new umbrella-type guide seal **(see illustration).**

15   Install the spring/damper assembly in position over the valve.

16   Install the valve spring retainer. Com-

6.8 Use a scraper to remove all traces of the intake manifold gaskets

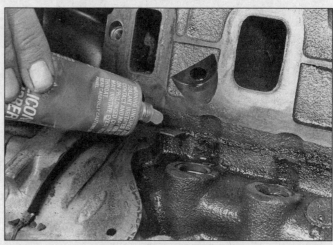

6.9 Apply RTV sealant to the corners where the ends of the cylinder heads meet the block ridges

press the valve spring assembly.

17 Position the keepers in the upper groove. Apply a small dab of grease to the inside of each keeper to hold it in place if necessary. Remove the pressure from the spring tool and make sure the keepers are seated.

18 Disconnect the air hose and remove the adapter from the spark plug hole. If a rope was used in place of air pressure, pull it out of the cylinder.

19 Refer to Section 4 and install the rocker arm(s) and pushrod(s).

20 Install the spark plug(s) and hook up the wire(s).

21 Refer to Section 3 and install the rocker arm cover(s).

22 Start and run the engine, then check for oil leaks and unusual sounds coming from the rocker arm cover area.

6   Intake manifold - removal and installation

Refer to illustrations 6.8, 6.9, 6.10 and 6.14

1 Relieve the fuel pressure (Chapter 4).

2 Disconnect the negative cable at the battery.

3 Drain the cooling system (Chapter 1).

4 Disconnect the fuel lines and remove the throttle body (Chapter 4).

5 Remove the upper radiator hose and water outlet heater hose.

6 Remove the distributor as described in Chapter 5.

7 Remove the manifold attaching bolts/studs. The fuel rails and injectors may remain in place. Lift the manifold out of the engine compartment.

8 Clean all traces of old gasket material from the mating surfaces **(see illustration)**. Lightly oil all threaded fasteners prior to assembly.

9 Apply RTV sealant to the intersections of the cylinder head and block **(see illustration)**.

10 Install the front and rear manifold seals **(see illustration)**.

11 Position the intake manifold gaskets in place over the tabs on the cylinder heads. **Note:** *Do not allow the sealer to dry prior to installation.*

12 Apply RTV sealer on top of the end seals.

13 Carefully lower the manifold into place. Avoid displacing the seals and gaskets.

14 Install the bolts and studs, tightening them in the sequence shown **(see illustration)** to the specified torque.

15 Install the thermostat housing with a new gasket (Chapter 3).

16 The remainder of the installation is the reverse of the removal procedure.

17 Change the engine oil and filter, then refill the cooling system (see Chapter 1). Reconnect the battery.

18 Run the engine and check for vacuum and fluid leaks.

7   Exhaust manifolds - removal and installation

1 Disconnect the negative cable at the battery.

2 Raise the vehicle and support it securely on jackstands.

3 Remove the engine cover under the dash (Chapter 11).

4 Unbolt the exhaust pipe(s) from the

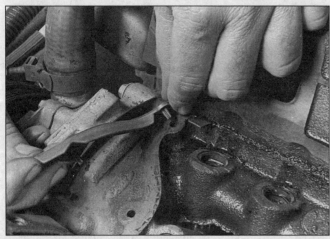

6.10 The end seals have locating pins which must be pressed into place

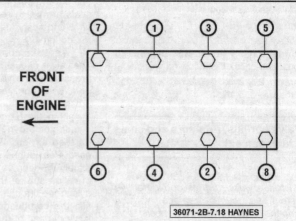

FRONT
OF
ENGINE
←

36071-2B-7.18 HAYNES

6.14 Intake manifold bolt tightening sequence

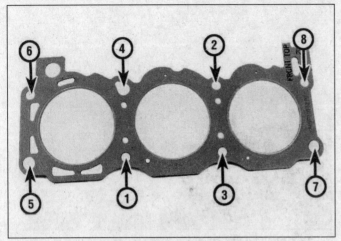

8.20 Head bolt tightening sequence

9.4 Remove the vibration damper with a puller

appropriate manifold(s) as described in Chapter 4.

Left side

5   Remove the oil dipstick tube bracket.
6   Remove the power steering hoses (when equipped).

Right side

7   Remove the heater hose support bracket.
8   Drain the cooling system (Chapter 1).
9   Remove the heater hoses and set aside.
10 Disconnect the oxygen sensor.
11 Remove the EGR tube.

Both sides

12 Unbolt and remove the manifold(s).
13 Clean the mating surfaces of the head and manifold.
14 Clean and lightly oil the threads of all fasteners.
15 Install the manifold and tighten all fasteners to the specified torque.
16 Reinstall all parts removed for access.
17 Refill the cooling system if it was drained.
18 Reconnect the battery.
19 Run the engine and check for leaks.
20 Reinstall the engine cover.

8   Cylinder heads - removal and installation

Refer to illustration 8.20

1   Disconnect the negative cable at the battery.
2   Drain the cooling system (Chapter 1), including removing engine block drain plugs.
3   Remove the air cleaner outlet tube (Chapter 4).
4   Remove the intake manifold (Section 6).
5   Remove the drivebelts (Chapter 1).

Left head

6   Remove the accessory drivebelt idler bracket.

7   If equipped with power steering, remove the pump mounting bracket bolts. Leave the pump hoses connected and tie the pump aside.
8   Remove the ignition coil bracket and dipstick tube.

Right head

9   Remove the alternator adjusting arm.
10 Remove the ground strap and throttle cable bracket.

Both heads

11 Remove the exhaust manifold(s) (Section 7).
12 Remove the rocker arm cover(s) (Section 3).
13 Remove the pushrods (Section 4).
14 Unbolt and remove the cylinder heads.
15 Thoroughly clean and inspect the gasket mating surfaces for cracks and warpage.
16 Service the head(s) as needed (Chapter 2 Part D).
17 Be sure all bolt and bolt hole threads are clean, dry and in good condition.
18 Lightly oil all bolt and stud threads prior to installation.
19 Carefully position the head gaskets on the block using the dowels for alignment.
20 Install the heads and head bolts and tighten them in two steps following the sequence shown **(see illustration)** to the specified torque. **Note:** *When head bolts have been tightened using the above procedure it is not necessary to retighten them after engine warm-up.*
21 Installation of the remaining parts is the reverse of removal.
22 Refill cooling system (Chapter 1).
23 Run the engine and check for leaks and proper operation.

9   Front crankshaft oil seal - replacement

Refer to illustrations 9.4 and 9.8

1   Disconnect the negative cable at the battery.
2   Remove the radiator (Chapter 3).
3   Remove the drivebelts (Chapter 1) and the crankshaft pulley.
4   Remove the vibration damper bolt. Install a puller **(see illustration)** and remove the vibration damper.
5   Carefully pry the old seal from the cover using a seal remover or a screwdriver. Do not nick or scratch the crankshaft or seal bore.
6   Thoroughly clean the seal area.
7   Apply engine oil to the seal lips. Slide the seal into place with the spring side facing the engine.
8   Press the seal into place using a seal installer **(see illustration)** or a large socket and hammer. Be sure the seal seats against the timing chain cover.
9   Reinstall the vibration damper and tighten the bolt to the specified torque.

9.8 A piece of pipe, a bolt and several large washers can be used to fabricate a seal installation tool

10.6 Timing chain cover bolt and stud locations

12.3 Mount a dial indicator on the pushrod to detect valve lifter movement

10   The remainder of the installation is the reverse of removal.

11   Refill the cooling system and run the engine, checking for leaks.

10   Timing chain cover - removal and installation

Refer to illustration 10.6

1   Remove the radiator and lower hose (Chapter 3).

2   Remove the idler pulley and bracket assembly.

3   Remove the drivebelts (Chapter 1) and water pump (Chapter 3).

4   Remove the crankshaft pulley and damper (Section 9).

5   Remove the oil pan-to-timing cover bolts (front of oil pan).

6   Unbolt the cover and remove it **(see illustration)**.

7   While the cover is off, replace the front oil seal. Drive the old one out from the rear. Place the cover on a block of wood and install a new seal using a hammer and a suitable driver.

8   Carefully cut and remove the exposed part of the old oil pan gasket. Clean off all traces of old gasket material from the cover and block.

9   Cut the required section of gasket from a new oil pan gasket and position it and the timing cover gasket with RTV sealer.

10   Install the cover and tighten all fasteners to the specified torque. **Note:** *Use sealant on all cover bolts which go into the water jacket.*

11   The remainder of the installation is the reverse of removal.

12   Refill the cooling system and run the engine, checking for leaks.

11   Top Dead Center (TDC) for number one piston - locating

1   Top Dead Center (TDC) is the highest point in the cylinder that each piston reaches as it travels up-and-down when the crankshaft turns. Each piston reaches TDC on the compression stroke and again on the exhaust stroke, but TDC generally refers to piston position on the compression stroke. The timing marks on the vibration damper installed on the front of the crankshaft are referenced to the number one piston at TDC on the compression stroke.

2   Positioning the piston(s) at TDC is an essential part of many procedures such as rocker arm removal, valve adjustment, timing chain and sprocket replacement and distributor removal.

3   In order to bring any piston to TDC, the crankshaft must be turned using one of the methods outlined below. When looking at the front of the engine, normal crankshaft rotation is clockwise. **Warning:** *Before beginning this procedure, be sure to place the transmission in Neutral and disconnect and ground the coil wire at the distributor cap to disable the ignition system.*

a) *The preferred method is to turn the crankshaft with a large socket and breaker bar attached to the vibration damper bolt that is threaded into the front of the crankshaft.*

b) *A remote starter switch, which may save some time, can also be used. Attach the switch leads to the switch and battery terminals on the solenoid. Once the piston is close to TDC, use a socket and breaker bar as described in the previous paragraph.*

c) *If an assistant is available to turn the ignition switch to the Start position in short bursts, you can get the piston close to TDC without a remote starter switch. Use a socket and breaker bar as described in Paragraph a to complete the procedure.*

4   Using a felt tip pen, make a mark on the distributor housing directly below the number one spark plug wire terminal on the distributor cap.

5   Remove the distributor cap as described in Chapter 1.

6   Turn the crankshaft (see Paragraph 3 above) until the zero on the vibration damper is aligned with the pointer on the timing cover. The timing pointer and vibration damper are located low on the front of the engine, near the pulley that turns the drivebelt.

7   The rotor should now be pointing directly at the mark on the distributor housing. If it is 180-degrees off, the piston is at TDC on the exhaust stroke.

8   To get the piston to TDC on the compression stroke, turn the crankshaft one complete turn (360-degrees) clockwise. The rotor should now be pointing at the mark. When the rotor is pointing at the number one spark plug wire terminal in the distributor cap (which is indicated by the mark on the housing) and the timing marks are aligned, the number one piston is at TDC on the compression stroke.

9   After the number one piston has been positioned at TDC on the compression stroke, TDC for any of the remaining cylinders can be located by turning the crankshaft 120-degrees at a time and following the firing order (refer to the Specifications).

12   Timing chain and sprockets - checking, removal and installation

Refer to illustrations 12.3 and 12.14

Checking

1   Remove the left rocker arm cover (Section 3).

2   Loosen the number 5 cylinder exhaust valve rocker arm (4th one back) and rotate it aside.

3   Mount a dial indicator on the end of the pushrod **(see illustration)**.

4   Turn the crankshaft clockwise until the number 1 piston is at Top Dead Center (Section 11). The damper timing mark should point to TDC on the timing degree indicator.

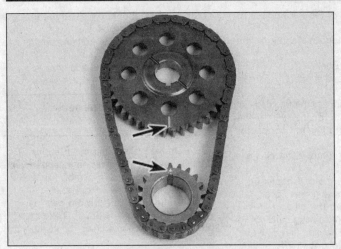

12.14 Align the timing marks on the camshaft and crankshaft sprockets

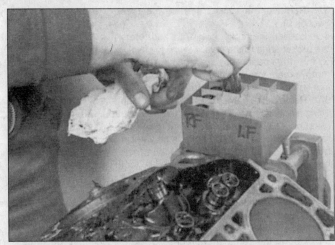

13.2 A clearly labeled box with subdividers is a handy way to keep the lifters in order so that they can be reinstalled in the same lifter bores

This will take up slack on the right side of the chain.

5   Zero the dial indicator.

6   Slowly turn the crankshaft *counterclockwise* until the slightest movement is seen on the dial indicator. Check to see how far the damper TDC timing mark has moved away from the pointer.

7   If the reading on the timing indicator exceeds 6 degrees, replace the timing chain and sprockets.

Removal

8   Turn the crankshaft clockwise until the number one piston is at Top Dead Center (Section 11).

9   Remove the timing chain cover (Section 10).

10   Stuff a clean rag into the oil pan opening.

11   Unbolt the camshaft sprocket. Be sure the engine does not turn during bolt removal. Carefully remove the sprocket with the chain still on it.

12   Pry the crankshaft sprocket off. Do not lose the small crankshaft key.

13   Check the timing chain and sprockets for visible wear and damage. Replace as a set if necessary.

Installation

14   Slide both sprockets and the chain onto the camshaft and crankshaft with the timing marks aligned **(see illustration)**. Tap the crankshaft sprocket into place with a brass drift and hammer.

15   Install the camshaft bolt and washer and tighten to the specified torque. **Caution:** *The camshaft bolt has a drilled oil passage. Be sure that passage is open before installation. DO NOT substitute any other type bolt.*

16   Apply oil to the timing chain and sprockets.

17   Install the timing chain cover (Section 10).

13   Valve lifters - removal, inspection and installation

Refer to illustrations 13.2 and 13.3

Removal and installation

1   Remove the intake manifold (Section 6) and pushrods (Section 4).

2   Remove the lifters with a magnet and place them in a rack in sequence so they can be replaced in their original order **(see illustration)**.

3   If stuck carefully use a lifter puller tool to remove them by rotating them back and forth to loosen them from gum or varnish deposits **(see illustration)**.

4   Clean the lifters and apply engine oil to both the lifters and bores before installing. Be sure to install the lifters in their original bores.

5   Install the manifold and pushrods.

Lifter inspection

6   The faces of the lifters which bear on the camshaft should show no signs of pitting, scoring or other forms of wear. They should also not be a loose fit in the lifter bore. Wear

is only normally encountered at very high mileages or in cases of neglected engine lubrication.

7   If the lifters show signs of wear they must be replaced as a complete set and the camshaft must be replaced as well.

8   Press down on the plunger using a phillips screwdriver to check for freeness of operation. Replace the entire lifter if the plunger is not free in the body.

9   See Chapter 2 Part D for further information.

14   Camshaft - removal and installation

Note: *Refer to Chapter 2, Part D and check the camshaft lobe lift as described there before removing the camshaft.*

Removal

1   Open the hood and remove the engine cover, then cover the seats to avoid damage to the upholstery.

2   Remove the intake manifold (see Section 6).

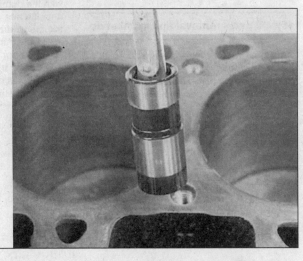

13.3 The lifters in an engine that has accumulated many miles may have to be removed with a special tool

3   Remove the rocker arm covers (see Section 3).

4   Loosen the rocker arm bolts and rotate the rocker arms to the side.

5   If the pushrods are to be reused, remove them from the engine and place them in a holder to keep them in order (Section 4).

6   Remove the valve lifters from the engine using a special tool designed for this purpose (Section 13). Sometimes they can be removed with a magnet if there is no varnish build-up or wear. If they are stuck in their bores you will have to obtain the special tool designed for grasping lifters internally and work them out. Do not use pliers as the surface will be damaged.

7   Remove the timing chain cover (Section 10).

8   Unbolt the camshaft sprocket. Be sure you do not turn the engine while doing so.

9   Remove the timing chain and sprockets (Section 12).

10   Remove the bolts securing the camshaft thrust plate to the engine block.

11   Slowly withdraw the camshaft from the engine, being careful not to nick, scrape or otherwise damage the bearings with the cam lobes.

12   See Chapter 2, Part D for camshaft and lifter inspection procedures.

Installation

13   Lubricate the camshaft journals with engine oil and apply engine assembly lube to the cam lobes.

14   Slide the camshaft into position, being careful not to scrape or nick the bearings.

15   Install the camshaft thrust plate. Tighten the thrust plate retaining bolts to the specified torque. Install the timing chain and sprockets.

16   Check the timing chain deflection as described in Section 12.

17   Install the hydraulic valve lifters in their original bores if the old ones are being used. Make sure they are coated with engine assembly lube. Never use old lifters with a new camshaft or new lifters with an old camshaft.

18   The remainder of the installation procedure is the reverse of removal.

15   Oil pan - removal and installation

Note: *The factory recommends removal of the transmission when removing the oil pan to assure proper pan sealing upon re-installation. However, if proper care is exercised dur-*

ing re-sealing, removal of the transmission is not necessary.

1   Disconnect the negative cable at the battery.

2   Remove the oil dipstick.

3   Raise the vehicle and support it securely on jackstands.

4   If equipped, disconnect the low oil level sensor by removing the retainer clip at the sensor and then separating the connector from the sensor.

5   Drain the oil and remove the oil filter (Chapter 1).

6   Remove the starter motor (Chapter 5).

7   Remove the bellhousing dust cover.

8   Unbolt and remove the oil pan. If it is stuck on, gently pry it off, taking care not to bend or distort the pan sealing flange. **Note:** *It may be necessary to unbolt the engine mounts and raise the engine with a jack to provide enough clearance for the oil pan removal. Be sure to safely block the engine in the raised position.*

9   Thoroughly clean the gasket surfaces of the block and pan.

10   Apply a 4 to 5 mm (1/4-inch) bead of RTV sealer to the junctions of the rear main bearing cap and engine block and also to the timing chain cover-to-block.

11   Attach the oil pan gasket to oil pan with sealer and position the pan on the engine.

12   Install the oil pan bolts and tighten to the specified torque.

13   Reinstall any parts removed and refill with engine oil.

14   Run the engine and check for leaks.

16   Oil pump and pickup - removal and installation

Note: *We recommend oil pump replacement any time the engine is rebuilt and when the pump has high mileage or is otherwise suspect.*

1   Remove the oil pan as described in Section 15.

2   While supporting the oil pump, remove the pump-to-rear main bearing cap bolt. The oil pan baffle may then be removed.

3   Lower the pump and remove it along with the pump driveshaft.

4   If a new oil pump is installed, make sure the pump driveshaft is mated with the shaft inside the pump.

5   Position the pump on the engine and make sure the hex in the upper end of the driveshaft is aligned with the lower end of the distributor shaft. The distributor drives the oil

pump, so it is absolutely essential that the components mate properly.

6   Install the mounting bolt and tighten it to the specified torque.

7   Install the oil pan and add oil.

17   Rear crankshaft oil seal - replacement

1   Remove the transmission (Chapter 7).

2   On manual transmission equipped vehicles, remove the clutch assembly (Chapter 8).

3   Remove the flywheel/driveplate.

4   Carefully pry out the old seal using a seal remover or screwdriver. Take care to prevent nicking or scratching the seal bore or crankshaft.

5   Clean and inspect the oil seal area.

6   Lubricate the new seal with engine oil.

7   Start the seal into its recess with the spring side facing the engine.

8   Press the seal into place with a rear seal installation tool, available at most auto parts stores or a large socket and hammer.

9   Install the flywheel/driveplate.

10   The remainder of installation is the reverse of removal.

11   Run the engine and check for leaks.

18   Front engine mounts - removal and installation

1   Remove the fan shroud attaching bolts.

2   Support the engine using a wood block and a jack placed under the oil pan.

3   Remove the nuts and washers attaching the insulators to the crossmember and lift the engine enough to disengage the insulator stud from the crossmember.

4   Remove the engine insulator assembly to cylinder block attaching bolts and lockwashers and remove the engine mount assembly.

5   To install, position the engine mount assembly to the cylinder block, install the bracket bolts and tighten to the specified torque.

6   Lower the engine until the insulator stud engages and bottoms in the slot on the crossmember pedestal. Tighten the nuts to the specified torque.

7   Remove the block of wood and jack from the engine oil pan.

8   Install the fan shroud attaching bolts and tighten.

Chapter 2 Part D
4.0L V6 engine

Contents

Specifications

General

Displacement	4.0 liters (244 cubic inches)
Cylinder numbers (front-to-rear)	
Left (driver's) side	4-5-6
Right side	1-2-3
Firing order*	1-4-2-5-3-6

Camshaft

Lobe lift (intake and exhaust)	0.2756 inch
Allowable lobe lift loss	0.005 inch
Endplay	
Standard	0.0008 to 0.004 inch
Service limit	0.009 inch
Thrust plate thickness	0.158 to 0.159 inch
Journal-to-bearing (oil) clearance	
Standard	0.001 to 0.0026 inch
Service limit	0.006 inch
Bearing inside diameter (standard)	
No. 1	1.954 to 1.955 inch
No. 2	1.939 to 1.940 inch
No. 3	1.919 to 1.920 inch
No. 4	1.924 to 1.925 inch

4.0L PUSHROD V6

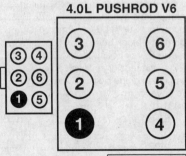

36024-1-specs.A HAYNES

Firing order and spark plug wire connection points - 4.0L V6 engine

Camshaft (continued)

Journal diameter (standard)

No. 1	1.951 to 1.952 inch
No. 2	1.937 to 1.938 inch
No. 3	1.922 to 1.923 inch
No. 4	1.907 to 1.908 inch
Front bearing location	0.040 to 0.060 inch below face of block

Oil pan-to-transmission spacer thicknesses

Coded yellow	0.010 inch
Coded blue	0.020 inch
Coded pink	0.030 inch

Crankshaft rear seal

Dimension from rear face of block	0.413 +0.008 -0.012 inch
Square to crankshaft centerline within	0.015 inch

Torque specifications

Ft-lbs (unless otherwise indicated)

Camshaft sprocket bolt	44 to 50
Camshaft thrust plate bolts	84 to 120 in-lbs
Crankshaft pulley bolt	
Step 1	30 to 37
Step 2	Turn an additional 80 to 90 degrees
Cylinder head bolts (use new bolts)	
Through 1994	
Step 1	Tighten head bolts to 44 ft-lbs
Step 2	Tighten intake manifold fasteners to 36 to 72 in-lbs
Step 3	Tighten head bolts to 59 ft-lbs
Step 4	Tighten intake manifold fasteners to 72 to 132 in-lbs
Step 5	Tighten head bolts an additional 80 to 85 degrees
Step 6	Tighten intake manifold fasteners to 15 to 18
1995 to 1997	
Step 1	Tighten head bolts to 22 to 26 ft-lbs
Step 2	Tighten intake manifold fasteners to 72 in-lbs
Step 3	Tighten head bolts to 52 to 56 ft-lbs
Step 4	Tighten intake manifold fasteners to 132 in-lbs
Step 5	Tighten head bolts an additional 90 degrees
Step 6	Tighten intake manifold fasteners to 16 ft-lbs
Flywheel/driveplate bolts	
Step 1	108 to 132 in-lbs
Step 2	50 to 55
Engine mount insulator-to-frame nuts	71 to 94
Engine mount insulator-to-bracket nuts	65 to 85
Engine mount bracket-to-block bolts	45 to 60
Exhaust manifold bolts	19
Exhaust pipe-to-manifold nuts	20
Timing chain cover bolts	13 to 15
Intake manifold fasteners	
Step 1	72 in-lbs
Step 2	132 in-lbs
Step 3	16
Intake manifold studs to block	72 to 84 in-lbs
Oil pump drive gear bolt	156 to 180 in-lbs
Oil pump pick-up tube-to-pump bolts	84 to 120 in-lbs
Oil pump-to-block bolts	156 to 180 in-lbs
Oil pan-to-block bolts	60 to 84 in-lbs
Timing chain tensioner bolts	84 to 96 in-lbs
Timing chain guide bolts	84 to 108 in-lbs
Valve cover bolts	
1991	
Step 1	30 to 60 in-lbs
Step 2	Wait 2 minutes
Step 3	Retighten to 30 to 60 in-lbs
1992 to 1994	53 to 70 in-lbs
1995 to 1996	71 to 89 in-lbs
1997	59 to 71 in-lbs
Rocker arm shaft stand bolts	46 to 52

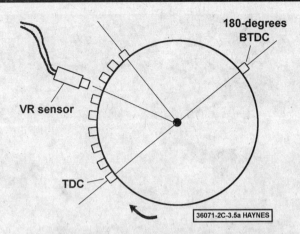

3.5a The crankshaft pulley has a gap at 60-degrees BTDC - TDC is located at the sixth tooth from the gap

3.5b The pointer on the front of the engine lines up with a notch in the crankshaft pulley to indicate TDC

1 General information

This Part of Chapter 2 is devoted to in-vehicle repair procedures for the 4.0L V6 engine. All information concerning engine removal and installation and engine block and cylinder head overhaul can be found in Part E of this Chapter.

The following repair procedures are based on the assumption that the engine is installed in the vehicle. If the engine has been removed from the vehicle and mounted on a stand, many of the steps outlined in this Part of Chapter 2 will not apply.

The Specifications included in this Part of Chapter 2 apply only to the procedures contained in this Part. Part E of Chapter 2 contains the Specifications necessary for cylinder head and engine block rebuilding.

2 Repair operations possible with the engine in the vehicle

Many major repair operations can be accomplished without removing the engine from the vehicle.

Clean the engine compartment and the exterior of the engine with some type of degreaser before any work is done. It will make the job easier and help keep dirt out of the internal areas of the engine.

Depending on the components involved, it may be helpful to remove the hood to improve access to the engine as repairs are performed (refer to Chapter 11 if necessary). Cover the fenders to prevent damage to the paint. Special pads are available, but an old bedspread or blanket will also work.

If vacuum, exhaust, oil or coolant leaks develop, indicating a need for gasket or seal replacement, the repairs can generally be made with the engine in the vehicle. The intake and exhaust manifold gaskets and cylinder head gaskets are all accessible with the engine in place.

Exterior engine components, such as the intake and exhaust manifolds, the water pump, the starter motor, the alternator, the distributor and the fuel system components can be removed for repair with the engine in place.

Since the cylinder heads can be removed without pulling the engine, valve component servicing can also be accomplished with the engine in the vehicle.

3 Top Dead Center (TDC) for number one piston - locating

Refer to illustrations 3.5a and 3.5b
Note: The 4.0L engine is not equipped with a distributor. Piston position must be determined by feeling for compression at the number one spark plug hole, then aligning the ignition timing marks as described in Step 5.
1   Top Dead Center (TDC) is the highest point in the cylinder that each piston reaches as it travels up-and-down when the crankshaft turns. Each piston reaches TDC on the compression stroke and again on the exhaust stroke, but TDC generally refers to piston position on the compression stroke.
2   Positioning the piston(s) at TDC is an essential part of many other repair procedures discussed in this manual.
3   Before beginning this procedure, be sure to place the transmission in Neutral and apply the parking brake or block the rear wheels. Remove the spark plugs (see Chapter 1). Disable the ignition system by disconnecting the wiring harness connector from the ignition coil pack, located above the left valve cover.
4   In order to bring any piston to TDC, the crankshaft must be turned using one of the methods outlined below. When looking at the front of the engine, normal crankshaft rotation is clockwise.
a) The preferred method is to turn the crankshaft with a socket and ratchet attached to the bolt threaded into the front of the crankshaft.

b) A remote starter switch, which may save some time, can also be used. Follow the instructions included with the switch. Once the piston is close to TDC, use a socket and ratchet as described in the previous paragraph.
c) If an assistant is available to turn the ignition switch to the Start position in short bursts, you can get the piston close to TDC without a remote starter switch. Make sure your assistant is out of the vehicle, away from the ignition switch, then use a socket and ratchet as described in Paragraph a) to complete the procedure.
5   The crankshaft pulley has 35 teeth, evenly spaced every 10-degrees around the pulley, and a gap where a 36th tooth would be. The gap is located at 60-degrees Before Top Dead Center (BTDC) (see illustration). Turn the crankshaft (see Paragraph 4 above) until you feel compression at the number one spark plug hole, then turn it slowly until the sixth tooth from the missing tooth is aligned with the Variable Reluctance (VR) sensor and the TDC notch is aligned with the pointer (located at the front of the engine) (see illustration).
6   After the number one piston has been positioned at TDC on the compression stroke, TDC for any of the remaining pistons can be located by turning the crankshaft and following the firing order.

4 Valve covers - removal and installation

Refer to illustrations 4.6a, 4.6b, 4.10, 4.14, 4.18 and 4.20

Removal

1   Disconnect the negative cable from the battery.
2   Remove the inside engine cover and remove the fresh air intake shield and tube.

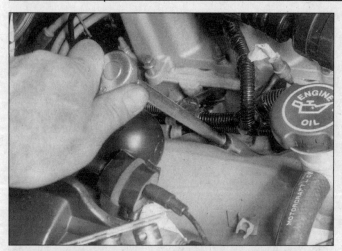

4.6a Remove the valve cover bolts with a ratchet and extension . . .

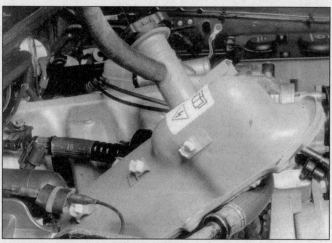

4.6b . . . and lift the cover off - you may have to angle it out, as shown here - tap it gently with a soft-face hammer if necessary to break the gasket seal

4.10 Disconnect the compressor electrical connector from the back of the compressor, then remove the mounting bolts and shift the compressor forward as shown - DO NOT disconnect any refrigerant lines

Right valve cover

3   Remove the oil fill pipe, oil dipstick tube and transmission fluid filler tube.
4   Remove the alternator and ignition coil pack (refer to Chapter 5).
5   Detach the spark plug wires from the valve cover clips.
6   Remove the valve cover bolts and rein-forcing plates **(see illustration)**. Lift the valve cover off **(see illustration)**. Tap it gently with a soft-face hammer if necessary to break the gasket seal.

Left valve cover

7   If you haven't already done so, remove the bolt from the air conditioning tube above the upper intake manifold.
8   Remove the upper intake manifold (refer to Chapter 4).
9   Disconnect the electrical connector for the air conditioning compressor clutch. Care-fully pry the wiring harness from the back of the compressor with a door panel clip remover or similar tool.

10   Remove the mounting bolts from the air conditioning compressor (see Chapter 3). Lift the air conditioning compressor, then posi-tion it out of the way **(see illustration)**. **Warn-ing:** *DO NOT disconnect any refrigerant lines!*
11   Disconnect the brake booster vacuum hose.
12   Label and disconnect the vacuum hoses from the tee on the plenum.
13   Detach the PCV hose. Remove the PCV hose and valve (see Chapter 1).
14   Carefully pry the wiring harnesses away from the valve cover with a door trim panel remover or similar tool **(see illustration)**. Place the harnesses out of the way.
15   Disconnect the driver's side spark plug wires from the plugs and detach them from the clips on the valve cover. Position the wires out of the way.
16   Carefully lift the engine wiring harness clip with your thumb to separate it from the valve cover. Don't pull on the harness.
17   Remove the bolt that secures the fuel line clip to the front of the engine. Move the fuel line just enough to provide access to the front valve cover bolt. DO NOT disconnect any fuel lines without first relieving fuel pres-

sure (see Chapter 4).
18   Remove the valve cover bolts and rein-forcing plates **(see illustration 4.6a)**. Lift the valve cover off **(see illustration)**. Tap it gently with a soft-face hammer if necessary to break the gasket seal.

Installation

19   Clean the gasket surfaces on the intake manifold, cylinder head and valve cover. Use a scraper to remove the pieces of old gasket material, then wipe off all residue with lacquer thinner or acetone.
20   Most valve cover gaskets are equipped with self-sticking sealant on the valve cover side. Pull the plastic film off the gasket **(see illustration)** and stick the gasket to the valve cover.
21   **Note:** *In this step, apply silicone sealant to one side of the engine at a time (if both valve covers were removed), then install the valve cover.* Apply silicone sealant to the seam where the cylinder head joins the intake manifold **(see illustration 4.6a)**. Apply a 1/8-inch ball of sealant to the valve cover bolt holes on the outer (exhaust) side of the cylin-der head.

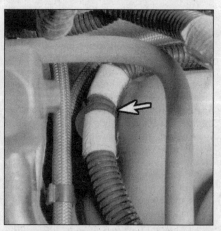

4.14 Wiring harnesses are secured to the valve cover by clips (arrow)

4.18 Move the wiring harnesses out of the way and lift the valve cover off

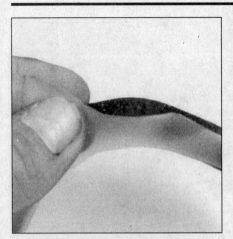

4.20 Just before installing the new gasket, peel the plastic film off the self-sticking sealant on the valve-cover side of the gasket

5.2 Remove the bolts evenly, two turns at a time, to prevent the shafts from being bent by valve spring pressure

5.3 Once the bolts are loose, lift the rocker assembly off the engine

22   The remainder of installation is the reverse of the removal Steps. Tighten the valve cover bolts evenly, starting with the center bolts and working out, to the torque listed in this Chapter's Specifications.

5 Rocker arms and pushrods - removal, inspection and installation

Removal

Refer to illustrations 5.2, 5.3, 5.4a and 5.4b

1   Refer to Section 4 and remove the valve cover(s).
2   Loosen the rocker arm shaft support bolts two turns at a time, starting with the center bolt and working out, until the bolts can be removed by hand **(see illustration)**.
3   Lift the rocker shaft assembly off the cylinder head **(see illustration)**. The pins will hold the components together. Mark each shaft assembly so it can be returned to the same side of the engine.
4   Lift the pushrods out of the engine **(see**

illustration). Place the pushrods in order in a holder **(see illustration)** so they can be returned to their original positions. Be sure to store them so you can reinstall them with the same end facing up.

Inspection

5   Remove the pins and disassemble the rocker arm assembly. Place the parts in order on a clean workbench. Be sure you don't mix up the parts - they must be reassembled in exactly the same order they were before disassembly.
6   Check each rocker arm for wear, cracks and other damage, especially where the pushrods and valve stems contact the rocker arm faces.
7   Make sure all oil holes are open and not plugged. Plugging can be cleared with a piece of wire.
8   Check each rocker arm bore, and its corresponding position on the rocker shaft, for wear, cracks and galling. If the rocker arms or shaft are damaged, replace them with new ones.
9   Inspect the pushrods for cracks and

excessive wear at the ends. Roll each pushrod across a piece of plate glass to see if it's bent (if it wobbles, it's bent).
10   If necessary, remove the plug from each end of the rocker shaft. Drill into one plug and insert a long steel rod through it to knock out the other plug. Knock out the first plug in the same manner.
11   If the rocker arm shaft plugs were removed, tap in new ones with a hammer and suitable drift.
12   Assemble the rocker assembly. Lubricate at friction points (pushrod ends, rocker arm bores and ends) with assembly lube.
13   Install new cotter pins in the ends of the rocker shaft. Be sure the rocker shaft oil holes will face down when the shaft is installed. The position of the oil holes is indicated by a notch on the front of each shaft.

Installation

14   Coat each end of each pushrod with assembly lube, then install them in the engine. If you are reinstalling the original pushrods, be sure to return them to their original positions.
15   Coat the rocker arm pads with assembly lube.

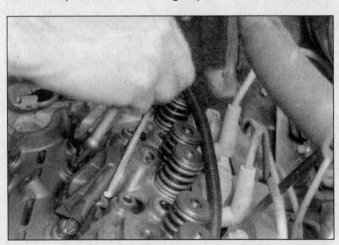

5.4a Pull the pushrods out of the lifters . . .

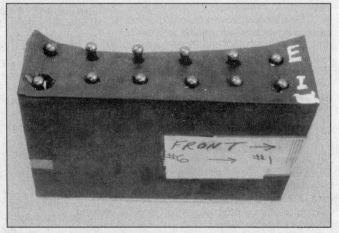

5.4b . . . and place them in a labeled holder so they can be returned to their original positions

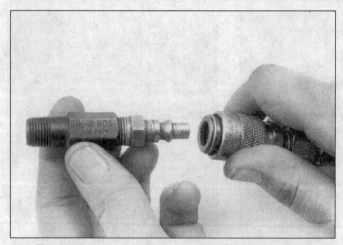

6.4 This is what the air hose adapter that threads into the spark plug hole looks like - they're commonly available from auto parts stores

6.9 Compress the valve spring, then remove the retainer locks

16   Install the rocker arm assembly on the engine. The notch on the front end of each shaft should face down.

17   Position the rocker arm ball ends in the pushrods.

18   Tighten the rocker shaft support bolts two turns at a time, working from the center bolt out, to the torque listed in this Chapter's Specifications.

19   The remainder of installation is the reverse of the removal Steps.

6   Valve springs, retainers and seals - replacement

Refer to illustrations 6.4 and 6.9

Note: *Broken valve springs and defective valve stem seals can be replaced without removing the cylinder heads. Two special tools and a compressed air source are normally required to perform this operation, so read through this Section carefully and rent or buy the tools before beginning the job. If compressed air isn't available, a length of nylon rope can be used to keep the valves from falling into the cylinder during this procedure.*

1   Refer to Section 4 and remove the valve cover from the affected cylinder head. If all of the valve stem seals are being replaced, remove both valve covers.

2   Remove the spark plug from the cylinder which has the defective component. If all of the valve stem seals are being replaced, all of the spark plugs should be removed.

3   Turn the crankshaft until the piston in the affected cylinder is at top dead center on the compression stroke (refer to Section 3 for instructions). If you're replacing all of the valve stem seals, begin with cylinder number one and work on the valves for one cylinder at a time. Move from cylinder-to-cylinder following the firing order sequence (see this Chapter's Specifications).

4   Thread an adapter into the spark plug hole **(see illustration)** and connect an air hose from a compressed air source to it. Most auto parts stores can supply the air hose adapter. **Note:** *Many cylinder compression gauges utilize a screw-in fitting that may work with your air hose quick-disconnect fitting.*

5   Remove the rocker assembly on the affected side of the engine (see Section 5). If all of the valve stem seals are being replaced, remove both rocker assemblies.

6   Apply compressed air to the cylinder. **Warning:** *The piston may be forced down by compressed air, causing the crankshaft to turn suddenly. If the wrench used when positioning the number one piston at TDC is still attached to the bolt in the crankshaft nose, it could cause damage or injury when the crankshaft moves.*

7   The valves should be held in place by the air pressure. If the valve faces or seats are in poor condition, leaks may prevent air pressure from retaining the valves - refer to the alternative procedure below.

8   If you don't have access to compressed air, an alternative method can be used. Position the piston at a point just before TDC on the compression stroke, then feed a long piece of nylon rope through the spark plug hole until it fills the combustion chamber. Be sure to leave the end of the rope hanging out of the engine so it can be removed easily. Use a large ratchet and socket to rotate the crankshaft in the normal direction of rotation until slight resistance is felt.

9   Stuff shop rags into the cylinder head holes above and below the valves to prevent parts and tools from falling into the engine, then use a valve spring compressor to compress the spring. Remove the retainer locks with small needle-nose pliers or a magnet. **Note:** *A couple of different types of tools are available for compressing the valve springs with the head in place. One type grips the lower spring coils and presses on the retainer as the knob is turned* **(see illustration)**, *while the other type, utilizes a bar installed in place of the rocker arm shaft for leverage. Both types work very well, although the knob type is more readily available.*

10   Remove the spring retainer and valve spring, then remove the valve stem seal. **Note:** *If air pressure fails to hold the valve in the closed position during this operation, the valve face or seat is probably damaged. If so, the cylinder head will have to be removed for additional repair operations.*

11   Wrap a rubber band or tape around the top of the valve stem so the valve won't fall into the combustion chamber, then release the air pressure. **Note:** *If a rope was used instead of air pressure, turn the crankshaft slightly in the direction opposite normal rotation.*

12   Inspect the valve stem for damage. Rotate the valve in the guide and check the end for eccentric movement, which would indicate that the valve is bent.

13   Move the valve up-and-down in the guide and make sure it doesn't bind. If the valve stem binds, either the valve is bent or the guide is damaged. In either case, the head will have to be removed for repair.

14   Reapply air pressure to the cylinder to retain the valve in the closed position, then remove the tape or rubber band from the valve stem. If a rope was used instead of air pressure, rotate the crankshaft in the normal direction of rotation until slight resistance is felt.

15   Lubricate the valve stem with engine oil and install a new stem seal.

16   Install the spring in position over the valve.

17   Install the valve spring retainer. Compress the valve spring and carefully position the retainer locks in the groove. Apply a small dab of grease to the inside of each lock to hold it in place.

18   Remove the pressure from the spring tool and make sure the retainer locks are seated.

19   Disconnect the air hose and remove the adapter from the spark plug hole. If a rope was used in place of air pressure, pull it out of the cylinder.

20   Refer to Section 5 and install the rocker

7.6 Remove the eight intake manifold bolts and nuts

7.8 Before installing the intake manifold gasket, apply a film of silicone sealant around the water jacket ports as well as the front sealing surface - be sure to apply extra beads to the corners (arrows) at the front and rear of the manifold

arm assembly(ies).

21 Install the spark plug(s) and hook up the wire(s).

22 Refer to Section 4 and install the valve cover(s).

23 Start and run the engine, then check for oil leaks and unusual sounds coming from the valve cover area.

7 Lower intake manifold - removal and installation

Refer to illustrations 7.6, 7.8, 7.9 and 7.11

Removal

1 Disconnect the negative cable from the battery.

2 Remove the upper intake manifold (see Chapter 4).

3 Remove the valve covers (see Section 4).

4 Disconnect the electrical connectors from the coolant temperature sender and EEC-IV system coolant temperature sensor at the front of the intake manifold.

5 Disconnect the heater hose from the front of the intake manifold. Either remove the thermostat and housing (see Chapter 3) or disconnect the upper radiator hose from the outlet fitting.

6 Remove the intake manifold nuts and bolts and lift the manifold off **(see illustrations)**. If it's stuck, tap it lightly with a soft-face hammer to break the gasket seal. If necessary, pry the manifold off, but pry between a casting protrusion and the engine - don't pry against gasket surfaces.

Installation

7 Clean away all traces of old gasket material. Remove oil and dirt with a cloth and solvent, such as acetone or lacquer thinner.

8 Apply silicone sealant around the water jacket ports (the smaller ports at the ends of the heads) and to the front sealing surface. Also apply a bead of sealant to the four corners where the manifold meets the engine **(see illustration)**. **Note:** *Install the manifold gasket immediately after applying the sealant. If allowed to set up (approximately 15 minutes), the sealant won't work properly.*

9 Install the manifold gasket **(see illustra-**

tion), then reapply sealant to the four corners.

10 Install the manifold over the studs. Install the nuts and bolts and tighten them finger-tight.

11 Tighten the nuts and bolts in the stages listed in this Chapter's Specifications, following the sequence **(see illustration)**.

12 The remainder of installation is the reverse of the removal steps.

13 Run the engine and check for oil, coolant and vacuum leaks.

8 Exhaust manifolds - removal and installation

Removal

Left manifold

1 Remove the oil dipstick tube bracket from the engine.

2 If the power steering pump hoses obstruct manifold removal, disconnect them (see Chapter 10). Cap the hoses and fittings to keep out dirt and place the hose ends out of the way.

7.9 As soon as the sealant is applied, install the gasket and place extra beads of sealant in the corners

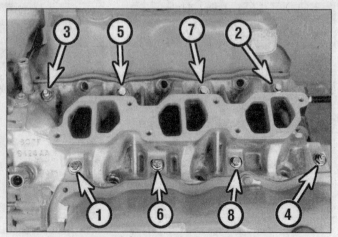

7.11 Intake manifold tightening sequence

8.5 Remove the exhaust manifold bolts (arrows)

9.15 Remove the head bolts with a T55 Torx bit - don't use an Allen wrench, since it may round out the bolt heads

Right manifold

3   Detach the heater hose bracket and disconnect the heater hoses (see Chapter 3).

Both manifolds

Refer to illustration 8.5

4   Detach the exhaust pipe from the manifold(s). **Note:** *For access to the exhaust pipe nuts, it may be necessary to raise the vehicle and support it securely on jackstands.*

5   Unbolt the manifold from the cylinder head and take it off **(see illustration)**.

Installation

6   Using a scraper, thoroughly clean the mating surfaces on cylinder head, manifold and exhaust pipe. Remove residue with a solvent such as acetone or lacquer thinner.

7   **Note:** *These engines were originally assembled without exhaust manifold gaskets and can be reassembled the same way so long as the mating surfaces are perfectly flat and not damaged in any way. Warped or damaged manifolds may require a gasket or machining. Apply some graphite grease to the cylinder head mating surface and place the manifold on the cylinder head. Tighten the*

bolts evenly to the torque listed in this Chapter's Specifications.

8   Connect the exhaust pipe to the manifold and tighten the nuts evenly to the torque listed in this Chapter's Specifications.

9   The remainder of installation is the reverse of the removal steps. If the heater hoses were disconnected, fill the cooling system (see Chapter 1).

10   Run the engine and check for exhaust leaks.

---

9   Cylinder heads - removal and installation

Removal

Note: *Head bolt removal requires a Torx bit. Obtain the necessary tool before starting. DO NOT try to use an Allen wrench since it may round out the bolt head.*

1   Disconnect the negative cable from the battery.

2   Drain the cooling system (see Chapter 3).

3   Remove the valve covers (see Section 4).

4   Remove the rocker arms and pushrods (see Section 5).

5   Remove the intake manifold (see Section 7).

Left head

6   Remove the drivebelt (see Chapter 1).

7   If not already done, detach the air conditioning compressor from the engine and place it out of the way **(see illustration 4.10)**. DO NOT disconnect any refrigerant lines!

8   Detach the power steering pump and bracket and set them out of the way (see Chapter 10). Don't disconnect the power steering hoses.

9   There's a wiring harness attached to the rear of the head. The harness is secured to the plastic retainer with tape. Unwrap or cut the tape to detach the harness from the retainer.

Right head

10   Remove the drivebelt (see Chapter 1).

11   Remove the alternator and bracket (see Chapter 5).

12   Remove the ignition coil pack and bracket (see Chapter 5).

Both heads

Refer to illustrations 9.15 and 9.16

13   Remove the spark plugs (see Chapter 1).

14   Remove the exhaust manifolds (see Section 8).

15   Remove and discard the cylinder head bolts **(see illustration)**. The bolts must be replaced with new ones whenever they are removed. Loosen the bolts in several stages.

16   Lift the cylinder head off the engine. If it's difficult to remove, carefully pry it off. Pry against a casting protrusion, not against gasket surfaces **(see illustration)**.

Installation

Refer to illustrations 9.17, 9.20 and 9.23

17   Thoroughly remove all traces of gasket material with a gasket scraper and clean all parts with solvent **(see illustration)**. Use a

9.16 If necessary, pry the head loose; pry against a casting protrusion so the gasket surfaces won't be damaged

9.17 Remove all traces of the old gasket with a scraper, taking care not to gouge the sealing surfaces

9.20 Be sure the FRONT TOP mark on the head gasket is positioned correctly

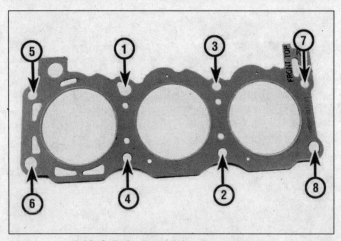

9.23 Cylinder head tightening sequence

rag and acetone or lacquer thinner to remove any traces of oil from the gasket mating surfaces. See Chapter 2 Part E for cylinder head inspection procedures.

18 Use a tap of the correct size to chase the threads in the head bolt holes.

19 Recheck all head bolt holes and cylinder bores for any traces of coolant, oil or other foreign matter. Remove as needed.

20 Position the new gaskets over the dowel pins on the block. Don't use sealant on the gaskets. Be sure the FRONT TOP marks are positioned correctly **(see illustration)**.

21 Install the heads and new head bolts finger tight.

22 Position the intake manifold on the engine (see Section 7).

23 Following the sequence shown **(see illustration)**, tighten the head bolts in three steps to the torque listed in this Chapter's Specifications. **Note:** *The head bolts and intake manifold bolts are tightened in alternate stages to align the manifold with the heads. Be sure to refer to this Chapter's Specifications for the correct tightening stages.*

24 The remainder of installation is the reverse of the removal steps.

25 Run the engine and check for oil, coolant and vacuum leaks.

10 Crankshaft pulley, front oil seal and timing chain cover removal and installation

Refer to illustrations 10.3a and 10.3b

Crankshaft pulley removal and installation

1 Remove the drivebelt (see Chapter 1).

2 Remove the crankshaft pulley bolt.

3 Insert a longer, smaller-diameter bolt into the crankshaft so the puller will have something to bottom against, then remove the crankshaft pulley with a puller **(see illustrations)**. DO NOT pry the pulley off or use an impact puller!

4 Using clean engine oil, lubricate the surface of the pulley where the front seal rides. Install the crankshaft pulley with a special tool, available at most auto parts stores, or equivalent. Don't hammer the pulley on.

Tighten the pulley bolt to the torque listed in this Chapter's Specifications.

Front oil seal replacement

Refer to illustration 10.5

5 Remove the crankshaft pulley (see Steps 1 through 3 above). Carefully pry the oil seal out with a screwdriver **(see illustration)**.

6 Clean the seal bore and check it for nicks or gouges.

7 Coat the lip of the new seal with clean engine oil and drive it into the bore with a socket or large piece of pipe slightly smaller in diameter than the seal. The open side of the seal faces into the engine. Install the crankshaft pulley (see Step 4 above).

Timing chain cover
Removal
Refer to illustration 10.13 and 10.14

8 Remove the crankshaft pulley (see Steps 1 through 3 above). Also remove the crankshaft timing sensor.

9 Drain the cooling system (see Chapter 1).

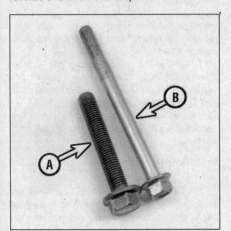

10.3a To remove the crankshaft pulley, remove the pulley bolt (A) and install a longer bolt that's smaller in diameter (B) into the crankshaft . . .

10.3b . . . so the puller screw (arrow) will have something to push against

10.5 To remove the front cover seal with the cover on the engine, pry it out with a screwdriver, taking care not to gouge the cover

10.13 Label the studs and brackets so they can be returned to their original locations

10.14 Remove the cover from the engine - if it's stuck, recheck to make sure all bolts and studs have been removed

10   Remove the engine cooling fan and radiator (see Chapter 3).

11   Detach the air conditioning compressor and bracket (if equipped) and set them aside (see Chapter 3). DO NOT disconnect any refrigerant lines! Remove the oil pan (see Section 14).

12   Remove the water pump (see Chapter 3). Remove the heater and radiator hoses as needed to provide removal access for the timing chain cover.

13   Remove the brackets, studs and cover bolts **(see illustration). Note:** *The bolts are different lengths, and bolts and studs go in different locations. Label them so they can be installed in the correct holes.*

14   Take the cover off **(see illustration).** If it's stuck, tap it lightly with a soft-face hammer or pry it carefully to break the gasket seal. Don't use excessive force or you'll crack the cover. If it's difficult to remove, check to make sure you've removed all the bolts.

Installation

15   Thoroughly clean and inspect all parts and use a scraper to remove all traces of gasket material. Remove oil film with a solvent such as lacquer thinner or acetone.

16   Apply RTV sealant to the gasket mating surfaces. Install the guide sleeves (if removed). Install the front cover and start the studs bolts two or three turns by hand. Note that the bolts are different lengths; be sure to install them in the correct holes.

17   Tighten the cover bolts evenly to the torque listed in this Chapter's Specifications.

18   Install the crankshaft timing sensor (see Chapter 5).

19   Install the crankshaft pulley (see Step 4 above).

20   The remainder of installation is the reverse of the removal steps.

21   Run the engine and check for oil or coolant leaks.

11   Timing chain and sprockets - inspection, removal and installation.

Camshaft endplay check

1   Remove the timing chain cover (see Section 10).

2   Remove the rocker arm shafts (see Section 5).

3   Push the camshaft to the rear as far as it will go.

4   Install a dial indicator with its pointer on the camshaft sprocket bolt. Set the indicator to zero.

5   Pry the camshaft forward with a large screwdriver or prybar between the camshaft sprocket and the block. Note the dial indicator reading and compare with the endplay listed in this Chapter's Specifications. Replace the camshaft thrust plate (see Section 13) if endplay is excessive.

Timing chain and tensioner inspection

Refer to illustration 11.7

6   Remove the timing chain cover (see Section 10).

7   Remove the timing chain tensioner **(see illustration).**

8   Rotate the crankshaft counterclockwise (as viewed from the front of the engine) to take up the slack in the right side of the chain.

9   Mark a reference point on the block approximately halfway along the chain and measure from that point to the right side of the chain **(see illustration 11.7).**

10   Turn the crankshaft clockwise to take up the slack on the left side of the chain.

11   Push the chain out and measure from the reference point to the chain.

12   The difference between the two measurements is deflection. If deflection is

excessive, replace the timing chain (see below).

13   Check the tensioner for wear and damage. If tensioner face wear is excessive, replace the tensioner. Before reinstalling the tensioner, it must be retracted (see Step 28). Tighten the tensioner bolts to the torque listed in this Chapter's Specifications.

Chain and sprocket

Removal

Refer to illustration 11.20

14   Drain the cooling system and engine oil (see Chapter 1).

15   Remove the oil pan (see Section 14).

16   Replace the oil filter (see Chapter 1).

17   Remove the drivebelt (see Chapter 1).

18   Remove the timing chain cover (see Section 10).

19   Turn the crankshaft (see Section 3, Step 4) until the timing marks are aligned. The mark on the crankshaft sprocket will be in the 12-o'clock position and the mark on the camshaft sprocket will be in the 6-o'clock position. Remove the camshaft sprocket Torx bolt and the crankshaft sprocket key. Also remove the timing chain tensioner **(see illustration 11.7).**

20   Remove the sprockets together with the chain **(see illustration).** Do not disturb the crankshaft or camshaft while the timing chain and sprockets are removed.

21   If necessary, remove the Torx bolts that secure the chain guide and remove it from the block.

Installation

Refer to illustration 11.28

22   Be sure the crankshaft and camshaft are positioned so the sprocket timing marks will align correctly after the sprockets are installed.

23   If the chain guide was removed, install it. Be sure its pin is inserted into the block oil hole, then tighten the guide bolts to the

11.7 Unbolt the chain tensioner from the block - when checking timing chain deflection, mark a reference point on the block (arrow)

11.20 Remove the timing chain and sprockets together

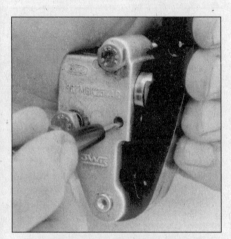

11.28 To retract the timing chain tensioner so it can be installed, squeeze the tensioner pad and use a sharp-tipped probe to push the ratchet mechanism down, then in, to release the latch and let the tensioner retract

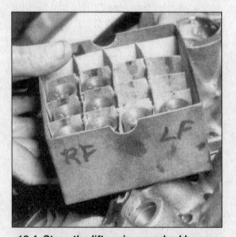

12.4 Store the lifters in a marked box so they can be returned to their original positions

torque listed in this Chapter's Specifications.
24 Place the sprockets in the chain with their timing marks aligned. The mark on the crankshaft sprocket will be in the 12-o'clock position and the mark on the camshaft sprocket will be in the 6-o'clock position. Install the sprockets and chain together on the crankshaft and camshaft.
25 Install the crankshaft key.
26 Make sure the sprocket timing marks are still aligned. The guide side of the chain must be straight, without any slack, for the marks to align accurately.
27 Install the camshaft sprocket bolt and tighten to the torque listed in this Chapter's Specifications.
28 Squeeze the tensioner pad with your fingers. At the same time, use a sharp-tipped probe to push the ratchet mechanism down, then in. This will release the latch and let the tensioner retract (see illustration). Retain the

tensioner pad in the retracted position by holding it or installing a clip or similar device.
29 Install the timing chain tensioner and tighten its bolts to the torque listed in this Chapter's Specifications. Remove the clip.
30 The remainder of installation is the reverse of the removal Steps.
31 Run the engine and check for oil or coolant leaks.

12 Valve lifters - removal, inspection and installation

Refer to illustration 12.4

Removal

1 Remove the intake manifold (see Section 7).
2 Remove the cylinder heads (see Section 9).
3 The lifters protrude from the bores, so if there isn't a lot of varnish buildup, simply pull

them out of their bores with your fingers. On engines with a lot of sludge and varnish, work the lifters up and down, using carburetor spray cleaner to loosen the deposits. If the lifters are particularly stubborn, special tools designed to grip and remove lifters are manufactured by many tool companies and are widely available.
4 Before removing the lifters, arrange to store them in a clearly labeled box to ensure that they're installed in their original locations (see illustration).
5 Remove the lifters and store them where they won't get dirty.

Inspection

Refer to Chapter 2E.

Installation

6 Install the lifters in the bores. Coat them with moly-based grease or engine assembly lube. Note that when the lifters are installed, the alignment tab must fit in the groove in the lifter bore.
7 The remainder of installation is the reverse of the removal Steps.

13 Camshaft - removal, inspection and installation

Endplay check

1 Refer to Section 11 for this procedure.

Lobe lift check

Refer to Chapter 2E.

Removal

Refer to illustrations 13.9a, 13.9b, 13.17 and 13.18
2 Disconnect the negative cable from the battery.
3 Drain the cooling system and engine oil (see Chapter 1).

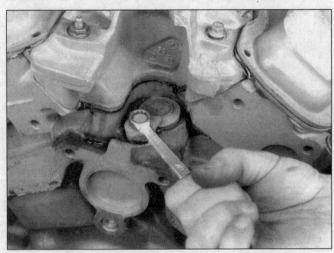

13.9a Remove the oil pump drive gear bolt . . .

13.9b . . . and lift the gear out of the block

4   Remove the radiator and fan (see Chapter 3).

5   Remove the drivebelt (see Chapter 1).

6   Remove the spark plug wires (see Chapter 1).

7   Remove the ignition coil pack and bracket (see Chapter 5).

8   Remove the alternator (see Chapter 5).

9   Remove the oil pump drive gear from the block **(see illustrations).**

10   Remove the intake manifold (see Section 7). The upper intake manifold can be left attached to the lower manifold.

11   Remove the valve covers (see Section 4).

12   Remove the rocker arms and pushrods (see Section 5).

13   Remove the valve lifters (see Section 12).

14   Remove the oil pan (see Section 14).

15   Remove the timing chain cover (see Section 10).

16   Remove the timing chain and sprockets (see Section 11).

17   Remove the camshaft thrust plate bolts with a Torx bit **(see illustration)** and take the thrust plate off the block.

18   Carefully pull the camshaft out of the block, rotating it as you pull **(see illustration).** Don't let the camshaft lobes or journals nick the camshaft bearings.

Inspection

Refer to Chapter 2E.

Installation

19   Lubricate the camshaft bearing journals and cam lobes with moly-base grease or engine assembly lube.

20   Slide the camshaft into the engine. Support the cam near the block and be careful not to scrape or nick the bearings.

21   Coat both sides of the thrust plate with assembly lube or SAE 50 engine oil, then install it. Tighten its bolts to the torque listed in this Chapter's Specifications.

22   The remainder of installation is the reverse of the removal Steps.

14   Oil pan and baffle - removal and installation

Removal

Refer to illustrations 14.4 and 14.5.

Note: *The factory recommends removal of the transmission when removing the oil pan to assure proper pan sealing upon re-installation. However, if proper care is exercised during re-sealing, removal of the transmission is not necessary.*

1   Disconnect the battery negative cable.

2   Remove the starter motor (refer to Chapter 5).

3   Remove the transmission-to-oil pan bolts and spacers at the two mounting pad locations.

4   Remove the oil pan-to-block fasteners along with the two Torx-type fasteners at rear of the crankcase on each side of the oil seal **(see illustration).** Remove the pan from the engine.

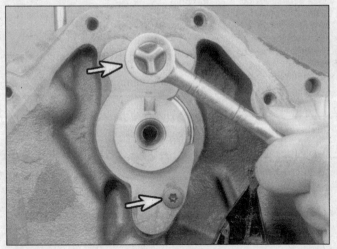

13.17 Remove the camshaft thrust plate bolts with a Torx bit and take the thrust plate off - note which end of the thrust plate is up so it can be installed correctly

13.18 As you're withdrawing it, support the camshaft near the block with both hands - be careful not to let the camshaft nick the bearings

14.4 Use a Torx driver to remove the two bolts (arrows) adjacent to the crankshaft rear oil seal

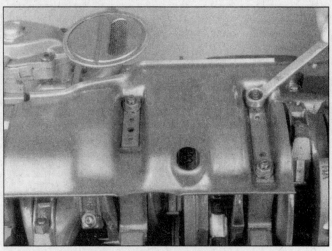

14.5 The oil baffle is secured to the main bearing caps

5   If necessary, remove the baffle fasteners and remove the baffle (see illustration).
6   Using a gasket scraper, thoroughly clean all old gasket material from the pan and its mounting surfaces. Remove residue and oil film with solvent such as acetone or lacquer thinner.

Installation

Refer to illustrations 14.12a, 14.12b and 14.13.
Caution: *Since two of the oil pan bolts are attached to the transmission, spacers are used between the oil pan and transmission to allow for standard manufacturing and installation tolerances. The spacer gap must be measured and the correct spacer selected and installed or oil pan damage and/or oil leaks may occur.*
7   Place the oil pan on the engine and install several fasteners finger tight. Measure the gap between the oil pan and the transmission at the spacer mounting pad locations.
8   If the engine is out of the vehicle, place a straightedge across the transmission mounting surface on the block and one of the mounting pad locations and measure the gap.
9   Record the gap measurement.
10   Move to the second mounting pad location and record the measurement. Remove the oil pan.
11   Verify that the original spacers are of the correct thickness or select new spacers to compensate for the gap. Spacers are available in three thicknesses from your dealer (see this Chapter's Specifications).
12   At the rear of the engine, apply a thin bead of RTV sealant to the outside edge of the gasket surface and another bead to the area around the crankshaft rear oil seal (see illustration). At the front of the engine, apply a bead of sealant to each of the engine block-to-timing-chain cover seams and to the corners of the front seal cover (see illustration).
13   Install a new oil pan gasket on the engine and apply sealant around the front and rear oil seal areas. Insert the spacers into the oil pan mounting pads and install the pan with the fasteners finger tight. Gently sung up the oil pan to transmission bolts (correct spacers should be installed). Next, tighten the oil pan to block bolts to the correct torque as listed in this Chapter's Specifications. The long screw next to the crankshaft rear oil seal passes through the boss on the pan (see illustration).
14   Apply final torque to the oil pan to transmission fasteners as listed in this Chapter's Specifications. If the engine is out of the vehicle, remember to install the spacers between the mounting pad and transmission before bolting the engine to the transmission.

15   Oil pump - removal and installation

Removal

Refer to illustrations 15.2 and 15.3
1   Remove the oil pan (see Section 14).

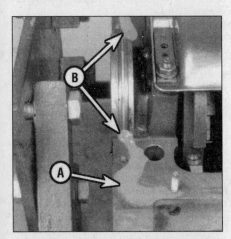

14.12a At the rear of the engine, apply thin beads of sealant to the outside edges of the gasket surface (A) and also to the crankshaft rear oil seal corners (B)

14.12b Apply beads of sealant to the corners of the front seal cover (arrows) and at the seams where the timing chain cover meets the engine block

14.13 The long Torx screw next to the crankshaft rear oil seal passes through a boss on the pan, then through the pan flange and into the block

15.2 Remove the oil pump bolts and lift the pump off the engine

15.3 As you remove the intermediate driveshaft, note the different shapes of its ends and the location of the retainer ring on the shaft

2   Remove the oil pump bolts and take the pump off the engine **(see illustration)**.
3   Pull the oil pump intermediate driveshaft out of the engine **(see illustration)**.
4   Unbolt the inlet tube and screen from the pump.

Installation

5   Fill one of the pump ports with clean engine oil and rotate the pump by hand to prime it.
6   Install the oil pump pick-up tube on the pump, using a new gasket.
7   Insert the intermediate driveshaft into the engine, pointed end first. The pressed-on retaining ring on the shaft should be positioned 4.2-inches from the oil pump end of the shaft.
8   Install the oil pump on the block, using a new gasket. Install the oil pump bolts and tighten to the torque listed in this Chapter's Specifications.
9   The remainder of installation is the reverse of removal.
10   Run the engine and make sure oil pressure comes up to normal quickly. If it doesn't, stop the engine and find out the cause. Severe engine damage can result from running an engine with insufficient oil pressure!

16 Crankshaft oil seals - replacement

Front seal - timing chain cover in place

1   Refer to Section 10 for this procedure.

Front seal - timing chain cover removed

2   Support the timing chain cover on wooden blocks, then drive out the seal with a punch and hammer.
3   Check the seal bore in the timing chain

cover for nicks or burrs.
4   Coat the seal lip and outer circumference with engine oil.
5   Position the new seal with its lip facing IN. Drive in the new seal with a seal driver. If a seal driver isn't available, use a socket or piece of pipe the same diameter as the seal.

Rear seal

6   Remove the transmission (see Chapter 7).
7   Remove the clutch (if equipped) (see Chapter 8).
8   Remove the flywheel or driveplate (see Section 17).
9   Remove the engine rear plate.
10   With a sharp awl or similar tool, punch two holes in the seal on opposite sides just above the point where the main bearing cap meets the cylinder block.
11   Thread a sheet metal screw into each hole. Pry against the sheet metal screws with two large screwdrivers or small prybars to remove the seal. **Note:** *If you need a fulcrum to pry against, use two small blocks of wood.*
Caution: *Don't scratch or gouge the crankshaft seal surface.*
12   Clean the oil seal bore in the block and main bearing cap. Check the seal bore and crankshaft sealing surface for nicks or burrs.
13   Apply a thin coat of clean engine oil to the outer diameter of the seal. Apply a thin coat of Lubriplate or equivalent to the contact surfaces of the seal and crankshaft.
14   Position the seal in the bore with its open end facing IN.
15   Drive the seal in with a seal driver until it is securely seated. Use a socket or piece of pipe the same diameter as the seal if a seal driver isn't available.
16   After installation, squareness of the rear seal with the crankshaft centerline and the dimension from the seal to the rear face of the block must be within the tolerances listed in this Chapter's Specifications.

17 Flywheel/driveplate - removal and installation

Refer to illustrations 17.3 and 17.4

1   Raise the vehicle and support it securely on jackstands, then refer to Chapter 7 and remove the transmission. If it's leaking, now would be a very good time to replace the front pump seal/O-ring (automatic transmission only).
2   Remove the pressure plate and clutch disc (see Chapter 8) (manual transmission equipped vehicles). Now is a good time to check/replace the clutch components and pilot bearing.
3   Paint alignment marks on the flywheel/driveplate and crankshaft to ensure correct alignment during reinstallation **(see illustration)**.
4   Remove the bolts that secure the flywheel/driveplate to the crankshaft **(see illustration)**. If the crankshaft turns, wedge a screwdriver through the starter opening to jam the flywheel.
5   Remove the flywheel/driveplate from the crankshaft. Since the flywheel is fairly heavy, be sure to support it while removing the last bolt.
6   Clean the flywheel to remove grease and oil. Inspect the surface for cracks, rivet grooves, burned areas and score marks. Light scoring can be removed with emery cloth. Check for cracked and broken ring gear teeth. Lay the flywheel on a flat surface and use a straightedge to check for warpage.
7   Clean and inspect the mating surfaces of the flywheel/driveplate and the crankshaft. If the crankshaft rear seal is leaking, replace it before reinstalling the flywheel/driveplate.
8   Position the flywheel/driveplate against the crankshaft. Be sure to align the marks made during removal. Note that some engines have an alignment dowel or staggered bolt holes to ensure correct installation. Before installing the bolts, apply thread

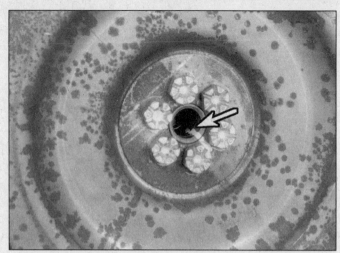

17.3 Paint alignment marks on the flywheel/driveplate and crankshaft - note how deep the paint mark must go to reach t he crankshaft

17.4 Remove the mounting bolts and take the flywheel/driveplate off the crankshaft - the manual transmission flywheel is heavy, so be sure to support it during removal

locking compound to the threads.

9   Wedge a screwdriver through the starter motor opening to keep the flywheel/drive-plate from turning as you tighten the bolts to the torque listed in this Chapter's Specifications.

10   The remainder of installation is the reverse of the removal procedure.

18   Engine mounts - check and replacement

1   Engine mounts seldom require attention, but broken or deteriorated mounts should be replaced immediately or the added strain placed on the driveline components may cause damage or wear.

Check and replacement

2   During the check, the engine must be raised slightly to remove the weight from the mounts.

3   Raise the vehicle and support it securely on jackstands, then position a jack under the engine oil pan. Place a large block of wood between the jack head and the oil pan, then carefully raise the engine just enough to take the weight off the mounts. **Warning:** *DO NOT place any part of your body under the engine when it's supported only by a jack!*

4   Check the mounts to see if the rubber is cracked, hardened or separated from the metal plates. Sometimes the rubber will split right down the center.

5   Check for relative movement between the mount plates and the engine or frame (use a large screwdriver or pry bar to attempt to move the mounts). If movement is noted, lower the engine and tighten the mount fasteners.

6   Rubber preservative should be applied to the mounts to slow deterioration.

Replacement

7   Disconnect the negative battery cable from the battery, then raise the vehicle and support it securely on jackstands (if not already done).

8   Remove the fasteners and detach the mount from the frame bracket.

9   Raise the engine slightly with a jack or hoist (make sure the fan doesn't hit the radiator or shroud). Remove the mount-to-block bolts and detach the mount.

10   Installation is the reverse of removal. Use thread locking compound on the mount bolts and be sure to tighten them securely. Cleaning the engine compartment and engine prior to removal will help keep tools clean and organized.

Notes

Chapter 2 Part E
General engine overhaul procedures

Contents

Specifications

2.3L four-cylinder engine

General

Oil pressure ... 40 to 60 psi @ 2000 rpm
Compression pressure ... Lowest reading cylinder must be within 75% of highest reading cylinder (100 psi minimum)

Valves and related components

Valve face angle ... 44°
Valve seat angle ... 45°
Valve margin width ... 1/32-in
Valve stem diameter
  Intake ... 0.3416 to 0.3423 in
  Exhaust ... 0.3411 to 0.3418 in
Valve stem-to-guide clearance
  Intake ... 0.0010 to 0.0027 in
  Exhaust ... 0.0015 to 0.0032 in
Valve seat width
  Intake ... 0.060 to 0.080 in
  Exhaust ... 0.070 to 0.090 in
Valve spring free length ... 1.877 in
Valve spring installed height ... 1.49 to 1.55 in
Head surface warpage limit ... 0.006 in over the length of the head

2.3L four-cylinder engine (continued)

Crankshaft and connecting rods

Crankshaft end play	0.004 to 0.008 in
Connecting rod side clearance	0.0035 to 0.0105 in
Main bearing journal diameter	2.399 to 2.3982 in
Out-of-round limit	0.0006 in maximum
Taper limit	0.0006 in per in
Journal runout limit	0.002 in maximum
Runout service limit	0.005 in
Main bearing oil clearance	
Nominal	0.0008 to 0.0015 in
Allowable	0.0008 to 0.0026 in
Rear oil seal installed depth	0.005 in maximum
Connecting rod bearing journal diameter	2.0462 to 2.0472 in
Connecting rod bearing oil clearance	
Nominal	0.0008 to 0.0015 in
Allowable	0.0008 to 0.0026 in

Engine block

Cylinder bore	3.7795 to 3.7825 in
Out-of-round limit	0.0015 in
Taper limit	0.010 in

Pistons and rings

Compression ring side clearance	0.0020 to 0.0040 in
Piston diameter	3.7780 to 3.7786 in
Piston-to-bore clearance (select fit)	0.0014 to 0.0022 in
Piston ring groove width	
Top and bottom compression	0.080 to 0.081 in
Oil ring	0.189 to 0.190 in
Piston ring end gap	
Compression	0.010 to 0.020 in
Oil ring	0.015 to 0.055 in

Camshaft

Lobe lift (intake and exhaust)	0.2381 in
Bearing journal diameter	1.7713 to 1.7720 in
Bearing oil clearance	0.001 to 0.003 in
Camshaft end play	0.001 to 0.007 in

Torque specifications*

	Ft-lbs
Main bearing cap bolts	
Step 1	50 to 60
Step 2	80 to 85
Connecting rod nuts	
Step 1	25 to 30
Step 2	30 to 36
Subframe-to-frame bolts	187 to 260

* Note: *Refer to Part A for additional torque specifications.*

2.8L V6 engine

General

Oil pressure	40 to 60 psi @ 2000 rpm
Compression pressure	Lowest reading cylinder must be within 75% of highest reading cylinder (100 psi minimum)

Valves and related components

Valve face angle	44 degrees
Valve seat angle	45 degrees
Valve margin width	1/32-in
Valve stem-to-guide clearance	
Intake	0.0008 to 0.0025 in
Exhaust	0.0018 to 0.0035 in
Valve seat width	0.060 to 0.079 in
Valve spring installed height	1-37/64 to 1-39/64-in
Valve spring free length	1.91 in
Head warpage limit	0.006 in over the length of the head

Crankshaft and connecting rods

Crankshaft end play	0.004 to 0.008 in
Connecting rod side clearance	0.004 to 0.011 in
Main bearing journal	
Diameter	2.2433 to 2.2441 in
Out-of-round limit	0.0006 in
Taper limit	0.0006 in
Runout limit	0.002 in
Main bearing oil clearance	
Nominal	0.0008 to 0.0015 in
Allowable	0.0005 to 0.0019 in
Connecting rod journal	
Diameter	2.1252 to 2.1260 in
Out-of-round limit	0.0006 in
Taper limit	0.0006 in
Connecting rod bearing oil clearance	
Nominal	0.0006 to 0.0016 in
Allowable	0.0005 to 0.0022 in

Engine block

Cylinder bore diameter	3.6614 to 3.6630 in
Out-of-round	
Nominal	0.0015 in
Service limit	0.005 in
Taper limit	0.010 in

Pistons and rings

Compression ring side clearance	0.0020 to 0.0033 in
Oil ring side clearance	Snug fit
Piston-to-bore clearance	0.0011 to 0.0019 in
Piston ring end gap	
Compression rings	0.015 to 0.023 in
Oil ring	0.015 to 0.055 in

Camshaft

Lobe lift	0.2555 in
Bearing journal diameter	
Number 1	1.7285 to 1.7293 in
Number 2	1.7135 to 1.7143 in
Number 3	1.6985 to 1.6992 in
Number 4	1.6835 to 1.6842 in
Bearing oil clearance	0.0017 in

Torque specifications*

	Ft-lbs
Connecting rod nuts	19 to 24
Main bearing cap bolts	65 to 75
Subframe-to-frame bolts	187 to 260

* **Note:** *Refer to Part B for additional torque specifications*

3.0L V6 engine

General

Oil pressure	40 to 60 psi @ 2500 rpm
Compression pressure	Lowest reading cylinder must be within 75% of highest reading cylinder (100 psi minimum)

Valves and related components

Valve face angle	44 degrees
Valve seat angle	45 degrees
Valve margin width	1/32-in
Valve stem-to-guide clearance	
Intake	0.001 to 0.0027 in
Exhaust	0.0015 to 0.0032 in
Valve seat width	0.060 to 0.080 in
Valve spring installed height	
1987	1.58 in
1988	1.85 in
Valve spring free length (1987 only)	1.84 in
Head warpage limit	0.006 in over the length of the head

3.0L V6 engine (continued)

Crankshaft and connecting rods

Crankshaft end play	0.004 to 0.008 in
Connecting rod side clearance	0.006 to 0.014 in
Main bearing journal	
Diameter	2.5190 to 2.5198 in
Out-of-round limit	0.0003 in
Taper limit	0.0006 in
Runout limit	0.002 in
Main bearing oil clearance	
Nominal	0.001 to 0.0014 in
Allowable	0.0005 to 0.0023 in
Connecting rod journal	
Diameter	2.1253 to 2.1261 in
Out-of-round limit	0.0003 in
Taper limit	0.0006 in
Connecting rod bearing oil clearance	
Nominal	0.001 to 0.0014 in
Allowable	0.00086 to 0.0027 in

Engine block

Cylinder bore diameter	3.504 in
Out-of-round	
Nominal	0.001 in
Service limit	0.002 in
Taper limit	0.002 in

Pistons and rings

Compression ring side clearance	0.0016 to 0.0037 in
Oil ring side clearance	Snug fit
Piston-to-bore clearance	0.0012 to 0.0023 in
Piston ring end gap	
Compression rings	0.010 to 0.020 in
Oil ring	0.010 to 0.049 in

Camshaft

Lobe lift	0.260 in
Bearing journal diameter	2.0074 to 2.0084 in
Bearing oil clearance	
Standard	0.001 to 0.0026 in
Service limit	0.006 in

Torque specifications*

	Ft-lbs
Connecting rod nuts	26
Main bearing cap bolts	
1988 and earlier	65 to 81
1989 through 1992	66
1993 and later	59
Subframe-to-frame bolts	187 to 260

* **Note:** *Refer to Part C for additional torque specifications.*

4.0L V6 engine

General

Displacement	4.0 liters
Cylinder compression pressure	Lowest reading cylinder must be within 75% of highest reading cylinder (100 psi minimum)
Oil pressure (engine warm at 2000 rpm)	40 to 60 psi

Cylinder head warpage limit

Cylinder head warpage limit	0.003 inch per 6 inches; 0.006 inch overall

Valves and related components

Minimum valve margin width	1/32 inch
Intake valve	
Seat angle	45-degrees
Seat width	0.060 to 0.079 inch
Seat runout limit	0.0015 inch total indicator reading
Stem diameter	0.3159 to 0.3167 inch
Stem-to-guide clearance	0.0008 to 0.0025 inch
Valve face runout limit	0.002 inch
Valve face angle	44-degrees

Exhaust valve
   Seat angle 45-degrees
   Seat width .... 0.060 to 0.079 inch
   Seat runout limit .... 0.0015 inch total indicator reading
   Stem diameter .... 0.3149 to 0.3156 inch
   Stem-to-guide clearance .... 0.0018 to 0.0035 inch
   Valve face runout limit .... 0.002 inch
   Valve face angle 44-degrees
Valve spring
   Pressure .... 60.0 to 68.0 lbs at 1.585 inch
   Service limit .... 10% pressure loss at specified length
   Free length (approximate) .... 1.91 inch
   Installed height .... 1-37/64 to 1-39/64 inch
   Maximum out-of-square .... 5/64 inch
Valve lifter
   Diameter (standard) .... 0.8742 to 0.8755 inch
   Lifter-to-bore clearance
      Standard .... 0.0005 to 0.0022 inch
      Service limit .... 0.005 inch

Crankshaft and connecting rods

Connecting rod journal
   Diameter (standard) .... 2.1252 to 2.1260 inches
   Out-of-round limit .... 0.0006 inch
   Taper limit .... 0.0006 inch per inch
   Bearing oil clearance
      Desired .... 0.0005 to 0.0022 inch
      Allowable .... 0.0003 to 0.0024 inch

Crankshaft and connecting rods

Connecting rod side clearance (endplay)
   Standard .... 0.0002 to 0.0025 inch
   Service limit .... 0.014 inch
Main bearing journal
   Diameter .... 2.2433 to 2.2441 inches
   Out-of-round limit .... 0.0006 inch
   Taper limit .... 0.0006 inch per inch
   Bearing oil clearance
      Desired .... 0.0008 to 0.0015 inch
      Allowable .... 0.0005 to 0.0019 inch
Crankshaft endplay
   Standard .... 0.0016 to 0.0126 inch
   Service limit .... 0.012 inch

Cylinder bore

Diameter .... 3.9527 to 3.9543 inches
Out-of-round service limit .... 0.005 inch
Taper service limit .... 0.010 inch

Pistons and rings

Piston diameter .... 3.9524 to 3.9531 inches
Piston-to-bore clearance .... 0.0008 to 0.0019 inch
Piston ring end gap
   Compression rings .... 0.015 to 0.023 inch
   Oil ring rails .... 0.015 to 0.055 inch
Piston ring side clearance
   Compression rings
      Standard .... 0.0020 to 0.0033 inch
      Service limit .... 0.006 inch
   Oil ring .... Snug fit

Torque specifications*   Ft-lbs

Main bearing cap bolts .... 66 to 77
Connecting rod cap nuts .... 18 to 24
Subframe-to-frame bolts .... 187 to 260

*Note: *Refer to Part D for additional torque specifications.*

1   General information

Included in this portion of Chapter 2 are the general overhaul procedures for the cylinder heads and internal engine components. The information ranges from advice concerning preparation for an overhaul and the purchase of replacement parts to detailed, step-by-step procedures covering removal and installation of internal engine components and the inspection of parts.

The following Sections have been written based on the assumption that the engine has been removed from the vehicle. For information concerning in-vehicle engine repair, as well as removal and installation of the external components necessary for the overhaul, see Part A, B, C or D of Chapter 2 (depending on engine type) and Section 7 of this Part.

The Specifications included here in Part E are only those necessary for the inspection and overhaul procedures which follow. Refer to Part A, B, C or D for additional Specifications related to the various engines covered in this manual.

2   Cylinder compression check

Refer to illustration 2.4
1   A compression check will tell you what mechanical condition the upper end (pistons, rings, valves, head gaskets) of the engine is in. Specifically, it can tell you if the compression is down due to leakage caused by worn piston rings, defective valves and seats or a blown head gasket. **Note:** *The engine must be at normal operating temperature for this check and the battery must be fully charged.*
2   Begin by cleaning the area around the spark plugs before you remove them (compressed air works best for this). This will prevent dirt from getting into the cylinders as the compression check is being done. Remove all of the spark plugs from the engine.
3   Block the throttle wide open. Disconnect the coil wire from the distributor cap and ground it. On the 4.0L engine, disable the ignition system by detaching the electrical connector from the coil pack above the left valve cover. The fuel pump circuit should also be disabled (see Chapter 4).
4   With the compression gauge in the number one spark plug hole **(see illustration)**, crank the engine over at least four compression strokes and watch the gauge. The compression should build up quickly in a healthy engine. Low compression on the first stroke, followed by gradually increasing pressure on successive strokes, indicates worn piston rings. A low compression reading on the first stroke, which does not build up during successive strokes, indicates leaking valves or a blown head gasket (a cracked head could also be the cause). Record the highest gauge reading obtained.
5   Repeat the procedure for the remaining

2.4 Access for compression check is through wheel wells on V6 engines (the gauge shows a low reading cylinder)

cylinders. The lowest reading cylinder should be within 75-percent of the highest reading cylinder.
6   Add some engine oil (about three squirts from a plunger-type oil can) to each cylinder, through the spark plug hole, and repeat the test.
7   If the compression increases after the oil is added, the piston rings are definitely worn. If the compression does not increase significantly, the leakage is occurring at the valves or head gasket. Leakage past the valves may be caused by burned valve seats and/or faces or warped, cracked or bent valves.
8   If two adjacent cylinders have equally low compression, there is a strong possibility that the head gasket between them is blown. The appearance of coolant in the combustion chambers or the crankcase would verify this condition.
9   If the compression is unusually high, the combustion chambers are probably coated with carbon deposits. If that is the case, the cylinder heads should be removed and decarbonized.
10   If compression is way down or varies greatly between cylinders, it would be a good idea to have a leak-down test performed by an automotive repair shop. This test will pinpoint exactly where the leakage is occurring and how severe it is.

3   Engine overhaul - general information

It's not always easy to determine when, or if, an engine should be completely overhauled, as a number of factors must be considered.

High mileage is not necessarily an indication that an overhaul is needed, while low mileage doesn't preclude the need for an overhaul. Frequency of servicing is probably the most important consideration. An engine that's had regular and frequent oil and filter changes, as well as other required maintenance, will most likely give many thousands of miles of reliable service. Conversely, a neglected engine may require an overhaul very early in its life.

Excessive oil consumption is an indication that piston rings, valve seals and/or valve guides are in need of attention. Make sure that oil leaks aren't responsible before deciding that the rings and/or guides are bad. Have a cylinder compression or leakdown test performed by an experienced tune-up mechanic to determine the extent of the work required.

If the engine is making obvious knocking or rumbling noises, the connecting rod and/or main bearings may be at fault. Check the oil pressure with a gauge installed in place of the oil pressure sending unit and compare it to the Specifications. If it's extremely low, the bearings and/or oil pump are probably worn out.

a) *On four-cylinder engines the oil pressure switch is located at the left rear of the engine*
b) *On the 2.8L V6 the oil pressure sending unit is located adjacent to the fuel pump*
c) *On the 3.0L V6 engine the oil pressure switch or sending unit (depending on type of gauges) is located at the right rear corner of the engine*
d) *On the 4.0L V6 the oil pressure sending unit (switch) is located on the driver's side of the engine block next to the water pump inlet*

Loss of power, rough running, excessive valve train noise and high fuel consumption rates may also point to the need for an overhaul, especially if they're all present at the same time. If a complete tune-up doesn't remedy the situation, major mechanical work is the only solution.

An engine overhaul involves restoring the internal parts to the specifications of a new engine. During an overhaul, the piston rings are replaced and the cylinder walls are reconditioned (rebored and/or honed). If a rebore is done, new pistons are required. The main bearings, connecting rod bearings and camshaft bearings are generally replaced with new ones and, if necessary, the crankshaft may be reground to restore the journals. Generally, the valves are serviced as well, since they're usually in less-than-perfect condition at this point. While the engine is being overhauled, other components, such as the distributor, starter and alternator, can be rebuilt as well. The end result should be a like new engine that will give many trouble free miles. **Note:** *Critical cooling system components such as the hoses, drivebelts, thermostat and water pump MUST be replaced with new parts when an engine is overhauled. The radiator should be checked carefully to ensure that it isn't clogged or leaking; if in doubt, replace it with a new one. Also, we don't recommend overhauling the oil pump - always install a new one when an engine is rebuilt.*

Before beginning the engine overhaul, read through the entire procedure to familiarize yourself with the scope and requirements of the job. Overhauling an engine isn't difficult, but it is time consuming. Plan on the vehicle being tied up for a minimum of two weeks, especially if parts must be taken to an automotive machine shop for repair or reconditioning. Check on availability of parts and make sure that any necessary special tools and equipment are obtained in advance. Most work can be done with typical hand tools, although a number of precision measuring tools are required for inspecting parts to determine if they must be replaced. Often an automotive machine shop will handle the inspection of parts and offer advice concerning reconditioning and replacement. **Note:** *Always wait until the engine has been completely disassembled and all components, especially the engine block, have been inspected before deciding what service and repair operations must be performed by an automotive machine shop.* Since the block's condition will be the major factor to consider when determining whether to overhaul the original engine or buy a rebuilt one, never purchase parts or have machine work done on other components until the block has been thoroughly inspected. As a general rule, time is the primary cost of an overhaul, so it doesn't pay to install worn or substandard parts.

As a final note, to ensure maximum life and minimum trouble from a rebuilt engine, everything must be assembled with care in a spotlessly clean environment.

4   Engine rebuilding alternatives

The do-it-yourselfer is faced with a number of options when performing an engine overhaul. The decision to replace the engine block, piston/connecting rod assemblies and crankshaft depends on a number of factors, with the number one consideration being the condition of the block. Other considerations are cost, access to machine shop facilities, parts availability, time required to complete the project and experience.

Some of the rebuilding alternatives include:

Individual parts - If the inspection procedures reveal that the engine block and most engine components are in reusable

condition, purchasing individual parts may be the most economical alternative. The block, crankshaft and piston/connecting rod assemblies should all be inspected carefully. Even if the block shows little wear, the cylinder bores should receive a finish hone.

Crankshaft kit - This rebuild package usually consists of a reground crankshaft, main and rod bearings, and a matched set of pistons with rings and connecting rods. The pistons will already be installed on the connecting rods. These kits are commonly available for standard cylinder bores, as well as for engine blocks which have been bored to a regular oversize.

Short block - A short block consists of an engine block with a crankshaft and piston/connecting rod assemblies already installed. All new bearings are incorporated and all clearances will be correct. Depending on where the short block is purchased, a guarantee may be included. The existing camshaft, valve train components, cylinder head and external parts can be bolted to the short block with little or no machine shop work necessary.

Long block - A long block consists of a short block plus an oil pump, oil pan, cylinder head, rocker arm cover, camshaft and valve train components, timing sprockets and chain and timing chain cover. All components are installed with new bearings, seals and gaskets incorporated throughout. The installation of manifolds and external parts is all that is necessary. Some form of guarantee is usually included with the purchase.

Give careful thought to which alternative is best for you and discuss the situation with local automotive machine shops, auto parts dealers and experienced rebuilders before ordering or purchasing replacement parts.

5   Engine removal - methods and precautions

If it has been decided that an engine must be removed for overhaul or major repair work, certain preliminary steps should be taken.

It is important to note that all engines installed in the Aerostar are removed from the bottom - dropping the engine/subframe assembly and lifting away the vehicle body. For this reason, the removal and installation process is somewhat unconventional and requires lifting devices and tools not common to a home mechanic. If it is decided to attempt this procedure, read through the instructions in Section 6 and ensure that the job can be performed safely before beginning.

Locating a suitable work area is extremely important. A shop is, of course, the most desirable place to work. Adequate work space, along with storage space for the vehicle, is very important.

Cleaning the engine compartment and

engine prior to removal will help keep tools clean and organized.

A vehicle hoist will also be necessary. Make sure that the equipment is rated in excess of the combined weight of the vehicle and its contents. Safety is of primary importance, considering the potential hazards involved in lifting the vehicle off of the engine and subframe.

If the engine is being removed by a novice, a helper should be available. Advice and aid from someone more experienced would also be helpful. There are many instances when one person cannot simultaneously perform all of the operations required when lifting the vehicle away from the engine and subframe.

Plan the operation ahead of time. Arrange for or obtain all of the tools and equipment you will need prior to beginning the job. Some of the equipment necessary to perform engine removal and installation safely and with relative ease are (in addition to a vehicle hoist) a heavy duty floor jack, complete sets of wrenches and sockets as described in the front of this manual, wooden blocks, a heavy duty dolly and plenty of rags and cleaning solvent for mopping up the inevitable spills. You will also need a way to lift the engine from the subframe to an engine stand.

Plan for the vehicle to be out of use for a considerable amount of time. A machine shop will be required to perform some of the work which the do-it-yourselfer cannot accomplish due to a lack of special equipment. These shops often have a busy schedule, so it would be wise to consult them before removing the engine in order to accurately estimate the amount of time required to rebuild or repair components that may need work.

Always use extreme caution when removing and installing the engine. Serious injury can result from careless actions. Plan ahead. Take your time and a job of this nature, although major, can be accomplished successfully.

6   Engine - removal and installation

Refer to illustrations 6.17, 6.21, 6.23, 6.29 and 6.32

Warning: *Have a dealer or air conditioning shop discharge the system prior to disconnecting any a/c components.*

1   On a/c equipped four-cylinder models only, have the refrigerant discharged by a professional.

2   Place the van on a hoist designed to lift vehicles for servicing. It must be configured to allow engine/subframe removal out the bottom of the vehicle. Drive-on ramp and axle lift (twin post) hoists won't work. Center post hoists may work if the transmission is removed first.

3   Relieve the fuel system pressure (Chapter 4).

6.17 Remove all ground wires, such as this one behind the dipstick bracket (2.8L V6) - hoses and wires can be difficult to see so search carefully

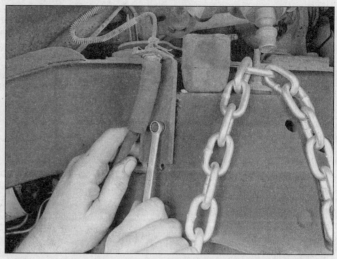

6.21 Unbolt the brake line brackets prior to subframe removal

4   Disconnect and remove the battery (Chapter 5).
5   Drain the cooling system (Chapter 1).
6   Remove the engine cover under the dash (Chapter 11).
7   Remove the right front seat for access (Chapter 11).
8   Remove the air cleaner assembly (Chapter 4).
9   Remove the upper and lower radiator hoses and disconnect the heater hoses.
10   Remove the radiator and shroud (Chapter 3).
11   Disconnect the optional auxiliary transmission cooler lines.
12   Remove the drivebelts (Chapter 1).
13   Remove the engine cooling fan (Chapter 3).
14   On a/c equipped four-cylinder models, disconnect the refrigerant lines from the compressor and plug the open fittings. On V6 models, unplug the wiring connector and unbolt the compressor from the engine without disconnecting the lines. Tie the compressor up out of the way.
15   Disconnect the throttle cable, transmission kickdown linkage and cruise control cable (as equipped), from the carburetor or throttle body and unbolt its retaining bracket (Chapter 4).
16   Label the vacuum lines, emission and coolant hoses, wiring connectors, ground straps and fuel lines to ensure correct reinstallation.
17   Carefully disconnect the vacuum lines, coolant and emission hoses, fuel lines, ground straps **(see illustration)** and electrical connectors which connect the engine to the vehicle. Refer to Chapters 4, 5 and 6 as needed.
18   Remove the starter motor (Chapter 5).
19   Set the front wheels and steering wheel in the straight ahead position (centered). Remove the bolt holding the intermediate steering shaft to the steering gear and sepa-

rate the shaft from the gear (Chapter 10).
20   Remove the front wheels.
21   Remove the front brake calipers (Chapter 9) without disconnecting the hoses. Unbolt the brake line brackets from the subframe **(see illustration)** and tie the calipers out of the way.
22   Remove the front stabilizer bar end nuts (Chapter 10).
23   Wrap the lower control arms with safety chains **(see illustration)** and separate the upper balljoints from the spindles (Chapter 10).
24   On manual transmission models, remove the floorshift (Chapter 7) and disconnect the hydraulic hose at the slave cylinder.
25   Remove the catalytic converter (Chapter 4).
26   Disconnect the transmission linkage, wiring harness and speedometer cable, cooling lines (automatic transmission only) and driveshaft. Refer to Chapter 7 or 8 as needed.
27   The engine may be removed with the transmission still attached, although you may want to reduce the overall weight by removing the transmission at this time.
28   Carefully check to be sure everything connecting the engine/subframe to the vehicle has been disconnected except the engine mounts (and crossmember, if transmission is still attached).
29   Position a wheeled dolly (or wooden cradle) under the engine/subframe and lower the vehicle until the subframe just rests on the dolly **(see illustration)**. Place wooden blocks between the engine (and transmission, if attached) and dolly to keep it level. Secure the subframe to the dolly with safety chains.
30   If the transmission is still attached, place a jack under it and remove the crossmember.
31   With the engine/subframe securely supported, remove the three nuts from the mounting bolts which attach the subframe to the frame on each side of the vehicle.

6.23 Wrap the lower control arm with a safety chain to keep the coil spring caged

32   Carefully raise the vehicle, making sure nothing is still connected or snagged **(see illustration)**.
33   With the vehicle fully raised, unbolt the power steering pump and remove the torque converter and transmission (Chapter 7), if applicable. Remove the flywheel/driveplate. Attach a shop crane or other safe lifting device to the engine.
34   Unbolt the engine mounts from the block and lift the engine from the subframe.
35   Attach the engine to an engine stand or place it on a sturdy workbench.
36   Upon installation, bolt the engine to the subframe and position the engine/subframe unit correctly under the raised vehicle. Slowly and carefully lower the vehicle and bolt the unit to the frame. Ensure that the engine/transmission/subframe assembly is secure before

6.29 Gently lower the vehicle until the subframe rests on a strong dolly or wooden platform

6.32 Carefully raise the vehicle off the engine/subframe, making sure that nothing is still connected

connecting and installing the various components to complete the installation.

37   Add coolant, oil, power steering and transmission fluids as needed.

38   Run the engine and check for proper operation and leaks. Correct as needed.

39   Reinstall the engine cover and seat.

40   If the a/c system was discharged, take the vehicle to an air conditioning shop or service station for evacuation, recharging and leak testing.

41   Have the wheel alignment checked.

7   Engine overhaul - disassembly sequence

1   It is much easier to disassemble and work on the engine if it is mounted on an engine stand. These stands can often be rented for a reasonable fee from an equipment rental yard. Before the engine is mounted on a stand, the flywheel/driveplate should be removed from the engine.

2   If a stand is not available, it is possible to disassemble the engine with it blocked up on a sturdy workbench or on the floor. Be extra careful not to tip or drop the engine when working without a stand.

3   If you are going to obtain a rebuilt engine, all external components must come off your old engine first in order to be transferred to the replacement engine (just as they will if you are doing a complete engine overhaul yourself). These include:

Alternator and brackets
Emissions control components
Distributor (except 4.0L), spark plug wires
  and spark plugs
Coil pack (4.0L only)
Thermostat and housing
Water pump
Carburetor/fuel injection components
Intake/exhaust manifolds
Oil filter

Fuel pump
Engine mounts
Flywheel/driveplate

Note: *When removing the external components from the engine, pay close attention to details that may be helpful or important during installation. Note the installed position of gaskets, seals, spacers, pins, washers, bolts and other small items.*

4   If you are obtaining a short block, which consists of the engine block, crankshaft, pistons and connecting rods all assembled, then the cylinder head, oil pan and oil pump will have to be removed as well. See Section 4, *Engine rebuilding alternatives,* for additional information regarding the different possibilities to be considered.

5   If you are planning a complete overhaul, the engine should be dismantled to the basic "short block" stage, which includes the removal of:

Four-cylinder engine

Cam cover
Timing belt/sprocket cover
Timing belt and sprockets
Cylinder head
Oil pan

V6 engines

Valve covers
Rocker assemblies and pushrods
Intake and exhaust manifolds
Cylinder heads
Oil pan
Oil pump
Timing chain cover and cover plate
Timing chain and sprockets
Valve lifters
Camshaft
Piston/connecting rod assemblies
Crankshaft and main bearings

6   Before beginning the disassembly and overhaul procedures, make sure the following items are available:

Common hand tools
Small cardboard boxes or plastic bags for
  storing parts
Gasket scraper
Ridge reamer
Vibration damper puller
Micrometers and/or vernier caliper
Telescoping (snap) gauge
Dial indicator set
Valve spring compressor
Cylinder surfacing hone
Electric drill motor
Tap and die set
Wire brushes
Cleaning solvent

8   Cylinder head - disassembly

Refer to illustrations 8.3, 8.4 and 8.5

Note: *New and rebuilt cylinder heads are commonly available for most engines at dealerships and auto parts stores. Due to the fact that some specialized tools are necessary for the disassembly and inspection procedures, and replacement parts may not be readily available, it may be more practical and economical for the home mechanic to purchase a replacement head rather than taking the time to disassemble, inspect and recondition the original.*

1   On four cylinder engines, the camshaft, followers and lash adjusters must be removed prior to cylinder head disassembly (see Part A). On V6 engines, the rocker arm shaft assemblies or rocker arms must be removed (see Part B, C and D).

2   Remove the deposits from the combustion chambers and valve heads with a scraper and a wire brush before removing the valves. **Caution:** *Be careful not to scratch the gasket surfaces.*

3   Compress the valve springs with a valve spring compressor. Remove the keepers and

8.3 Use a valve spring compressor to compress the spring, then remove the keepers from the valve stem

8.4 If the tips of the valves are mushroomed, they must be filed down prior to valve removal

release the springs **(see illustration)**.

4   Remove the retainer, spring assembly and seal from the valve. The valve can now be removed through the bottom of the head. If the valve binds in the guide (will not pull through), push it back into the head and deburr the area around the head of the stem with a fine file or whetstone **(see illustration)**.

5   Repeat the procedure for the remaining valves. Remember to keep all the parts for each valve together so they can be reinstalled in the same locations **(see illustration)**.

6   Once the valves have been removed and stored in an organized manner, the head should be thoroughly cleaned and inspected. If a complete engine overhaul is being done, finish the engine disassembly procedures before beginning the cylinder head cleaning and inspection process.

9   Cylinder head - cleaning and inspection

Refer to illustrations 9.12a, 9.12b, 9.14, 9.21, 9.22a and 9.22b

1   Thorough cleaning of the cylinder head(s) and related valve train components, followed by a detailed inspection, will enable you to decide how much valve service work must be done during the engine overhaul.

Cleaning

2   Scrape all traces of old gasket material and sealing compound off the head gasket, intake manifold and exhaust manifold sealing surfaces.

3   Remove built-up scale from the coolant passages.

4   Run a stiff wire brush through the oil holes to remove any deposits that may have formed in them.

5   Run a tap into each of the threaded holes to remove corrosion and thread sealant that may be present. If compressed air is

available, use it to clear the holes of debris produced by this operation.

6   Clean the exhaust and intake manifold stud threads (V6 engine) with a die. Clean the rocker arm pivot bolt or stud threads, if applicable, with a wire brush.

7   Clean the cylinder head with solvent and dry it thoroughly. Compressed air will speed the drying process and ensure that all holes and recessed areas are clean. **Note:** *Decarbonizing chemicals are available and may prove very useful when cleaning cylinder heads and valve train components. They are very caustic and should be used with caution. Be sure to follow the instructions on the container.*

8   Clean the rocker arms and pushrods (V6 engine) with solvent and dry them thoroughly. Compressed air will speed the drying process and can be used to clean out the oil passages.

9   Clean all the valve springs, keepers, retainers, shields and shims with solvent and dry them thoroughly. Do the components from one valve at a time to avoid mixing up the parts.

10   Scrape off any heavy deposits that may have formed on the valves, then use a motorized wire brush to remove deposits from the valve heads and stems. Again, make sure the valves do not get mixed up.

Inspection

Note: *Be sure to perform all of the following inspection procedures before concluding that machine shop work is required. Make a list of the items that need attention.*

Cylinder head

11   Inspect the head very carefully for cracks, evidence of coolant leakage and other damage. If cracks are found, a new cylinder head should be obtained.

12   Using a straightedge and feeler gauge, check the head gasket mating surfaces for warpage **(see illustrations)**. If the warpage exceeds specifications, the head can be

8.5 A small plastic bag, with an appropriate label, can be used to store the valve train components so they can be kept together and reinstalled in the correct guide

resurfaced at an automotive machine shop.

13   Examine the valve seats in each of the combustion chambers. If they are pitted, cracked or burned, the head will require valve service that is beyond the scope of the home mechanic.

14   Check the valve stem-to-valve guide clearance. Use a dial indicator to measure the lateral movement of each valve stem with the valve in the guide and approximately 1/16-inch off the seat **(see illustration)**. The valve stem-to-guide clearance is one-half the dial indicator reading. Compare the reading to the desired stem-to-guide clearance in the Specifications. If, after this check, there is still some doubt as to the condition of the valve guides, the exact clearance and condition of the guides can be checked by an automotive machine shop, usually for a very small fee.

Cam followers (four-cylinder engine)

15   Clean all the parts thoroughly. Make sure that all oil passages are open.

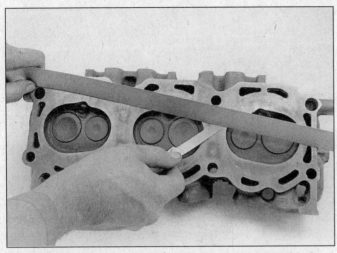

9.12a Check for head warpage with a straightedge and feeler gauge both diagonally and straight across

9.12b Check for head-to-manifold mating surface warpage with a straightedge and feeler gauge

16   Inspect the pad at the valve end of the follower and the camshaft end pad for indications of scuffing or abnormal wear. If the pad is grooved replace the follower. Do not attempt to true the surface by grinding.

Rocker arm components (V6 engine)

17   Check the rocker arm faces for pits, wear and rough spots. Check the pivot contact areas as well.
18   Inspect the pushrod ends for scuffing and excessive wear. Roll the pushrod on a flat surface, such as a piece of glass, to determine if it is bent.
19   Any damaged or excessively worn parts must be replaced with new ones.

Valves

20   Carefully inspect each valve face for cracks, pits and burned spots. Check the valve stem and neck for cracks. Rotate the valve and check for any obvious indication that it is bent. Check the end of the stem for pits and excessive wear. The presence of any

of these conditions indicates the need for valve service by a machine shop.
21   Measure the width of the valve margin on each valve (see illustration) and compare it to the Specifications. Any valve with a margin narrower than specified will have to be replaced with a new one.

Valve train components

22   Check each valve spring for wear on the ends and pits. Measure the free length (see illustration) and compare it to the Specifications. Any springs that are shorter than specified have sagged and should not be reused. Check each valve spring for squareness (see illustration). The tension of all springs should be checked with a special fixture before deciding that they're suitable for use in a rebuilt engine (take the springs to an automotive machine shop for this check).
23   Check the spring retainers and keepers for obvious wear and cracks. Any questionable parts should be replaced with new ones, as extensive damage will occur in the event of failure during engine operation.

9.14 To measure valve guide clearance with a dial indicator, rock the valve back and forth (arrows)

24   If the inspection process indicates that the valve components are in generally poor

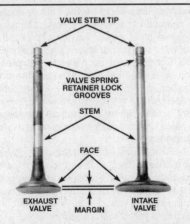

9.21 The margin width on each valve must be as specified (if no margin exists, the valve cannot be reused)

9.22a Measure the free length of each valve spring and replace any that have sagged or collapsed

9.22b Check each valve spring for squareness

11.3a Using a deep socket and hammer to install a valve guide seal (V6 engines)

11.3b Make sure each seal is seated properly on the guide

condition and worn beyond the limits specified, which is usually the case in an engine that is being overhauled, reassemble the valves in the cylinder head and refer to Section 10 for valve servicing recommendations.
25   If the inspection turns up no excessively worn parts, and if the valve faces and seats are in good condition, the valve train components can be reinstalled in the cylinder head without major servicing. Refer to the appropriate Section for cylinder head reassembly procedures.

10   Valves - servicing

1   Because of the complex nature of the job and the special tools and equipment needed, servicing of the valves, the valve seats and the valve guides, commonly known as a valve job, is best left to a shop.
2   The home mechanic can remove and disassemble the head, do the initial cleaning and inspection, then reassemble and deliver

the head to a dealer service department or an automotive machine shop for the actual valve servicing.
3   The dealer service department, or automotive machine shop, will remove the valves and springs, recondition or replace the valves and valve seats (four-cylinder engine), recondition the valve guides, check and replace the valve springs, spring retainers and keepers (as necessary), replace the valve seals with new ones (four-cylinder engines require a plastic installation cap), reassemble the valve components and make sure the installed spring height is correct. The cylinder head gasket surface will also be resurfaced if it is warped.
4   After the valve job has been performed by a shop the head will be in like new condition. When the head is returned, be sure to clean it again before installation on the engine to remove any metal particles and abrasive grit that may still be present from the valve service or head resurfacing operations. Use compressed air, if available, to blow out all the oil holes and passages.

11   Cylinder head - reassembly

Refer to illustrations 11.3a, 11.3b, 11.6 and 11.7

1   Regardless of whether or not the head was sent to an automotive repair shop for valve servicing, make sure it is clean before beginning reassembly.
2   If the head was sent out for valve servicing, the valves and related components will already be in place.
3   On V6 engines, lubricate and install the valves, then install new seals on each of the valve guides. Using a hammer and a deep socket, gently tap each seal into place until it is properly seated on the guide **(see illustrations)**. Do not twist or cock the seals during installation or they will not seal properly on the valve stems.
4   On four-cylinder engines install the valves, then install the valve seals. Install the plastic installation cap (usually included with the seals) over the end of the valve stem to allow the seal to pass over the ridges on valve stem. After installation remove the plastic cap and push the seal down until it is fully seated.
5   Install the valve spring shim(s) (if required) around the valve guide boss and set the valve spring, cap and retainer in place.
6   Compress the spring with a compressor and install the keepers. Release the compressor, making sure the keepers are seated properly in the valve stem groove(s). If necessary, grease can be used to hold the keepers in place until the compressor is released **(see illustration)**.
7   Check the installed valve spring height **(see illustration)**. If it was correct before reassembly it should still be within the specified limits. If it is not, install additional valve spring shims (available from a dealer) to bring the height to within the specified limit.
8   On V6 engines, install the rocker arms (and shafts on 2.8L and 4.0L engines). Be sure to lubricate the ball pivots with moly-base grease or engine assembly lube.

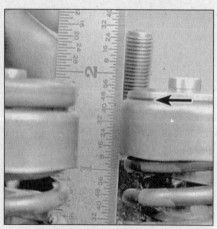

11.6 Apply a small dab of grease to each keeper before installation - it will hold them in place on the valve stem until the spring is released

11.7 Be sure to check the valve spring installed height (the distance from the top of the seat/shims to the top of the shield)

12.1 A ridge reamer is required to remove the ridge from the top of each cylinder - do this before removing the pistons!

12.2a Some manufacturers mark the connecting rods and caps with numbers

9   On four-cylinder engines, install the camshaft, lash adjusters and followers.

12 Piston/connecting rod assembly - removal

Refer to illustrations 12.1, 12.2a, 12.2b, 12.3 and 12.4

1   Using a ridge reamer, completely remove the ridge at the top of each cylinder (follow the manufacturer's instructions provided with the ridge reaming tool) **(see illustration)**. Failure to remove the ridge before attempting to remove the piston/connecting rod assemblies may result in piston breakage.
2   Check the connecting rods and connecting rod caps for identification marks **(see illustration)**. If they are not plainly marked, identify each rod and cap, using a small punch to make the appropriate number of indentations to indicate the cylinders they are associated with **(see illustration)**.

3   Before the connecting rods are removed, check the endplay with feeler gauges. Slide them between the first connecting rod and the crankshaft throw until the play is removed **(see illustration)**. The endplay is equal to the thickness of the feeler gauge(s). If the endplay exceeds the service limit, new connecting rods will be required. If new rods (or a new crankshaft) are installed, the endplay may fall under the specified minimum (if it does, the rods will have to be machined to restore it - consult an automotive machine shop for advice if necessary). Repeat the procedure for the remaining connecting rods.
4   Loosen each of the connecting rod cap nuts 1/2-turn. Remove the number one connecting rod cap and bearing insert. Do not drop the bearing insert out of the cap. Slip a short length of plastic or rubber hose over each connecting rod cap bolt to protect the crankshaft journal and cylinder wall when the piston is removed **(see illustration)** and push the connecting rod/piston assembly out

12.2b If the rods and caps are not identified, use a hammer and punch to mark them

through the top of the engine. Use a wooden tool to push on the upper bearing insert in the connecting rod. If resistance is felt, double-check to make sure that all of the ridge was removed from the cylinder.

12.3 Connecting rod side clearance is measured with feeler gauges inserted between the rod cap and the crankshaft

12.4 To prevent damage to the crankshaft journals and cylinder walls, slip sections of hose over the rod bolts before removing the pistons

13.1 Checking crankshaft endplay with a dial indicator

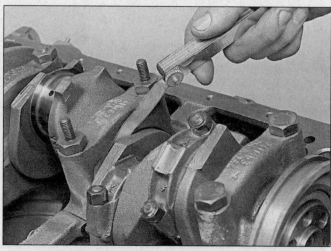

13.3 Checking crankshaft endplay with a feeler gauge

5   Repeat the procedure for the remaining cylinders. After removal, reassemble the connecting rod caps and bearing inserts in their respective connecting rods and install the cap nuts finger tight. Leaving the old bearing inserts in place until reassembly will help prevent the connecting rod bearing surfaces from being accidentally nicked or gouged.
6   Don't separate the pistons from the connecting rods (see Section 17 for additional information).

13   Crankshaft - endplay check and removal

Refer to illustrations 13.1, 13.3, 13.5 and 13.6
Note: *The crankshaft can be removed only after the engine has been removed from the vehicle. It's assumed that the flywheel or driveplate, vibration damper, timing chain, oil pan, oil pump and piston/connecting rod assemblies have already been removed.*

1   Before the crankshaft is removed, check the endplay. Mount a dial indicator with the stem in line with the crankshaft and just touching one of the crank throws **(see illustration)**.
2   Push the crankshaft all the way to the rear and zero the dial indicator. Next, pry the crankshaft to the front as far as possible and check the reading on the dial indicator. The distance that it moves is the endplay. If it's greater than the limit listed in this Chapter's Specifications, check the crankshaft thrust surfaces for wear. if no wear is evident, new main bearings should correct the endplay.
3   If a dial indicator isn't available, feeler gauges can be used. Gently pry or push the crankshaft all the way to the front of the engine. Slip feeler gauges between the crankshaft and the front face of the thrust main bearing (number 3 bearing) to determine the clearance **(see illustration)**.
4   Loosen each of the main bearing cap bolts 1/4-turn at a time, until they can be

removed by hand.
5   Check the main bearing caps to see if they are marked as to their locations. They are usually numbered consecutively (beginning with 1) from the front of the engine to the rear **(see illustration)**.
6   If they are not, mark them with number stamping dies or a centerpunch **(see illustration)**.
7   Gently tap the caps with a soft face hammer, then separate them from the engine block. If necessary, use the main bearing cap bolts as levers to remove the caps. Try not to drop the bearing insert if it comes out with the cap.
8   Carefully lift the crankshaft out of the engine. You may want to have an assistant available, since the crankshaft is quite heavy. With the bearing inserts in place in the engine block and in the main bearing caps, return the caps to their respective locations on the engine block and tighten the bolts finger tight.

13.5 Main bearing caps usually have a cast-in number to indicate position in the block and an arrow which should point to the front of the engine when the cap is installed

13.6 If the main bearing caps aren't numbered, use a hammer and punch to mark them

14.1a Use a hammer and a large punch to knock the soft plugs into the block

14.1b Grab the plug with pliers, turn it sideways and pull it out of the block

14 Engine block - cleaning

Refer to illustrations 14.1a, 14.1b, 14.8 and 14.10

1   Remove the soft plugs from the engine block. To do this, knock the plugs into the block (using a hammer and punch), then grasp them with large pliers and pull them back through the holes **(see illustrations)**.

2   Using a gasket scraper, remove all traces of gasket material from the engine block. Be very careful not to nick or gouge the gasket sealing surfaces.

3   Remove the main bearing caps and separate the bearing inserts from the caps and the engine block. Tag the bearings according to which journal they removed from (and whether they were in the cap or the block) and set them aside.

4   Remove the threaded oil gallery plugs from the block.

5   If the engine is extremely dirty it should be taken to an automotive machine shop to be steam cleaned or hot tanked. Any bearings left in the block (such as the camshaft bearings in V6 engines) will be damaged by the cleaning process, so plan on having new ones installed while the block is at the machine shop.

6   After the block is returned, clean all oil holes and oil galleries one more time. Brushes for cleaning oil holes and galleries are available at most auto parts stores. Flush the passages with warm water until the water runs clear, dry the block thoroughly and wipe all machined surfaces with a light, rust preventative oil. If you have access to compressed air, use it to speed the drying process and to blow out all the oil holes and galleries.

7   If the block is not extremely dirty or sludged up, you can do an adequate cleaning job with warm soapy water and a stiff brush. Take plenty of time and do a thorough job.

Regardless of the cleaning method used, be sure to thoroughly clean all oil holes and galleries, dry the block completely and coat all machined surfaces with light oil.

8   The threaded holes in the block must be clean to ensure accurate torque readings during reassembly. Run the proper size tap into each of the holes to remove any rust, corrosion, thread sealant or sludge and to restore any damaged threads **(see illustration)**. If possible, use compressed air to clear the holes of debris produced by this operation. Now is a good time to thoroughly clean the threads on the head bolts and the main bearing cap bolts.

9   Reinstall the main bearing caps and tighten the bolts finger tight.

10   After coating the sealing surfaces of the new soft plugs with a good quality gasket sealer, install them in the engine block **(see illustration)**. Make sure they are driven in straight and seated properly or leakage could

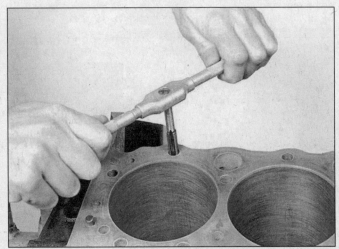

14.8 All bolt holes in the block - particularly the main bearing cap and head bolt holes - should be cleaned and restored with a tap (be sure to remove debris from the holes after this is done)

14.10 A large socket on an extension and a hammer can be used to install the new soft plugs in the block

result. Special tools are available for this purpose, but good results can be obtained using a large socket (with an outside diameter that will just slip into the soft plug) and a hammer.

11   If the engine is not going to be reassembled right away, cover it with a large plastic trash bag to keep it clean.

15 Engine block - inspection

Refer to illustrations 15.4a, 15.4b and 15.4c

1   Before the block is inspected, it should be cleaned as described in Section 14. Double-check to make sure the ridge at the top of each cylinder has been completely removed.

2   Visually check the block for cracks, rust and corrosion. Look for stripped threads in the threaded holes. It's also a good idea to have the block checked for hidden cracks by an automotive machine shop that has the special equipment to do this type of work. If defects are found, have the block repaired, if possible, or replaced.

3   Check the cylinder bores for scuffing and scoring.

4   Measure the diameter of each cylinder at the top (just under the ridge area), center and bottom of the cylinder bore, parallel to the crankshaft axis **(see illustrations)**. Next, measure each cylinder's diameter at the same three locations across the crankshaft axis. Compare the results to the Specifications. If the cylinder walls are badly scuffed or scored, or if they're out-of-round or tapered beyond the limits given in the Specifications, have the engine block rebored and honed at an automotive machine shop. If a rebore is done, oversize pistons and rings will be required.

5   If the required precision measuring tools aren't available, the piston-to-cylinder clearance can be obtained, though not as accurately, using feeler gauges. Feeler gauge stock comes in 12-inch lengths and various thicknesses and is generally available at auto parts stores.

6   To check the clearance, select a feeler gauge and slip it into the cylinder along with the matching piston. The piston must be positioned exactly as it normally would be. The feeler gauge must be between the piston and the cylinder on one side of the thrust face (90-degrees to the piston pin bore).

7   The piston should slip through the cylinder with moderate pressure.

8   If the piston falls through or slides through easily, the clearance is greater then that indicated on the feeler gauge and a new piston will be required. If the piston binds at the lower end of the cylinder and is loose at the top, the cylinder is tapered. If a tight spot is encountered as the piston is rotated in the bore, the cylinder is out-of-round.

9   Repeat this procedure for the remaining pistons and cylinders.

10   If the cylinders are in reasonably good condition and not worn to the outside of the limits, and if the piston-to-cylinder clearances can be maintained properly, then they don't have to be rebored. Honing is all that's necessary (Section 16).

16 Cylinder honing

Refer to illustrations 16.3a and 16.3b

1   Prior to engine reassembly, the cylinder bores must be honed so the new piston rings will seat correctly and provide the best possible combustion chamber seal. **Note:** *If you don't have the tools or don't want to tackle the honing operation, most automotive machine shops will do it for a reasonable fee.*

2   Before honing the cylinders, install the main bearing caps and tighten the bolts to the specified torque.

3   Two types of cylinder hones are commonly available - the flex hone or "bottle brush" type and the more traditional surfacing hone with spring-loaded stones. Both will do the job, but for the less experienced mechanic the "bottle brush" hone will probably be easier to use. You'll also need plenty of light oil or honing oil, some rags and an

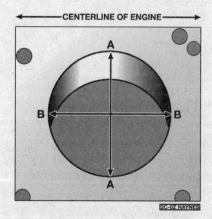

15.4a Measure the diameter of each cylinder at a right angle to the engine centerline (A), and parallel to the engine centerline (B) - out-of-round is the difference between A and B; taper is the difference between the diameter at the top of the cylinder and the diameter at the bottom of the cylinder

electric drill motor. Proceed as follows:

a) Mount the hone in the drill motor, compress the stones and slip it into the first cylinder **(see illustration)**. Be sure to wear safety goggles or a face shield!

b) Lubricate the cylinder with plenty of oil, turn on the drill and move the hone up-and-down in the cylinder at a pace that will produce a fine crosshatch pattern on the cylinder walls. Ideally, the crosshatch lines should intersect at approximately a 60 degrees angle **(see illustration)**. Be sure to use plenty of lubricant and don't take off any more material than is absolutely necessary to produce the desired finish. **Note:** Piston ring manufacturers may specify a smaller crosshatch angle than the traditional 60 degrees - read and follow any instructions included with the new rings.

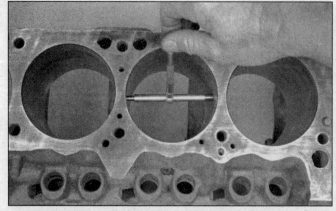

15.4b The ability to "feel" when the telescoping gauge is at the correct point will be developed over time, so work slowly and repeat the check until you're satisfied that the bore measurement is accurate

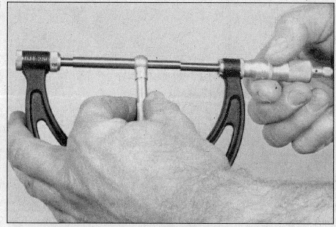

15.4c The gauge is then measured with a micrometer to determine the bore size

16.3a A "bottle-brush" hone is the easiest type of hone to use

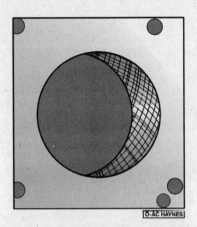

16.3b The cylinder hone should leave a smooth, crosshatch pattern with the lines intersecting at approximately a 60-degree angle

17.2 Use a piston ring tool to remove the old rings without scratching the pistons

c) *Don't withdraw the hone from the cylinder while it's running. Instead, shut off the drill and continue moving the hone up-and down in the cylinder until it comes to a complete stop, then compress the stones and withdraw the hone. If you're using a "bottle brush" type hone, stop the drill motor, then turn the chuck in the normal direction of rotation while withdrawing the hone from the cylinder.*

d) *Wipe the oil out of the cylinder and repeat the procedure for the remaining cylinders.*

4   After the honing job is complete, chamfer the top edges of the cylinder bores with a small file so the rings won't catch when the pistons are installed. **Be very careful not to nick the cylinder walls with the end of the file.**

5   The entire engine block must be washed again very thoroughly with warm, soapy water to remove all traces of the abrasive grit produced during the honing operation. **Note:** *The bores can be considered clean when a white cloth - dampened with clean engine oil*

- used to wipe them out doesn't pick up any more honing residue, which will show up as gray areas on the cloth. Be sure to run a brush through all oil holes and galleries and flush them with running water.

6   After rinsing, dry the block and apply a coat of light rust preventive oil to all machined surfaces. Wrap the block in a plastic trash bag to keep it clean and set it aside until reassembly.

17 Piston/connecting rod assembly - inspection

Refer to illustrations 17.2, 17.4a, 17.4b, 17.5, 17.10 and 17.11

1   Before the inspection process can be carried out, the piston/connecting rod assemblies must be cleaned and the original piston rings removed from the pistons. **Note:** *Always use new piston rings when the engine is reassembled.*

2   Using a piston ring installation tool, carefully remove the rings from the pistons **(see illustration)**. Do not nick or gouge the pistons in the process.

3   Scrape all traces of carbon from the top (or crown) of the piston. A hand held wire brush or a piece of fine emery cloth can be used once the majority of the deposits have been scraped away. Do not, under any circumstances, use a wire brush mounted in a drill motor to remove deposits from the pistons. The piston material is soft and may be eroded away by the wire brush.

4   Use a piston ring groove cleaning tool to remove any carbon deposits from the ring grooves **(see illustration)**. If a tool is not available, a piece broken off an old ring will do the job **(see illustration)**. Be very careful to remove only the carbon deposits. Do not remove any metal and do not nick or scratch the sides of the ring grooves.

5   Once the deposits have been removed, clean the piston/rod assemblies with solvent and dry them thoroughly. Make sure that the oil return holes or slots in the back sides of

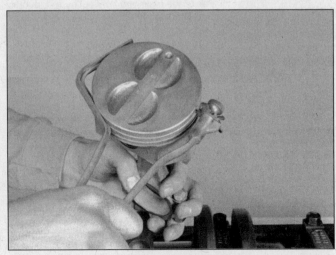

17.4a The piston ring grooves can be cleaned with a special tool, as shown here . . .

17.4b . . . or a section of broken ring

17.5 Make sure the oil return slots (arrow) are clear

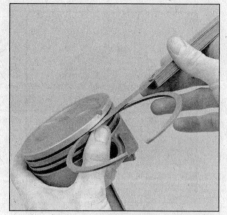

17.10 Check the ring side clearance with a feeler gauge at several points around the groove

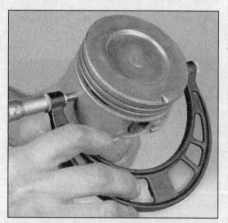

17.11 Measure the piston diameter at a 90 degrees angle to the piston pin and in line with it

the ring grooves are clear **(see illustration)**.

6   If the pistons are not damaged or worn excessively, and if the engine block is not rebored, new pistons will not be necessary. Normal piston wear appears as even vertical wear on the piston thrust surfaces and slight looseness of the top ring in its groove. New piston rings should always be used when an engine is rebuilt.

7   Carefully inspect each piston for cracks around the skirt, at the pin bosses and at the ring lands.

8   Look for scoring and scuffing on the thrust faces of the skirt, holes in the piston crown and burned areas at the edge of the crown. If the skirt is scored or scuffed, the engine may have been suffering from overheating and/or abnormal combustion, which caused excessively high operating temperatures. The cooling and lubrication systems should be checked thoroughly. A hole in the piston crown is an indication that abnormal combustion (preignition) was occurring. Burned areas at the edge of the piston crown are usually evidence of spark knock (detonation). If any of the above problems exist, the causes must be corrected or the damage will occur again.

9   Corrosion of the piston (evidenced by pitting) indicates that coolant is leaking into the combustion chamber and/or the crankcase. Again, the cause must be corrected or the problem may persist in the

rebuilt engine.

10   Measure the piston ring side clearance by laying a new piston ring in each ring groove and slipping a feeler gauge in between the ring and the edge of the ring groove **(see illustration)**. Check the clearance at three or four locations around each groove. If the side clearance is greater than specified, new pistons will have to be used.

11   Check the piston-to-bore clearance by measuring the bore (see Section 15) and the piston diameter. Make sure that the pistons and bores are correctly matched. Measure the piston across the skirt **(see illustration)**. Subtract the piston diameter from the bore diameter to obtain the clearance. If it is greater than specified, the block will have to be rebored and new pistons and rings installed. Check the piston-to-rod clearance by twisting the piston and rod in opposite directions. Any noticeable play indicates that there is excessive wear, which must be corrected. The piston/connecting rod assemblies should be taken to an automotive machine shop to have new piston pins installed and the pistons and connecting rods rebored.

12   If the pistons must be removed from the connecting rods, such as when new pistons must be installed, or if the piston pins have too much play in them, they should be taken to an automotive machine shop. While they are there, have the connecting rods checked

for bend and twist, as automotive machine shops have special equipment for this purpose. Unless new pistons or connecting rods must be installed, do not disassemble the pistons from the connecting rods.

13   Check the connecting rods for cracks and other damage. Temporarily remove the rod caps, lift out the old bearing inserts, wipe the rod and cap bearing surfaces clean and inspect them for nicks, gouges or scratches. After checking the rods, replace the old bearings, slip the caps into place and tighten the nuts finger tight.

18 Crankshaft - inspection

Refer to illustration 18.2

1   Clean the crankshaft with solvent and dry it thoroughly. Be sure to clean the oil holes with a stiff brush and flush them with solvent. Check the main and connecting rod bearing journals for uneven wear, scoring, pitting or cracks. Check the crankshaft for cracks and damage.

2   Using a micrometer, measure the diameter of the main and connecting rod journals **(see illustration)** and compare the results to the Specifications. By measuring the diameter at a number of points around the journal's circumference, you will be able to determine whether or not the journal is out of round. Take the measurement at each end of the journal, near the crank counterweights, to determine whether the journal is tapered.

3   If the crankshaft journals are damaged, tapered, out of round or worn beyond the limits given in the Specifications, have the crankshaft reground by an automotive machine shop. Be sure to use the correct oversize bearing inserts if the crankshaft is reconditioned.

4   Refer to Section 19 and examine the main and rod bearing inserts.

5   New main and connecting rod bearings should be installed whenever the engine is disassembled.

18.2 Measure the diameter of each crankshaft journal at several points to detect taper and out-of-round conditions

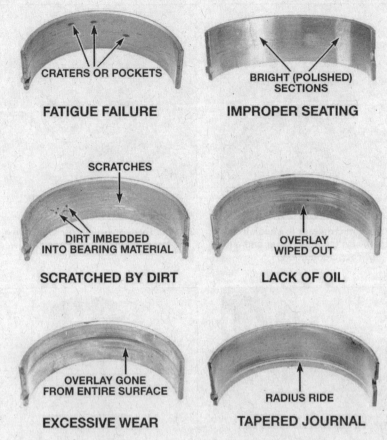

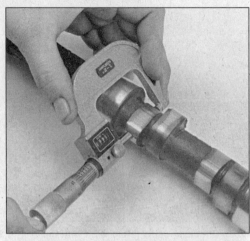

20.1 To verify camshaft lobe lift, measure the major and minor diameters of each lobe with a micrometer or vernier caliper - subtract each minor diameter form the major diameter to arrive at the lobe lift

19.2 Six common types of bearing problems

19 Main and connecting rod bearings - inspection

Refer to illustration 19.2

1   Even though the main and connecting rod bearings should be replaced with new ones during the engine overhaul, the old bearings should be retained for close examination, as they may reveal valuable information about the condition of the engine.

2   Bearing failure occurs mainly because of lack of lubrication, the presence of dirt or other foreign particles, overloading the engine and corrosion **(see illustration)**. Regardless of the cause of bearing failure, it must be corrected before the engine is reassembled to prevent it from happening again.

3   When examining the bearings, remove them from the engine block, the main bearing caps, the connecting rods and the rod caps and lay them out on a clean surface in the same general position as their location in the engine. This will enable you to match any noted bearing problems with the corresponding crankshaft journal.

4   Dirt and other foreign particles get into the engine in a variety of ways. It may be left in the engine during assembly, or it may pass through filters or breathers. It may get into the oil, and from there into the bearings.

Metal chips from machining operations and normal engine wear are often present. Abrasives are sometimes left in engine components after reconditioning, especially when parts are not thoroughly cleaned using the proper cleaning methods. Whatever the source, these foreign objects often end up embedded in the soft bearing material and are easily recognized. Large particles will not embed in the bearing and will score or gouge the bearing and shaft. The best prevention for this cause of bearing failure is to clean all parts thoroughly and keep everything spotlessly clean during engine assembly. Frequent and regular engine oil and filter changes are also recommended.

5   Lack of lubrication or lubrication breakdown has a number of interrelated causes. Excessive heat, which thins the oil, overloading, which squeezes the oil from the bearing face, and oil leakage or throw off from excessive bearing clearances, worn oil pump or high engine speeds, all contribute to lubrication breakdown. Blocked oil passages, which usually are the result of misaligned oil holes in a bearing shell, will also oil starve a bearing and destroy it. When lack of lubrication is the cause of bearing failure, the bearing material is wiped or extruded from the steel backing of the bearing. Temperatures may increase to the point where the steel backing turns blue from overheating.

6   Driving habits can have a definite effect on bearing life. Full throttle, low speed operation in too high a gear (or lugging the engine) puts very high loads on bearings, which tends to squeeze out the oil film. These loads cause the bearings to flex, which produces fine cracks in the bearing face (fatigue failure). Eventually the bearing material will loosen in pieces and tear away from the steel backing. Short trip driving leads to corrosion of bearings because insufficient engine heat is produced to drive off the condensed water and corrosive gases. These products collect in the engine oil, forming acid and sludge. As the oil is carried to the engine bearings, the acid attacks and corrodes the bearing material.

7   Incorrect bearing installation during engine assembly will lead to bearing failure as well. Tight fitting bearings leave insufficient bearing oil clearance and will result in oil starvation. Dirt or foreign particles trapped behind a bearing insert result in high spots on the bearing which lead to failure.

20 Camshaft, lifters and bearings - inspection and replacement

Camshaft inspection

Four-cylinder engine

Refer to illustration 20.1

1   Measure the major (A) and minor (B) diameters **(see illustration)** of each cam lobe and compare the readings to the Specifications. The difference between the readings is the lobe lift.

2   If the lobe lift is not as specified, replace the camshaft and all the followers.

3   After the camshaft is removed from the engine, cleaned with solvent and dried, inspect the bearing journals for uneven wear, pitting and evidence of seizure. If the journals

20.6 Checking camshaft lobe lift with the camshaft installed in the engine - always make sure the dial indicator plunger is directly in-line with the pushrod or lifter

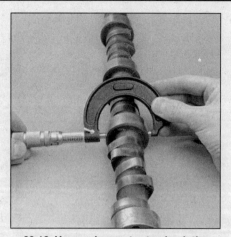

20.12 Use a micrometer to check the camshaft journals for wear, taper and out-of-round conditions

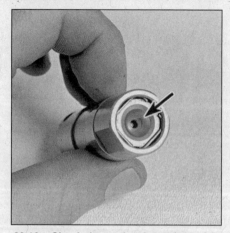

20.16a Check the pushrod seat (arrow) in the top of each lifter for excessive wear

20.16b A concave wear pattern on the base of this lifter indicates the need for a new camshaft and a complete set of lifters

20.16c The foot of each lifter should be slightly convex - the side of another lifter can be used as a straightedge to check it; if it appears flat, it's worn and must not be reused

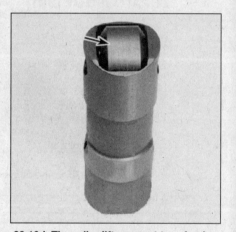

20.16d The roller lifters must turn freely - check for wear and excessive play as well

are damaged, the bearing inserts are probably damaged as well. Both the camshaft and bearings will have to be replaced with new ones.

V6 engines

Refer to illustrations 20.6 and 20.12

4   The lobe lift measurement on the V6 can be made with the camshaft still installed in the engine or as in Step 1 above.

5   Remove the rocker arm cover and rocker arms.

6   Beginning with the number one cylinder, mount a dial indicator with the stem resting on and directly in line with the pushrod **(see illustration)**.

7   Rotate the crankshaft very slowly in the direction of normal rotation (clockwise) until the lifter is on the heel of the cam lobe. At this point the pushrod will be at its lowest position.

8   Zero the dial indicator, then very slowly rotate the crankshaft in the direction of rotation until the pushrod is at its highest position. Note and record the reading on the dial indicator, then compare it to the lobe lift specifications.

9   Repeat the procedure for each of the remaining valves.

10   If the lobe lift measurements are not as specified, a new camshaft and lifters should be installed.

11   After the camshaft has been removed from the engine, cleaned with solvent and dried, inspect the bearing journals for uneven wear, pitting or evidence of seizure. If the journals are damaged, the bearing inserts are probably damaged as well. Both the camshaft and bearings will have to be replaced with new ones.

12   Measure the camshaft bearing journals with a micrometer **(see illustration)** to determine if they are excessively worn or out-of-round. If they are more than 0.001-inch out-of-round, the camshaft should be replaced with a new one.

13   Check the camshaft lobes for heat discoloration, score marks, chipped areas, pitting and uneven wear.

14   If the lobes are in good condition and if the lobe lift measurements are as specified, the camshaft can be reused.

Lifter inspection

Refer to illustrations 20.16a, 20.16b, 20.16c, and 20.16d

15   Clean the lifters with solvent and dry them thoroughly without mixing them up.

16   Check each lifter wall, pushrod seat and foot for scuffing, score marks and uneven wear **(see illustration)**. Each lifter foot (the surface that rides on the cam lobe) must be slightly convex - if they are concave **(see illustrations)** the lifters and camshaft must be replaced with new ones. If the lifter walls are damaged or worn (which is not very likely), inspect the lifter bores in the engine block as well. On 4.0L engines with roller lifters, make sure the roller turns freely without excessive play or wear **(see illustration)**.

17   If new lifters are being installed, a new camshaft must also be installed. If a new camshaft is installed, then use new lifters as well. Never install used lifters unless the original camshaft is used and the lifters can be installed in their original locations.

Bearing replacement

18   Camshaft bearing replacement requires special tools and expertise that place it out-

22.3a When checking piston ring end gap, the ring must be square in the cylinder bore (this is done by pushing the ring down with the top of a piston as shown)

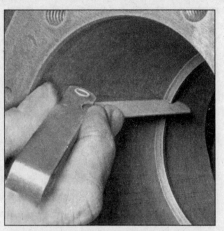

22.3b With the ring square in the cylinder, measure the end gap with a feeler gauge

22.5 If the end gap is too small, clamp a file in a vise and file the ring ends (from the outside in only) to enlarge the gap slightly

side the scope of the do-it-yourselfer. Take the block (V6) or head (four-cylinder) to an automotive machine shop to ensure that the job is done correctly.

21 Engine overhaul - reassembly sequence

1   Before beginning engine reassembly, make sure you have all the necessary new parts, gaskets and seals as well as the following items on hand:

Common hand tools
Torque wrench
Piston ring installation tool
Piston ring compressor
Short lengths of rubber or plastic hose to fit over connecting rod bolts
Plastigage
Feeler gauges
A fine-tooth file
New engine oil
Engine assembly lube or moly-base grease
RTV-type gasket sealant
Anaerobic-type gasket sealant
Thread locking compound

2   In order to save time and avoid problems, engine reassembly must be done in the following general order:

New camshaft bearings (must be done by automotive machine shop)
Piston rings
Crankshaft and main bearings
Piston/connecting rod assemblies
Oil pump
Camshaft and lifters
Cylinder heads, pushrods and rocker arms
Camshaft drive components
Oil pan
Camshaft drive cover
Intake and exhaust manifolds
Rocker arm covers
Flywheel/driveplate

22 Piston rings - installation

Refer to illustrations 22.3a, 22.3b, 22.5, 22.9a, 22.9b and 22.12

1   Before installing the new piston rings, the ring end gaps must be checked. It is assumed that the piston ring side clearance has been checked and verified correct (Section 17).

2   Lay out the piston/connecting rod assemblies and the new ring sets so the ring sets will be matched with the same piston and cylinder during the end gap measurement and engine assembly.

3   Insert the top (number one) ring into the first cylinder and square it up with the cylinder walls by pushing it in with the top of the piston (see illustration). To measure the end gap, slip a feeler gauge between the ends of the ring (see illustration). Compare the measurement to the Specifications.

4   If the gap is larger or smaller than specified, double-check to make sure that you have the correct rings before proceeding.

5   If the gap is too small, it must be enlarged or the ring ends may come in contact with each other during engine operation, which can cause serious damage to the

engine. The end gap can be increased by filing the ring ends very carefully with a fine file. Mount the file in a vise, slip the ring over the file with the ends contacting the file face and slowly move the ring to remove material from the ends. When performing this operation, file only from the outside in (see illustration).

6   Excess end gap is not critical unless it is greater than 0.040-inch (1 mm). Again, double-check to make sure you have the correct rings for your engine.

7   Repeat the procedure for each ring that will be installed in the first cylinder and for each ring in the remaining cylinders. Remember to keep rings, pistons and cylinders matched up.

8   Once the ring end gaps have been checked/corrected, the rings can be installed on the pistons.

9   The oil control ring (lowest one on the piston) is installed first. It is composed of three separate components. Slip the spacer/expander into the groove (see illustration), then install the side rails. Do not use a piston ring installation tool on the oil ring side rails, as they may be damaged. Instead, place one end of the side rail into the groove

22.9a Installing the spacer/expander in the oil control ring groove

22.9b DO NOT use a piston ring installation tool when installing the oil ring side rails

22.12 When installing the compression rings, always use a ring expander tool, don't mix up the top and second rings, and make sure the mark (if applicable) is on top of the ring when installed - arrow visible on top of piston points to front of engine when installed

23.10 Lay the Plastigage strips (arrow) on the main bearing journals, parallel to the crankshaft centerline

23.14 Compare the width of the crushed Plastigage to the scale on the container to determine the main bearing oil clearance (always take the measurement at the widest point of the Plastigage); be sure to use the correct scale - standard and metric scales are included

between the spacer/expander and the ring land, hold it firmly in place and slide a finger around the piston while pushing the rail into the groove **(see illustration)**.

10   After the three oil ring components have been installed, check to make sure that both the upper and lower side rails can be turned smoothly in the ring groove.

11   The number two (middle) ring is installed next. It should be stamped with a mark so it can be readily distinguished from the top ring. **Note:** *Always follow the instructions printed on the ring package or box - different manufacturers may require different approaches.* Do not mix up the top and middle rings, as they have different cross sections.

12   Use a piston ring installation tool and make sure that the identification mark is facing the top of the piston, then slip the ring into the middle groove on the piston **(see illustration)**. Do not expand the ring any more than is necessary to slide it over the piston.

13   Finally, install the number one (top) ring in the same manner. Make sure the identifying mark is facing up.

14   Repeat the procedure for the remaining pistons and rings.

23 Crankshaft - installation and main bearing oil clearance check

Refer to illustrations 23.10 and 23.14

1   Crankshaft installation is generally one of the first steps in engine reassembly. It is assumed at this point that the engine block and crankshaft have been cleaned, inspected and repaired or reconditioned.

2   Remove the main bearing cap bolts and lift out the caps. Lay them out in the proper order to help ensure that they are installed correctly.

3   If they are still in place, remove the old

bearing inserts from the block and the main bearing caps. Wipe the main bearing surfaces of the block and caps with a clean, lint-free cloth. They must be kept spotlessly clean.

4   Clean the back sides of the new main bearing inserts and lay one bearing half in each main bearing saddle in the block. Lay the other bearing half from each bearing set in the corresponding main bearing cap. Make sure the tab on the bearing insert fits into the recess in the block or cap.

5   The oil hole in the block must line up with the oil hole in the bearing insert. Do not hammer the bearing into place and do not nick or gouge the bearing faces. No lubrication should be used at this time.

6   The flanged thrust bearing must be installed in the number three cap main bearing position on both the 4 and 6 cylinder engines (third position back from front of engine).

7   Clean the faces of the bearings in the block and the crankshaft main bearing journals with a clean, lint-free cloth. Check or clean the oil holes in the crankshaft, as any dirt here can go only one way - straight through the new bearings.

8   Once you are certain that the crankshaft is clean, carefully lay it in position (an assistant would be very helpful here) in the main bearings.

9   Before the crankshaft can be permanently installed, the main bearing oil clearance must be checked.

10   Trim several pieces of the appropriate size of Plastigage (so they are slightly shorter than the width of the main bearings) and place one piece on each crankshaft main bearing journal, parallel with the journal axis **(see illustration)**.

11   Clean the faces of the bearings in the caps and install the caps in their respective positions (do not mix them up) with the arrows pointing toward the front of the engine. Do not disturb the Plastigage.

12   Starting with the center main and working out toward the ends, tighten the main

bearing cap bolts, in three steps, to the specified torque. Do not rotate the crankshaft at any time during this operation.

13   Remove the bolts and carefully lift off the main bearing caps. Keep them in order. Do not disturb the Plastigage or rotate the crankshaft. If any of the main bearing caps are difficult to remove, tap them gently from side-to-side with a soft face hammer to loosen them.

14   Compare the width of the crushed Plastigage on each journal to the scale printed on the Plastigage container to obtain the main bearing oil clearance **(see illustration)**. Check the Specifications to make sure it is correct.

15   If the clearance is not correct, double-check to make sure you have the right size bearing inserts. Also, make sure that no dirt or oil is between the bearing inserts and the main bearing caps or the block.

16   Carefully scrape all traces of the Plastigage material off the main bearing journals and/or the bearing faces. Do not nick or scratch the bearing faces. A fingernail or edge of a credit card works well.

17   Carefully lift the crankshaft out of the engine. Clean the bearing faces in the block, then apply a thin, uniform layer of clean, high quality moly base grease or engine assembly lube to each of the bearing surfaces. Be sure to coat the thrust flange faces as well as the journal face of the thrust bearing.

18   Make sure the crankshaft journals are clean, then lay the crankshaft back in place in the block. Clean the faces of the bearings in the caps, then apply a thin, uniform layer of clean, moly base grease to each of the bearing faces.

19   Install the caps in their respective positions with the arrows pointing toward the front of the engine. On 4.0L engines, install a new rear seal, referring to Section 24.

20   Install and tighten all main bearing bolts finger tight. Then tighten all bolts by starting

24.8 The wedge seal (arrow) must be installed on the rear main bearing cap before the cap is installed on the engine

24.9 Before installing the rear main cap, apply sealant to the cap mating surfaces and in the corners where the cap meets the block

with the center main and working outwards. Work up to the final torque in three steps.

21   Rotate the crankshaft a number of times by hand and check for any obvious binding.

22   The final step is to check the crankshaft end play with a feeler gauge or a dial indicator. This procedure is covered in Section 30.

24   Rear main oil seal - installation

Four-cylinder engine

1   Position a seal on a seal installer tool, available at most auto parts stores, and start the bolts into the crank-shaft flange.

2   To seat the seal properly, alternate the tightening of the bolts until the seal is positioned at the proper depth.

2.8L and 3.0L V6 engines

3   Make sure the engine block is free of all dirt, lint or foreign matter to ensure a proper fit.

4   To install coat the seal-to-cylinder block surface of the oil seal with engine oil.

5   Coat the seal surface and crankshaft with engine oil.

6   Start the seal in the recess and install it with a special tool available at most auto parts stores. Alternatively, use a hammer and socket the same diameter as the seal to drive the seal into the seal bore.

7   Drive the seal into position until it is firmly seated.

4.0L V6 engine

Refer to illustrations 24.8 and 24.9

8   Clean the face of the rear bearing cap, then apply lubricant to its insert. Install the wedge seal **(see illustration)** into the rear bearing cap. Also, install the rear main bearing oil seal over the end of the crankshaft. Be sure it is installed to the correct depth (see Specifications in Chapter 2D).

9   Install the cap with the arrow pointing towards the front of the engine. Apply sealant into the corners and on the mating surfaces where the rear main cap meets the cylinder block **(see illustration).**

10   Install the rear main bearing cap bolts and tighten them to the torque listed in this Chapter's Specifications.

25   Piston/connecting rod assembly - installation and rod bearing oil clearance check

Refer to illustrations 25.5, 25.8a, 25.8b, 25.9, 25.12, 25.13, 25.14 and 25.20

1   Before installing the piston/connecting rod assemblies the cylinder walls must be perfectly clean, the top edge of each cylinder must be chamfered, and the crankshaft must be in place.

2   Remove the connecting rod cap from the end of the number one connecting rod. Remove the old bearing inserts and wipe the bearing surfaces of the connecting rod and cap with a clean, lint-free cloth. They must be kept spotlessly clean.

3   Clean the back side of the new upper bearing half, then lay it in place in the connecting rod. Make sure that the tab on the bearing fits into the recess in the rod. Do not hammer the bearing insert into place and be very careful not to nick or gouge the bearing face. Do not lubricate the bearing at this time.

4   Clean the back side of the other bearing insert and install it in the rod cap. Again, make sure the tab on the bearing fits into the recess in the cap, and do not apply any lubricant. It is critically important that the mating surfaces of the bearing and connecting rod are perfectly clean and oil free when they are assembled.

5   Position the piston ring gaps **(see illustration),** then slip a section of plastic or rubber hose over the connecting rod cap bolts.

6   Lubricate the piston and rings with clean engine oil and attach a piston ring compressor to the piston. Leave the skirt protruding about 1/4-inch to guide the piston into the cylinder. The rings must be compressed as far as possible.

7   Rotate the crankshaft until the number one connecting rod journal is as far from the number one cylinder as possible (bottom

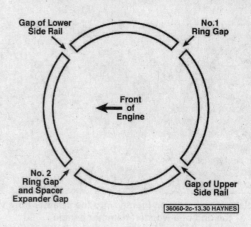

25.5 Compression and oil control ring gap positioning

Gap of Lower Side Rail

No.1 Ring Gap

Front of Engine

No. 2 Ring Gap and Spacer Expander Gap

Gap of Upper Side Rail

36060-2c-13.30 HAYNES

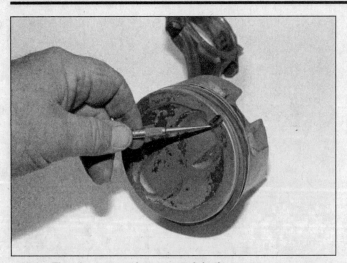

25.8a The pistons may have a notch in the crown, or an arrow, which should point to the front of the engine when the piston is installed

25.8b With the ring compressor installed, slide the piston and rod assembly into the cylinder

dead center), and apply a coat of engine oil to the cylinder walls.

8   With the notch on top of the piston facing to the front of the engine **(see illustration)**, gently place the piston/connecting rod assembly into the number one cylinder bore and rest the bottom edge of the ring compressor on the engine block **(see illustration)**. Tap the top edge of the ring compressor to make sure it is contacting the block around its entire circumference. On 4.0L engines, if the oil squirt hole faces the wrong way when the arrow on the piston points towards the front, the rod and piston have been assembled incorrectly.

9   Carefully tap on the top of the piston with the end of a wooden hammer handle **(see illustration)** while guiding the end of the connecting rod into place on the crankshaft journal.

10   The piston rings may try to pop out of the ring compressor just before entering the cylinder bore, so keep some downward pressure on the ring compressor. Work slowly,

and if any resistance is felt as the piston enters the cylinder, stop immediately. Find out what is hanging up and fix it before proceeding. Do not, for any reason, force the piston into the cylinder, as you will break a ring and/or the piston.

11   Once the piston/connecting rod assembly is installed, the connecting rod bearing oil clearance must be checked before the rod cap is permanently bolted in place.

12   Cut a piece of the appropriate size Plastigage slightly shorter than the width of the connecting rod bearing and lay it in place on the number one connecting rod journal, parallel with the journal axis (it must not cross the oil hole in the journal) **(see illustration)**.

13   Pull the connecting rod firmly against the crankshaft **(see illustration)**, remove the protective hoses from the connecting rod bolts and install the rod cap. Make sure the mating mark on the cap is on the same side as the mark on the connecting rod. Install the nuts and tighten them to the specified torque, working up to it in three steps. Do not rotate

the crankshaft at any time during this operation.

14   Remove the rod cap, being very careful not to disturb the Plastigage. Compare the width of the crushed Plastigage to the scale printed on the Plastigage container to obtain the oil clearance **(see illustration)**. Compare it to the Specifications to make sure the clearance is correct. If the clearance is not correct, double-check to make sure that you have the correct size bearing inserts. Also, recheck the crankshaft connecting rod journal diameter and make sure that no dirt or oil was between the bearing inserts and the connecting rod or cap when the clearance was measured.

15   Carefully scrape all traces of the Plastigage material off the rod journal and/or bearing face. Be very careful not to scratch the bearing- use your fingernail or a piece of hardwood. Make sure the bearing faces are perfectly clean, then apply a layer of moly base grease or engine assembly lube to both of them. You will have to push the piston into

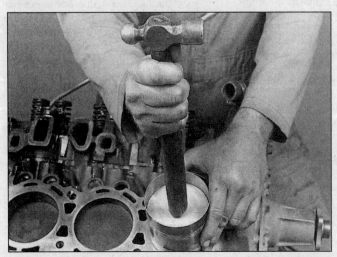

25.9 The piston can be driven (gently) into the cylinder bore with the end of a wooden hammer handle

25.12 Lay the Plastigage strips on each rod bearing journal, parallel to the crankshaft centerline

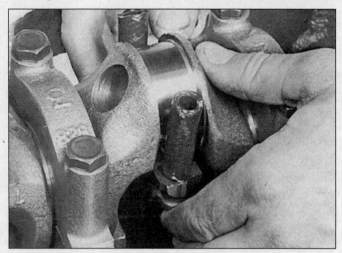

25.13 Pull the connecting rod firmly against the crankshaft

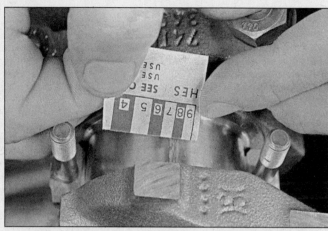

25.14 Measuring the width of the crushed Plastigage to determine the rod bearing oil clearance (be sure to use the correct scale - standard and metric scales are included)

the cylinder to expose the face of the bearing insert in the connecting rod. Be sure to replace the protective hoses over the rod bolts first.

16   Slide the connecting rod back into place on the journal, remove the protective hoses from the rod cap bolts, install the rod cap and tighten the nuts to the specified torque. Again, work up to the torque in three steps.

17   Repeat the entire procedure for the remaining piston/connecting rod assemblies. Keep the back sides of the bearing inserts and the inside of the connecting rod and cap perfectly clean when assembling them. Make sure you have the correct piston for the cylinder and that the notch on the piston faces to the front of the engine when the piston is installed. Remember, use plenty of oil to lubricate the piston before installing the ring compressor. Also, when installing the rod caps for the final time, be sure to lubricate the bearing faces.

18   After all the piston/connecting rod assemblies have been properly installed, rotate the crankshaft a number of times by hand and check for any obvious binding.

19   As a final step, the connecting rod side clearance must be checked. Compare the measured side clearance to the Specifications to make sure it is correct. Refer to Section 12 for the procedure.

20   Pry the connecting rod (4-cylinder) or rod pair (V6) all the way forward on the journal, then insert progressively thicker feeler gauges between the rod and crankshaft until all the space is taken up **(see illustration)**. The thickness of feeler gauge required is the side clearance.

26   Pre-oiling engine after overhaul

Refer to illustration 26.3

1   After an overhaul it is a good idea to pre-oil the engine before it is installed in the vehicle and started for the first time. Pre-oiling will reveal any problems with the lubrication system at a time when corrections can be made easily and will prevent major engine damage. It will also allow the internal engine parts to be lubricated thoroughly in the normal fashion without the heavy loads associated with combustion placed on them.

2   The engine should be completely assembled with the exception of the distributor and rocker arm covers. The oil filter and oil pressure sending unit must be in place and the specified amount of oil must be in the crankcase (see Chapter 1).

3   A modified Ford distributor will be needed for this procedure - a junkyard should be able to supply one for a reasonable price. In order to function as a pre-oil tool, the distributor must have the gear on the lower end of the shaft ground off **(see illustration)** and, if equipped, the advance weights on the upper end of the shaft removed.

4   Install the pre-oil distributor in place of the original distributor and make sure the lower end of the shaft mates with the upper end of the oil pump driveshaft. Turn the distributor shaft until they are aligned and the distributor body seats on the block. Install the distributor hold-down clamp and bolt.

25.20 Check the connecting rod side clearance with a feeler gauge as shown

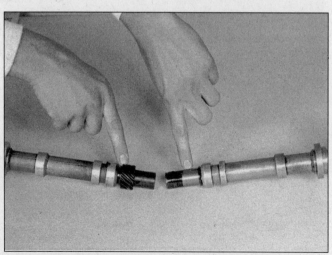

26.3 A typical pre-oil distributor (right) has the gear ground off and the advance weights (if equipped) removed

5   Mount the upper end of the shaft in the chuck of an electric drill and use the drill to turn the pre-oil distributor shaft, which will drive the oil pump and circulate the oil throughout the engine. **Note:** *The drill must turn in a clockwise direction.*

6   It may take two or three minutes, but oil should soon start to flow out of all of the rocker arm holes, indicating that the oil pump is working properly. Let the oil circulate for several seconds, then shut off the drill motor.

7   Remove the pre-oil distributor, then install the rocker arm covers. The distributor should be installed after the engine is installed in the vehicle, so plug the hole with a clean cloth.

27   Initial start-up and break-in after overhaul

1   Once the engine has been properly installed in the vehicle, double-check the engine oil and coolant levels.

2   With the spark plugs out of the engine and the coil high tension lead grounded to the engine block, crank the engine over until oil pressure registers on the gauge (if so equipped) or until the oil light goes off.

3   Install the spark plugs, hook up the plug wires and the coil high tension lead.

4   Make sure the carburetor choke plate is closed, then start the engine. It may take a few moments for the gasoline to reach the carburetor, but the engine should start without a great deal of effort.

5   As soon as the engine starts it should be set at a fast idle to ensure proper oil circulation and allowed to warm up to normal operating temperature. While the engine is warming up, make a thorough check for oil and coolant leaks.

6   Shut the engine off and recheck the engine oil and coolant levels. Restart the engine and check the ignition timing and the engine idle speed (refer to Chapter 1). Make any necessary adjustments.

7   Drive the vehicle to an area with mini-mum traffic, accelerate at full throttle from 30 to 50 mph, then allow the vehicle to slow to 30 mph with the throttle closed. Repeat the procedure 10 or 12 times. This will load the piston rings and cause them to seat properly against the cylinder walls. Check again for oil and coolant leaks.

8   Drive the vehicle gently for the first 500 miles (no sustained high speeds) and keep a constant check on the oil level. It is not unusual for an engine to use oil during the break-in period.

9   At approximately 500 to 600 miles, change the oil and filter and retorque the cylinder head bolts.

10   For the next few hundred miles, drive the vehicle normally. Do not either pamper it or abuse it.

11   After 2000 miles, change the oil and filter again and consider the engine fully broken in.

Chapter 3
Cooling, heating and air conditioning systems

Contents

Specifications

General

Radiator cap pressure rating	See Chapter 1
Coolant capacity	See Chapter 1
Coolant type	See Chapter 1
Refrigerant capacity	3.5 lbs

Torque specifications

Ft-lbs (unless otherwise indicated)

Thermostat housing	144 to 180 in-lbs
Fan-to-clutch	54 to 72 in-lbs
Clutch-to-water pump	
Four-cylinder engine	12 to 18
2.8L V6 engine	15 to 25
3.0L V6 engine	30 to 100 in-lbs
4.0L V6 engine	30 to 100 in-lbs
Temperature sending unit	8 to 18
Water pump-to-front cover	
Four-cylinder engine	14 to 21
2.8L V6 engine	84 to 108 in-lbs
3.0L V6 engine	72 to 96 in-lbs
4.0L V6 engine	72 to 108 in-lbs
A/C condenser-to-body bolts	96 to 144 in-lbs
A/C compressor-to-bracket	25 to 35
A/C hose manifold-to-compressor	13 to 27

3.6 Removing the thermostat housing (2.8L V6)

**3.9 The thermostat properly located inside the housing -
DO NOT install it backwards!**

1   General information

Engine cooling system

These vehicles employ a pressurized engine cooling system with thermostatically controlled coolant circulation. An impeller-type water pump mounted on the front of the block pumps coolant through the engine. The pump is driven by the timing belt on 4-cylinder engines and by a drivebelt on V6 engines.

The coolant flows around each cylinder and toward the rear of the engine. Cast-in coolant passages direct coolant around the intake and exhaust ports, near the spark plug areas and in close proximity to the exhaust valve guide inserts.

A wax pellet type thermostat is located in the front of the intake manifold, just ahead of the throttle body. During warm up, the closed thermostat prevents coolant from circulating through the radiator. When the engine reaches normal operating temperature, the thermostat opens and allows hot coolant to travel through the radiator, where it is cooled before returning to the engine.

The aluminum radiator is the crossflow type, with tanks on either side of the core.

The cooling system is sealed by a pressure type radiator cap. This raises the boiling point of the coolant and the higher boiling point of the coolant increases the cooling efficiency of the radiator.

If the system pressure exceeds the cap pressure relief value, the excess pressure in the system forces the spring-loaded valve inside the cap off its seat and allows the coolant to escape through the overflow tube into a coolant reservoir. When the system cools the excess coolant is automatically drawn from the reservoir back into the radiator.

The coolant reservoir maintains the proper radiator fluid level and acts as a holding tank for overheated coolant. This type of cooling system is known as a closed design because coolant that escapes past the pressure cap is saved and reused.

Heating system

The heating system consists of a blower fan and heater core located under the hood, the inlet and outlet hoses connecting the heater core to the engine cooling system and the heater/air conditioning control head on the dashboard. Hot engine coolant is circulated through the heater core at all times. When the heater mode is activated, a flap door opens to expose the heater box to the passenger compartment. A fan switch on the control head activates the blower motor, which forces air through the core, heating the air.

Air conditioning system

The air conditioning system consists of a condenser mounted in front of the radiator, an evaporator mounted under the hood, a compressor mounted on the engine, a filter-drier (accumulator) which contains a high pressure cycling switch and the plumbing connecting all of the above.

A blower fan forces the warmer air of the passenger compartment through the evaporator core (sort of a radiator-in-reverse), transferring the heat from the air to the refrigerant. The liquid refrigerant boils off into low pressure vapor, taking the heat with it when it leaves the evaporator.

2   Antifreeze - general information

Warning: *Do not allow antifreeze to come in contact with your skin or painted surfaces of the vehicle. Rinse off spills immediately with plenty of water. Antifreeze is highly toxic if ingested. Never leave antifreeze lying around in an open container or in puddles on the floor; children and pets are attracted by it's sweet smell and may drink it. Check with local authorities about disposing of used antifreeze. Many communities have collection centers which will see that antifreeze is disposed of safely.*

The cooling system should be drained, flushed and refilled at least every other year (see Chapter 1). The use of antifreeze solutions for periods of longer than two years is likely to cause damage and encourage the formation of rust and scale in the system.

Before adding antifreeze to the system, check all hose connections, because antifreeze tends to leak through very minute openings.

The exact mixture of antifreeze to water which you should use depends on the relative weather conditions. The mixture should contain at least 50 percent antifreeze, but should never contain more than 70 percent antifreeze. Read carefully all instructions printed on the antifreeze container.

Due to the aluminum radiator, use an antifreeze that is compatible with aluminum components. Coolant may be diluted with tap water except in areas where it has high alkaline content, in which case distilled water should be used.

3   Thermostat - replacement

Refer to illustrations 3.6 and 3.9
Warning: *The engine must be completely cool before beginning this procedure.*
1   Refer to the Warning in Section 2.
2   Drain the cooling system so that the coolant level is below the thermostat (Chapter 1).
3   On V6 engines the thermostat is mounted at the front of the intake manifold in an extension housing.
4   On 4-cylinder engines, disconnect the heater return hose at the thermostat housing, located on the front of the head.
5   Remove the upper radiator hose from the thermostat housing.
6   Remove the coolant outlet housing retaining bolts **(see illustration)**. Be prepared for additional coolant to flow from the engine, then pull the housing out sufficiently to provide access to the thermostat.
7   Remove the thermostat by turning counterclockwise and remove the gasket. **Caution:** *Do not pry on the thermostat to remove.*
8   Clean the thermostat housing surface and the engine surface thoroughly. All traces

of gasket material must be scraped clean before proceeding.

9   Install the thermostat with the bridge section in the outlet casting **(see illustration)**. Turn the thermostat clockwise to lock it in position on the flats cast into the outlet elbow.

10   On 4-cylinder models which have a heater hose connection, make sure that the thermostat is aligned properly to allow full flow to the heater; no part of the thermostat can be allowed to block the entrance to the heater tube.

11   Coat both sides of a new gasket with water resistant gasket sealer and position the gasket on the thermostat housing mating surface.

12   Position the thermostat housing carefully against the block or head, making sure the gasket is not disturbed.

13   Install and tighten the retaining bolts to the specified torque.

14   On 4-cylinder models, connect the heater hose to the thermostat housing.

15   Fill the cooling system with the recommended antifreeze and water mixture as described in Chapter 1.

16   Check for leaks and proper heater operation after the engine has reached normal operation temperature. After the engine has completely cooled remove the radiator cap and check for proper coolant level (see Chapter 1).

4   Radiator - removal and installation

Refer to illustrations 4.9 and 4.10
Warning: *Some models covered by this manual are equipped with Supplemental Restraint Systems (SRS), more commonly known as airbags. Always disable the airbag system before working in the vicinity of any airbag system component to avoid the possibility of accidental deployment of the airbag(s), which could cause personal injury.*

1   Disconnect the negative cable at the battery.

2   Drain the cooling system (Chapter 1).

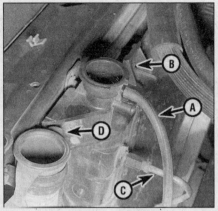

4.9 Radiator components

A   *Overflow hose*
B   *Fan shroud mounting bolt*
C   *Transmission cooling line (automatic only)*
D   *Radiator mounting bolt*

3   Disconnect the overflow hose near the filler neck.

4   Remove the two upper fan shroud bolts, lift up on the shroud until it pops out of the bottom locating pockets, then lay it over the fan.

5   Disconnect the upper and lower radiator hoses.

6   Disconnect the two transmission cooling lines from the radiator (automatic transmission only). Be sure to place a drain pan underneath. Cover the ends of the hoses with a plastic bag to prevent additional transmission fluid loss and contamination.

7   Unbolt the cooler tube bracket under the radiator (automatic transmission only).

8   Remove the hoses from the optional auxiliary transmission cooler (if equipped).

9   Remove the two radiator mounting bolts **(see illustration)**.

10   Tilt the radiator back about one inch and lift up, clear of the radiator support and fan. Push the A/C hoses aside, if so equipped **(see illustration)**, and carefully lift the radiator out of the vehicle. **Caution:** *Do not disconnect*

any of the air conditioning hoses or components without first having the system depressurized by a qualified dealer or specialist.

11   Following radiator removal, the shroud may be lifted out if desired.

12   Do not lose the rubber insulators from the bottom of the radiator.

13   Installation is the reverse of removal.

14   Refill the radiator with coolant (Chapter 1). Fill the system slowly to ensure the radiator is completely full. After driving, allow the system to cool and then check again that the level inside the radiator is full.

15   Check and correct the transmission fluid level (automatic transmission only).

5   Engine cooling fan and clutch - check and replacement

Refer to illustrations 5.6a and 5.6b
Warning: *The engine must be completely cool before beginning this procedure.*

Check

Note: *The following checks should be done with the engine off.*

1   Check fan clutch for fluid leakage. Slight seepage is acceptable. Substantial leaking requires replacement.

2   Rock the fan blades to check for bearing play in the hub. Replace the clutch assembly if excessive play or noise when turning the fan is noted.

3   If the fan completely freewheels (no resistance) when the engine is warm, or locks up completely regardless of temperature, replace the clutch assembly with a new one.

V6 engines

4   Unbolt the fan shroud. Place the shroud over the fan, against the engine.

5   Remove the fan to clutch bolts from the inside.

6   Remove the large nut which attaches the viscous clutch to the hub of the water pump shaft, using special tools, available at most auto parts stores, or their equivalents **(see illustration)**. **Caution:** *The 3.0L engine*

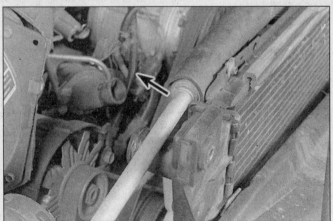

4.10 On A/C equipped vehicles, the refrigerant line must be pushed aside (arrow) to clear the hose neck

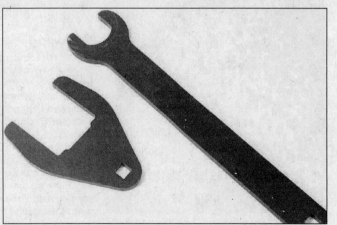

5.6a Two special wrenches are used for fan clutch removal - the shorter one holds the water pump pulley, while the longer one loosens the nut on the fan clutch

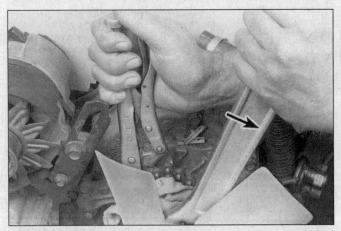

5.6b If the factory tools are unavailable, remove the lower pulley for clearance. Wrap a rag around the water pump pulley, attach a chain wrench or similar tool, and turn the hex-nut on fan hub with an adjustable wrench clock-wise to loosen (left hand threads). You may have to unbolt the fan from the clutch hub for clearance.

6.8a The coolant temperature sending unit (arrow) is located in left side of thermostat housing (2.8L V6). Coolant hose removed for access

shaft has left-hand threads and must be rotated clockwise for removal. In place of these special tools, we were able to use a chain wrench to hold the assembly stationary and adjustable wrench to loosen the nut **(see illustration)**.

7   Install the clutch/fan assembly on the water pump shaft and install the nut using the special tools or chain wrench/adjustable wrench. Remember that these are left-hand threads.

8   Attach the fan to the clutch (if separated).

9   Install the fan shroud.

4-cylinder engines

10   Disconnect the negative cable from the battery.

11   Remove the fan shroud.

12   Disconnect the overflow tube and lift the shroud off the lower retaining clips.

13   Move the shroud back over the fan in order to get to the fan bolts.

14   Remove the four bolts retaining the

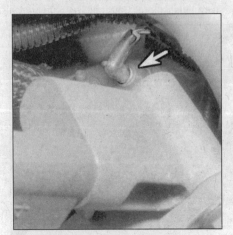

6.8b The temperature sending unit (arrow) on 4.0L engines is located on the intake manifold - it's on the left when viewed from the front of the vehicle

clutch and fan assembly to the pulley and remove the assembly from the vehicle.

15   If necessary, the fan and clutch can be separated by removing the four bolts which join them.

16   Prior to installation, attach the fan to the clutch with the four bolts (if separated).

17   Position the fan and clutch assembly to the pulley and secure with the four bolts.

18   Install the fan shroud.

19   Connect the battery cable.

---

6   Coolant temperature sending unit - check and replacement

Refer to illustrations 6.8a and 6.8b

Warning: *The engine must be completely cool before removing the sending unit. Antifreeze/coolant is toxic, keep it away from children and pets.*

Note: *The following procedure applies only to the standard analog instruments. The optional digital panel diagnosis requires special equipment the home mechanic is not likely to have. We recommend you take these to a dealer.*

Checking

1   If the coolant temperature gauge is inoperative, check the fuses first (Chapter 12).

2   If the temperature gauge indicates excessive temperature after running, see the Troubleshooting Section in the front of the manual.

3   If the temperature gauge indicates "Hot" as soon as the engine is started cold, disconnect the wire at the coolant temperature sensor. If the gauge reading drops, replace the sending unit. If the reading remains high, the wire to the gauge may be shorted to ground or the gauge is faulty.

4   If the coolant temperature gauge fails to show any indication after the engine has been warmed up, (approximately 10 minutes)

and the fuses checked out OK, shut off the engine. Disconnect the wire at the sending unit and, using a jumper wire, connect it to a clean ground on the engine. Turn on the ignition without starting the engine. If the gauge now indicates "Hot", replace the sending unit.

5   If the gauge still does not work, the circuit may be open or the gauge may be faulty - see Chapter 12 for additional information.

Replacement

6   Disconnect the negative cable at the battery.

7   With the engine completely cool, remove the cap from the radiator to relieve any pressure and then replace the cap. Place a drain pan under the sender.

8   Disconnect the temperature sending unit wire at the sending unit **(see illustrations)**.

9   Prepare the new temperature sending unit for installation by applying thread sealer to the threads.

10   Remove the temperature sending unit. Immediately install the new sending unit to minimize coolant loss.

11   Connect the temperature sending unit wire.

12   Refill the cooling system to replace lost coolant.

13   Connect the battery cable and start the engine. Check for leaks.

---

7   Coolant reservoir - removal and installation

Refer to illustration 7.6

Warning: *Store coolant/antifreeze in a tightly closed, properly labeled container, out of reach of children and pets. Immediately clean up any spills and observe precautions listed on the container.*

1   The coolant reservoir is located under the battery. It is combined with the wind-

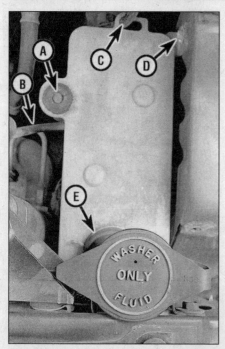

7.6 The combination coolant/windshield washer reservoir

A   *Coolant filler plug*
B   *Coolant overflow hose*
C   *Windshield washer hose*
D   *Mounting screw*
E   *Windshield washer filler tube clamp*

shield washer reservoir in a common housing separated by a bulkhead.

2   Remove the battery as described in Chapter 5.

3   Unbolt the starter solenoid as described in Chapter 5.

4   Remove the four bolts retaining the battery tray and lift it out.

5   Disconnect the hoses and wiring connectors from the reservoir.

6   Remove the retaining screw from the rear of the reservoir **(see illustration)** and lift it out while disengaging the locating flange at the front of the reservoir.

7   Installation is the reverse of removal.

8   Water pump - check

Refer to illustration 8.4

1   A failure of the water pump can cause overheating and serious engine damage (the pump will not circulate coolant through the engine).

2   There are three ways to check the operation of the water pump while it is installed on the engine. If the pump is defective, it should be replaced with a new or rebuilt unit.

3   With the engine running and at normal operating temperature, squeeze the upper radiator hose. If the water pump is working properly, a pressure surge will be felt as the hose is released.

4   Water pumps are equipped with a 'weep' or vent hole. If a pump seal failure occurs, small amounts of coolant will leak from the weep hole **(see illustration)**.

5   If the water pump shaft bearing fails, there may be a squealing sound emitted from the front of the engine while it is running. Shaft or bearing wear can be felt as free play if the water pump pulley is rocked up and down by hand.

9   Water pump - removal and installation

Refer to illustrations 9.27a, 9.27b, 9.30 and 9.31

Warning: *The engine must be completely cool before beginning this procedure. Also, see the Warning in Section 2 pertaining to antifreeze.*

4-cylinder engine

1   Disconnect the negative cable at the battery.

2   Drain the cooling system (refer to Chapter 1, if necessary).

3   Remove the shroud retaining bolts and move the shroud back over the fan.

4   Remove the four bolts retaining the fan assembly to the water pump shaft and then remove the fan (Section 5) and shroud from the engine compartment.

5   If equipped with air conditioning, loosen

the air conditioner compressor idler pulley and remove the drivebelt.

6   Loosen the alternator and power steering mounting bolts and remove the alternator and power steering belts.

7   Remove the water pump pulley.

8   Disconnect the heater hose at the water pump.

9   Remove the timing belt cover.

10   Disconnect the lower radiator hose from the water pump.

11   Remove the water pump retaining bolts and remove the water pump. Pay close attention to the various types and locations of the bolts for reinstallation.

12   Clean all old gasket and sealant from the gasket surfaces.

13   Coat both sides of the new gasket with sealant and place the gasket on the water pump surface.

14   Install the water pump on the cylinder block and install the retaining bolts. Apply water resistant sealer to the bolt threads prior to installing.

15   Install the lower radiator hose.

16   Install the heater hose.

17   Install the timing belt cover and tighten the bolts evenly.

18   Position the water pump pulley on the pump shaft.

19   Position the power steering (if so equipped) and alternator drivebelts to their respective pulleys.

20   Position the air conditioner compressor drivebelt (if so equipped) to its pulley.

21   Put the shroud around the fan assembly and install the fan assembly onto the water pump shaft (Section 5).

22   Position the shroud on the radiator and install the retaining bolts.

23   Adjust all drivebelts (Chapter 1).

24   Fill the cooling system (Chapter 1).

25   Start the engine and check for leaks.

V6 engines

26   Disconnect the negative cable at the battery.

27   To provide adequate clearance, remove the radiator (Section 4) and then the coolant hoses from the pump **(see illustrations)**.

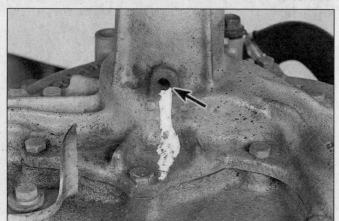

8.4 Coolant seepage stains from the water pump "weep hole" (arrow), viewed from below

9.27a Remove the coolant hose (arrow) by loosening the clamp and tapping the screw with a small hammer (2.8L V6 shown)

28 Remove the fan and clutch assembly (Section 5).

29 Loosen the alternator mounting bolts and remove the drive belt. On some models, it may be necessary to move the alternator completely to one side to gain access to the water pump bolts. Also, it may be necessary to remove the accessory drivebelt pulley or tensioner and the air conditioning compressor (if equipped) for the same reason. **Note:** *Do not disconnect the refrigerant line from the compressor.*

30 Remove the water pump pulley **(see illustration)**.

31 Remove the water pump assembly attaching bolts **(see illustration)** making note of the positions of the various length bolts. Remove the water pump (some models) from the front cover.

32 Remove all gasket material from the sealing surfaces on the front cover and water pump assembly.

33 Install the water pump assembly on the front cover with two bolts. Tighten the bolts only finger-tight.

34 Working around the pump, tighten all the water pump bolts to the specified torque.

35 Install the water pump pulley and the alternator and bracket if removed.

36 Install the radiator and connect all hoses.

37 Install all belts and adjust to the proper tension (Chapter 1).

38 Install the fan and clutch assembly (Section 6).

39 Connect the battery.

40 Fill the cooling system as described in Chapter 1.

41 Start the engine and check for leaks.

10 Heater blower motor - removal and installation

Refer to illustrations 10.6, 10.7, 10.15 and 10.16

Standard front blower motor

1 Disconnect the negative cable at the battery.

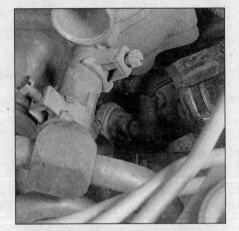

9.27b Remove the short coolant hose from the bottom of the pump (2.8L V6 shown)

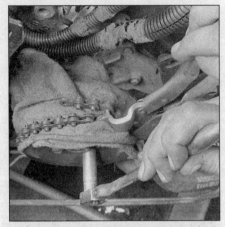

9.30 After fan removal, use the chain wrench to hold the pulley during bolt removal

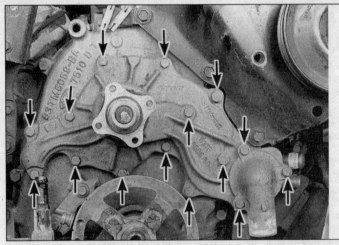

9.31 Arrows show locations of typical water pump mounting bolts (2.8L V6 shown)

2 The blower motor is located on the right side of the engine compartment firewall. Depending on the engine installed, the air cleaner and/or duct may have to be removed for access to the motor.

3 Disconnect the wiring harness at the blower motor. This is done by pushing down on the tab and then pulling the connector away.

4 On models so equipped, remove the two screws that attach the vacuum reservoir to the heater blower assembly and move the reservoir out of the way.

5 Disconnect the small rubber cooling tube at the blower motor.

6 Remove the screws which attach the motor to the heater blower assembly **(see illustration)**.

10.6 Remove the three blower retaining screws (arrows)

10.7 During blower removal you may have to twist the unit to clear obstructions

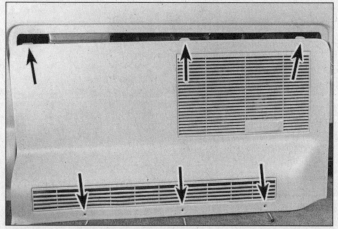

10.15 To remove the rear air conditioner/heater panel, remove the three screws at bottom (arrows) and pull out until clips at top (arrows) disengage

10.16 Rear mounted blower

A Hose
B Connector

C Hold-down screws (one is hidden)

7   Being very careful not to damage the gasket, pull the blower motor assembly out of the heater **(see illustration)**. You may have to turn the motor to clear the accumulator.

8   At this point the blower wheel can be separated from the motor by removing the clamp on the motor shaft.

9   Installation is essentially the reverse of removal, however note the following points:

10   The gasket between the blower motor and the heater assembly must be in good condition. If it was damaged on removal, use a new gasket.

11   Make sure the electrical connector is fully seated.

12   Reconnect the battery.

13   Test for proper operation upon completion.

Optional rear blower

14   Disconnect the negative cable at the battery.

15   Remove the access panel **(see illustration)**.

16   The remainder of the procedure is essentially the same as for the front mounted blower - follow Steps 3 to 13 described above **(see illustration)**.

11   Heater core and hoses - removal and installation

Warning: *Some models covered by this manual are equipped with Supplemental Restraint Systems (SRS), more commonly known as airbags. Always disable the airbag system before working in the vicinity of any airbag system component to avoid the possibility of accidental deployment of the airbag(s), which could cause personal injury.*

1   **Warning:** *Allow the engine to completely cool before beginning this procedure.*

Heater core

2   Using a thick cloth, turn the radiator cap

slowly to the first stop.

3   Step back and let the pressure release.

4   Once the pressure has been released, tighten the radiator cap.

5   Disconnect the heater hoses at the engine compartment firewall and plug the ends to prevent further draining of the system.

6   In the passenger compartment, remove the six screws attaching the heater core access cover to the plenum assembly. Remove the access cover.

7   Depress the retainer bracket at the top of the heater core. Pull the heater core rearward and down, removing it from the plenum assembly.

8   Position the new heater core and seal in the plenum assembly.

9   Install the heater core access cover to the plenum assembly.

10   Install the heater hoses.

11   Check the coolant level and add coolant as required (see Chapter 1).

12   Check the system for proper operation and for leaks.

Heater hoses

13   Drain the coolant from the cooling system (Chapter 1).

14   Disconnect the hoses by depressing the plastic tabs using a special tool, available at most auto parts stores. If you are unable to obtain this tool, you may be able to squeeze the retainer together with pliers or your fingers.

15   Pull the hoses apart until the plastic connector clears the locking tabs.

16   Damaged sections of hose may be repaired in an emergency by cutting out the damaged section and splicing in a new section using a short piece of 5/8-inch diameter tubing and hose clamps.

17   If a coupler clip is damaged, they may be purchased separately and installed.

18   Installation consists of pushing the two hose ends together until they click in, using new O-rings and silicone lubricant. Try to pull

them apart to ensure that they are properly seated.

19   Refill the cooling system (Chapter 1).

20   Check for leaks and proper system operation.

12   Air conditioning system - check and maintenance

Warning: *The air conditioning system is pressurized at the factory and depressurization requires special equipment. Discharging should be left to your dealer or air conditioning shop. Do not, under any circumstances, disconnect air conditioning fittings while the system is under pressure.*

Caution: *If you own a 1993 or later model vehicle, it is equipped with one of two different refrigerants (R-12 or R-134a). The refrigerants are not compatible and, although the system components may look identical, they are not interchangeable. The label stating which refrigerant is used is attached in the engine compartment. Use of the incorrect refrigerant will result in improper performance and damage to components.*

1   The following maintenance steps should be performed on a regular basis to ensure that the air conditioner continues to operate at peak efficiency.

a) *Check the tension of the drivebelt and adjust if necessary (Chapter 1).*

b) *Inspect the condition of the hoses. Check for cracking, hardening or other deterioration.* **Warning:** *Do not replace A/C hoses until the system has been discharged by a dealer or air conditioning specialist.*

c) *Inspect the fins of the condenser for leaves, bugs and any other foreign material. A soft brush and compressed air can be used to remove them.*

d) *Maintain the correct refrigerant charge.*

2   The A/C compressor should be run for about 10 minutes at least once a month. This

is particularly important during the winter months because long-term non-use can cause hardening of the internal seals.

3 Because of the complexity of the air conditioning system and the special equipment required to effectively work on it, accurate troubleshooting and repair of the system should be left to a professional mechanic. One probable cause for poor cooling that can be determined by the home mechanic is low refrigerant charge. Should the system lose its cooling ability, the following procedure will help you pinpoint the cause.

4 Warm the engine to normal operating temperature.

5 The hood and doors should be open.

6 Set the control mode selector lever to the A/C position.

7 Set the temperature selector lever to the Cold position.

8 Set the blower control selector to the Hi position.

9 With the compressor engaged, feel the evaporator inlet pipe between the orifice and the evaporator. Put your other hand on the surface of the accumulator can.

10 If both surfaces feel about the same temperature and if both feel a little cooler than the ambient temperature, the freon level is probably okay. The problem is elsewhere.

11 If the inlet pipe has frost accumulation or feels cooler than the accumulator surface, the freon charge is low.

13 Air conditioning system accumulator - removal and installation

Refer to illustration 13.2
Warning: *The A/C system is under high pressure. Have the system discharged by a dealer or A/C specialist prior to repairs.*
Note: *A number of special tools are necessary to properly disconnect the A/C lines and fittings at the accumulator. For this reason, this job may be better left to a dealer or spe-*

cialist equipped with these tools.

1 Disconnect negative cable at the battery.

2 Unplug the pressure switch connector **(see illustration).**

3 Disconnect the refrigerant line couplings using the special tool available at most auto parts stores.

4 Remove the mounting screws.

5 Installation is the reverse of removal.

6 Reconnect the refrigerant lines.

7 Have the system evacuated, recharged and leak tested by an air conditioning specialist.

14 Air conditioning system compressor - removal and installation

Refer to illustrations 14.3 and 14.5
Warning: *The A/C system is under high pressure. Have the system discharged prior to repairs by a dealer or A/C specialist.*

1 Disconnect the negative cable at the battery.

2 Remove the compressor drivebelt (Chapter 1).

3 Disconnect the refrigerant lines **(see illustration)** and cap them.

4 Disconnect the compressor wiring.

5 Remove the mounting bolts **(see illustration)** and lift the compressor out of the engine compartment.

6 Installation is the reverse of removal.

7 Have the system evacuated, recharged and leak tested by an air conditioning specialist.

15 Air-conditioning system condenser - removal and installation

Warning 1: *The A/C system is under high pressure. Have the system discharged prior*

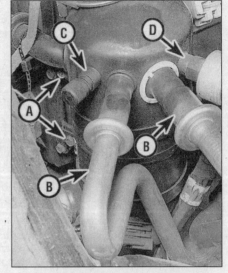

13.2 Air conditioning accumulator

A   Mounting screws
B   Refrigerant lines
C   Test gauge fitting
D   Pressure switch

to repairs by a dealer or A/C specialist.
Warning 2: *Some models covered by this manual are equipped with Supplemental Restraint Systems (SRS), more commonly known as airbags. Always disable the airbag system before working in the vicinity of any airbag system component to avoid the possibility of accidental deployment of the airbag(s), which could cause personal injury.*

1 Remove the radiator (Section 4).

2 After discharging system, disconnect the refrigerant lines from the condenser.

3 Remove the bolts from the upper corners of the condenser and lift it out of the vehicle.

4 Installation is the reverse of removal.

5 Have the system evacuated, recharged and leak tested by an air conditioning specialist.

14.3 Top view of compressor shows location of refrigerant line hold-down bolts (arrows) (typical side-mounted compressor)

14.5 View of A/C compressor on 2.8L V6 shows mounting bolt locations (arrows). Wrench is on right rear bolt

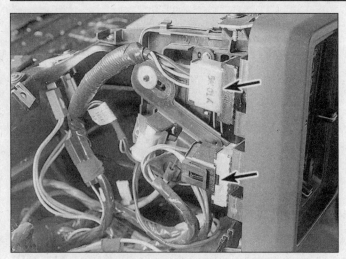

16.3 Unplug the electrical connectors (arrows)

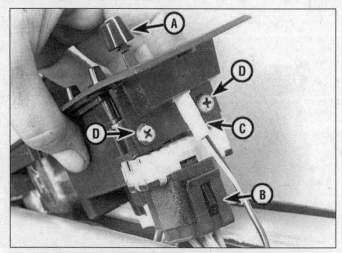

16.8 To replace the blower switch, pull off the control knob (A), unclip and remove the wiring connector (B), disconnect the light plug (C), and remove the screws (D)

16 Air conditioning and heater control assembly - removal and installation

Refer to illustrations 16.3 and 16.8
Note: *Early production vehicles were equipped with manual (cable) operated controls. Later models have vacuum operated air doors.*

Dash mounted control

Warning: *Some models covered by this manual are equipped with Supplemental Restraint Systems (SRS), more commonly known as airbags. Always disable the airbag system before working in the vicinity of any airbag system component to avoid the possibility of accidental deployment of the airbag(s), which could cause personal injury.*

1   Remove the instrument cluster cowl (Chapter 11).
2   Remove the four screws attaching the face plate to the control assembly. Remove the rear screw holding the control to the dash.
3   Pull the control assembly out towards you and unplug the electrical (and vacuum, if equipped) connectors **(see illustration)**.

4   Using pliers, carefully release the black function control cable and blue temperature control cable, as applicable.
5   Installation is the reverse of removal.
6   See Section 17 for adjustment procedures.

Optional rear mounted control

7   Remove the three cover mounting screws.
8   Lift the control out of the panel. Remove the knob, wiring and screws. Lift the switch out **(see illustration)**.
9   Installation is the reverse of removal.

17 Heater control cables - check and adjustment

Warning: *Some models covered by this manual are equipped with Supplemental Restraint Systems (SRS), more commonly known as airbags. Always disable the airbag system before working in the vicinity of any airbag system component to avoid the possibility of accidental deployment of the airbag(s), which could cause personal injury.*
Note: *Early production vehicles were equipped with manual (cable) operated con-*

trols. Later models have vacuum operated air doors.
1   Remove the heater control assembly (Section 16).
2   To check the function control cable, turn the blower on High and move the function control lever to its various positions and check for airflow at the various outlets.
3   To adjust or replace either cable, remove the upper finish panel.
4   To replace a broken cable, disconnect it at both ends and remove it from the vehicle. Install the cable at the control end first and adjust it using the following steps.
5   To adjust the function control cable, unclip the cable near the mode cam, place the control lever in the OFF position and align the holes in the mode cam with a Phillips screwdriver. Holding this position, snap the cable fastening clip into place.
6   To adjust the temperature control cable, unclip the cable at the temperature control cam and rotate the temperature control cam clockwise until the control lever is in the Cool position. Holding the cable in this position, snap the retaining clip into place, securing the cable housing.
7   Reinstall the control assembly and panels removed for access.

Notes

Chapter 4
Fuel and exhaust systems

Contents

Specifications

Fuel pump (carburetor-equipped models)
Fuel pressure (at idle)	4.5 to 6.5 psi
Fuel pump pushrod length	2.43 in (minimum)

Fuel pressure (EFI models)
Key on engine off	35 to 45 psi
At idle (vacuum disconnected)	35 to 45 psi
At idle (vacuum connected)	30 to 35 psi

Torque specifications
Ft-lbs (unless otherwise indicated)

Four-cylinder engine
Air bypass valve mounting bolts	71 to 102 in-lbs
Throttle body to upper intake manifold bolts/nuts	12 to 15
Upper intake manifold to lower intake manifold bolts	14 to 21
Throttle position sensor to throttle body	14 to 15 in-lbs
Fuel pressure relief valve	48 to 84 in-lbs
Fuel pressure regulator to fuel supply manifold	27 to 40 in-lbs
Fuel supply manifold to lower intake manifold bolts	15 to 22

2.8L V6 engine
Carburetor to intake manifold	14 to 16
Fuel pump mounting bolts	11 to 19

3.0L V6 engine
Air intake throttle body assembly bolts and studs	15 to 22

4.0L V6 engine
Upper intake manifold nuts	15 to 18 ft-lbs
Throttle body bolts	76 to 106 in-lbs
Air bypass valve screws	72 to 96 in-lbs
Fuel pressure regulator mounting screws	72 to 96 in-lbs
Fuel rail mounting bolts (studs)	84 to 120 in-lbs

2.6 The inertia switch has a reset button (arrow) which should be pushed in after reinstalling the switch

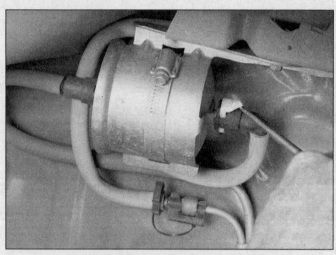

3.5 A hairpin clip type push-connect fitting

1   General information

Fuel system

The fuel system consists of the fuel tank, the fuel pump(s), an air cleaner assembly, either a carburetor or a fuel injection system and the various steel, plastic and/or nylon lines and fittings connecting everything together. 1986 Aerostars powered by the 2.8L V6 are equipped with a carburetor. 1986 and 1987 vehicles with 2.3L OHC fours and all 3.0L and 4.0L V6s are equipped with electronic fuel injection (EFI).

The fuel pump on carburetor equipped vehicles is a mechanical type mounted on the block and driven off the camshaft by a pushrod. The pumps on fuel injected vehicles are electric. A low pressure pump is mounted inside the fuel tank and a high pressure pump is mounted on the left frame rail, forward of the tank.

Exhaust system

All vehicles are equipped with a single exhaust manifold (2.3L OHC Four), or a pair of manifolds (V6), connecting pipe between the exhaust manifold and the catalytic converter, a catalytic converter, an exhaust pipe and a muffler. Any component of the exhaust system can be replaced.

2   Fuel pressure relief procedure (fuel injected vehicles)

Refer to illustration 2.6

Warning: *Gasoline is extremely flammable, so take extra precautions when you work on any part of the fuel system. Don't smoke or allow open flames or bare light bulbs near the work area, and don't work in a garage where a gas-type appliance (such as a water heater or clothes dryer) is present. Since gasoline is carcinogenic, wear latex gloves when there's*

a possibility of being exposed to fuel, and, if you spill any fuel on your skin, rinse it off immediately with soap and water. Mop up any spills immediately and do not store fuel-soaked rags where they could ignite. The fuel system on fuel-injected models is under constant pressure, so, if any fuel lines are to be disconnected, the fuel pressure in the system must be relieved first (see Chapter 4 for more information). When you perform any kind of work on the fuel system, wear safety glasses and have a Class B type fire extinguisher on hand.

1   The inertia switch, which shuts off fuel to the engine in the event of a collision, affords a simple and convenient means by which fuel pressure can be relieved before servicing fuel injection components. The switch is located on the toe-board to the right of the transmission hump, on the kick panel to the right of the passenger seat, or in the engine compartment.

2   Locate the switch on your vehicle, remove the mounting screws and remove the switch.

3   Unplug the inertia switch electrical connector.

4   Start the engine and allow it to idle until it eventually stops running on its own.

5   The fuel system pressure is now relieved. When finished working on the fuel system, check for fuel leaks, then simply plug the electrical connector back into the switch. **Warning:** *If you see or smell gasoline, do not reset the switch until you have located the source of the leak and fixed it.*

6   Reinstall the inertia switch and tighten the two mounting nuts securely. Insure the switch is operational by depressing the reset button **(see illustration)**.

3   Fuel lines and fittings - general information

Warning: *Gasoline is extremely flammable,*

so take extra precautions when you work on any part of the fuel system. See the **Warning** in Section 2.

Note: *The majority of fuel line fittings on these vehicles require special disconnect tools to allow removal of the fuel lines. These tools are available at most auto part stores.*

Push-connect fittings - disassembly and reassembly

1   The manufacturer uses two different push-connect fitting designs. Fittings used with 3/8 and 5/16-inch diameter lines have a "hairpin" type clip; fittings used with 1/4-inch diameter lines have a "duck bill" type clip. The procedure used for releasing each type of fitting is different. The clips should be replaced whenever a connector is disassembled.

2   Disconnect all push-connect fittings from fuel system components such as the fuel filter, the fuel charging assembly, the fuel tank, etc. before removing the assembly.

3/8 and 5/16-inch fittings (hairpin clip)

Refer to illustration 3.5

3   Inspect the internal portion of the fitting for accumulations of dirt. If more than a light coating of dust is present, clean the fitting before disassembly.

4   Some adhesion between the seals in the fitting and the line will occur over a period of time. Twist the fitting on the line, then push and pull the fitting until it moves freely.

5   Remove the hairpin clip from the fitting by bending the shipping tab down until it clears the body. Then spread each leg about 1/8-inch to disengage the body and push the legs through the fitting. If possible, don't use any tools to do this, to prevent damage to the plastic fitting on the fuel line. Finally, pull lightly on the triangular end of the clip and work it clear of the line and fitting **(see illustration)**.

6   Grasp the fitting and hose and pull it straight off the line.

3.10 A push-connect fitting with a duck bill clip

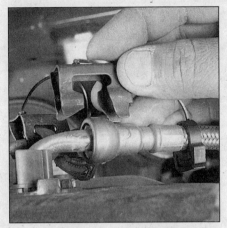

3.13 Remove the safety clip

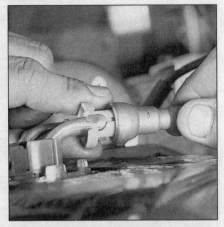

3.14 Duck bill clip fitting disassembly using the special tool

7   Do not reuse the original clip in the fitting. A new clip must be used.

8   Before reinstalling the fitting on the line, wipe the line end with a clean cloth. Inspect the inside of the fitting to ensure that it's free of dirt and/or obstructions.

9   To reinstall the fitting on the line, align them and push the fitting into place. When the fitting is engaged, a definite click will be heard. Pull on the fitting to ensure that it's completely engaged. To install the new clip, insert it into any two adjacent openings in the fitting with the triangular portion of the clip pointing away from the fitting opening. Using your index finger, push the clip in until the legs are locked on the outside of the fitting.

1/4-inch fittings (duck bill clip)

Refer to illustrations 3.10, 3.13 and 3.14

10   The duck bill clip type fitting consists of a body, spacers, O-rings and the retaining clip **(see illustration)**. The clip holds the fitting securely in place on the line. One of the two following methods must be used to disconnect this type of fitting.

11   Before attempting to disconnect the fitting, check the visible internal portion of the fitting for accumulations of dirt. If more than a light coating of dust is evident, clean the fitting before disassembly.

12   Some adhesion between the seals in the fitting and line will occur over a period of time. Twist the fitting on the line, then push and pull the fitting until it moves freely.

13   Remove the safety clip from the fuel line **(see illustration)**.

14   The preferred method used to disconnect the fitting requires a special tool available at most auto part stores. To disengage the line from the fitting, align the slot in the push-connect disassembly tool with either tab on the clip (90-degrees from the slots on the side of the fitting) and insert the tool **(see illustration)**. This disengages the duck bill from the line. **Note:** *Some fuel lines have a secondary bead which aligns with the outer surface of the clip. The bead can make tool insertion difficult. If necessary, use the alternative disassembly method described in*

Step 16. Holding the tool and the line with one hand, pull the fitting off. **Note:** *Only moderate effort is necessary if the clip is properly disengaged. The use of anything other than your hands should not be required.*

15   After disassembly, inspect and clean the line sealing surface. Also inspect the inside of the fitting and the line for any internal parts that may have been dislodged from the fitting. Any loose internal parts should be immediately reinstalled (use the line to insert the parts).

16   The alternative disassembly procedure requires a pair of small adjustable pliers. The pliers must have a jaw width of 3/16-inch or less.

17   Align the jaws of the pliers with the openings in the side of the fitting and compress the portion of the retaining clip that engages the body. This disengages the retaining clip from the body (often one side of the clip will disengage before the other - both sides must be disengaged).

18   Pull the fitting off the line. **Note:** *Only moderate effort is required if the retaining clip has been properly disengaged. Do not use any tools for this procedure.*

19   Once the fitting is removed from the line end, check the fitting and line for any internal parts that may have been dislodged from the fitting. Any loose internal parts should be immediately reinstalled (use the line to insert the parts).

20   The retaining clip will remain on the line. Disengage the clip from the line bead to remove it. Do not reuse the retaining clip - install a new one!

21   Before reinstalling the fitting, wipe the line end with a clean cloth. Check the inside of the fitting to make sure that it's free of dirt and/or obstructions.

22   To reinstall the fitting, align it with the line and push it into place. When the fitting is engaged, a definite click will be heard. Pull on the fitting to ensure that it's fully engaged.

23   Install the new replacement clip by inserting one of the serrated edges on the duck bill portion into one of the openings. Push on the other side until the clip snaps into place.

Spring lock couplings - disassembly and reassembly

Refer to illustrations 3.26a, 3.26b and 3.26c

24   The fuel supply and return lines used on SEFI engines utilize spring lock couplings at the engine fuel rail end instead of plastic push-connect fittings. The male end of the spring lock coupling, which is girded by two O-rings, is inserted into a female flared end engine fitting. The coupling is secured by a garter spring which prevents disengagement by gripping the flared end of the female fitting. A cup-tether assembly provides additional security.

25   To disconnect the 1/2-inch spring lock coupling supply fitting or the 3/8-inch return fitting, you will need to obtain the appropriate spring lock coupling tool from your local auto parts store. The tools come in two different sizes - 3/8 and 1/2 inch - which correspond with the diameter of the fuel supply and return lines.

26   Study the accompanying illustrations carefully before detaching either spring lock coupling fitting **(see illustrations)**.

3.26a If the spring lock couplings are equipped with safety clips, pry them off with a small screwdriver

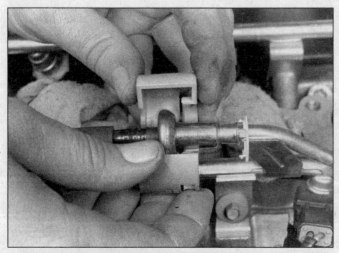

3.26b Open the spring-loaded halves of the spring lock coupling tool and place it in position around the coupling, then close it

3.26c To disconnect the coupling, push the tool into the cage opening to expand the garter spring and release the female fitting, then pull the male and female fittings apart

4   Fuel pump - check

Warning: *Gasoline is extremely flammable, so take extra precautions when you work on any part of the fuel system. Don't smoke or allow open flames or bare light bulbs near the work area, and don't work in a garage where a gas-type appliance (such as a water heater or clothes dryer) is present. Since gasoline is carcinogenic, wear latex gloves when there's a possibility of being exposed to fuel, and, if you spill any fuel on your skin, rinse it off immediately with soap and water. Mop up any spills immediately and do not store fuel-soaked rags where they could ignite. The fuel system on fuel-injected models is under constant pressure, so, if any fuel lines are to be disconnected, the fuel pressure in the system must be relieved first (see Chapter 4 for more information). When you perform any kind of work on the fuel system, wear safety glasses and have a Class B type fire extinguisher on hand.*

Mechanical fuel pump

General check

1   On carburetor-equipped engines, the fuel pump is located on the left side of the engine.
2   The fuel pump is sealed and no repairs are possible other than component replacement. However, the fuel pump can be inspected and tested on the vehicle as follows.
3   Make sure there is fuel in the fuel tank.
4   With the engine running, check for leaks at all fuel line connections between the fuel tank and the carburetor.
5   Tighten any loose connections and inspect all lines for flat spots and kinks which could restrict fuel flow. Air leaks or restrictions on the suction side of the fuel pump will greatly affect pump output.
6   Check for leaks at the fuel pump diaphragm flange.
7   Disconnect the coil wire from the distributor and fuel inlet line from the carburetor.

Direct the fuel line from the pump into a metal container.
8   Crank the engine several revolutions and make sure that well defined spurts of fuel are ejected from the line. If not, the fuel line is clogged or the fuel pump is defective.
9   Disconnect the fuel line at both ends of the line and blow through it with compressed air to determine if the fuel line is clogged.

Pressure test

10   Connect a fuel pressure gauge (0 to 15 psi) to the fuel filter end of the line.
11   Start the engine - it should be able to run for over 30 seconds on the fuel in the carburetor bowl - and read the pressure after ten seconds. Compare your reading to the specified pressure.
12   If pump pressure is not as specified, install a new fuel pump.
13   Reconnect the fuel lines and install the air cleaner.

Electric fuel pump (fuel-injected vehicles)

General check

14   An electric fuel pump malfunction will usually result in loss of fuel flow and/or pressure that is often reflected by a corresponding drop in performance (or a no-run condition).
15   Verify the pump actually runs. Remove the fuel filler cap and have an assistant turn the ignition switch to On - you should hear a brief whirring noise as the pump comes on and pressurizes the system. Have the assistant start the engine (if possible). This time you should hear a constant whirring sound from the pump (but it's more difficult to hear with the engine running).
16   If the pump doesn't run (makes no sound), check the fuel pump fuse (see Chapter 12). If the fuse is okay, check the continuity of the inertia switch with an ohmmeter (see Section 2 for the location of the switch). If the inertia switch is good, proceed to the

next Step. If the fuse is blown, replace it and see if the pump works (if it blows again, trace the fuel pump circuit for a short to ground). If the fuse doesn't blow but the fuel pump still doesn't work, proceed to the next Step.
17   With the ignition off, unplug the electrical connector from the fuel pump. Using a jumper wire, ground the negative terminal of the fuel pump. Using a fused jumper wire, apply 12-volts to the fuel pump feed terminal.
18   If the fuel pump doesn't run, replace it (see Section 7). If it does run, the problem lies somewhere in the electrical circuit to the fuel pump.

Pressure check

19   Relieve the fuel pressure (see Section 2).
20   Connect a fuel pressure gauge to the Schrader valve on the fuel rail.
21   Turn the ignition switch to the On position. The fuel pump should run for about two seconds, then the pressure should hold steady. Note the reading on the gauge and compare it to the pressure listed in this Chapter's Specifications.
22   Start the engine (if possible) and allow it to idle. The pressure should be lower by 3 to 10 psi (see this Chapter's specifications). Disconnect the vacuum hose from the fuel pressure regulator and compare your reading with the pressure listed in this Chapter's Specifications. If all the pressure readings are within specifications, the system is operating properly.
23   If the pressure did not drop by 3 to 10 psi after starting the engine, apply 10-inches Hg of vacuum to the pressure regulator. If the pressure drops, repair the vacuum source to the regulator. If the pressure doesn't drop, replace the regulator.
24   If the pressure is higher than specified, check for a faulty regulator or a pinched or clogged fuel return hose or pipe.
25   If the pressure is lower than specified:

a) *Inspect the fuel filter - make sure it's not clogged.*

b) *Look for a pinched or clogged fuel hose or line between the fuel tank and the fuel rail.*

c) *Check the pressure regulator for a malfunction. With the engine running, pinch the fuel return line (if possible). If the pressure rises, replace the fuel pressure regulator.*

d) *Look for leaks in the fuel feed line.*

e) *Check for leaking injectors.*

26 After the testing is done, relieve the fuel pressure (see Section 2) and remove the fuel pressure gauge.

5 Fuel tank - removal and installation

Note: *Don't begin this procedure until the gauge indicates that the tank is empty or nearly empty. If the tank must be removed when it's full (for example, if the fuel pump malfunctions), siphon any remaining fuel from the tank prior to removal.*

Warning: *Gasoline is extremely flammable, so take extra precautions when you work on any part of the fuel system. Don't smoke or allow open flames or bare light bulbs near the work area, and don't work in a garage where a gas-type appliance (such as a water heater or clothes dryer) is present. Since gasoline is carcinogenic, wear latex gloves when there's a possibility of being exposed to fuel, and, if you spill any fuel on your skin, rinse it off immediately with soap and water. Mop up any spills immediately and do not store fuel-soaked rags where they could ignite. The fuel system on fuel-injected models is under constant pressure, so, if any fuel lines are to be disconnected, the fuel pressure in the system must be relieved first (see Chapter 4 for more information). When you perform any kind of work on the fuel system, wear safety glasses and have a Class B type fire extinguisher on hand.*

1 On fuel injected vehicles, relieve the fuel pressure (refer to Section 2).

2 Detach the cable from the negative terminal of the battery.

3 Raise the vehicle and support it securely on jackstands.

4 Unless the vehicle has been driven far enough to completely empty the tank, it's a good idea to siphon the residual fuel out before removing the tank from the vehicle. Siphon or pump the fuel out through the fuel filler pipe. On US vehicles, a small diameter hose may be necessary because of the small trap door installed in the fuel filler pipe to prevent vapors from escaping during refueling. Fuel injected vehicles have reservoirs inside the tank to maintain fuel near the pump pick-up during vehicle cornering maneuvers and when the fuel level is low. The reservoirs could prevent siphon tubes or hoses from reaching the bottom of the fuel tank. This situation can be overcome by repositioning the siphon hose several times.

5 Loosen the hose clamps securing the

fuel filler neck hose and detach the hose.

6 Place a floor jack under the tank and position a block of wood between the jack pad and the tank. Raise the jack until it's supporting the tank.

7 Remove the bolt from the front and rear straps. (The straps are hinged at the other end so you can swing them out of the way.)

8 Lower the tank far enough to unplug the wiring harness and to disconnect the fuel feed return and vapor hoses. Note that each of these hoses has a different diameter, making it impossible to mix them up during reassembly.

9 Slowly lower the jack while steadying the tank. Remove the tank from the vehicle and remove the shield from the tank.

10 If you're replacing the tank, or having it cleaned or repaired, remove the sender unit (carburetor equipped vehicles) or fuel pump/sending unit (fuel injected vehicles) (see Section 7). For information regarding tank cleaning and repair, see Section 6.

11 When the tank is ready to be installed, install the fuel pump/sending unit (see Section 7).

12 Install the shield.

13 Installation is the reverse of removal.

6 Fuel tank - cleaning and repair

1 Repairs to the fuel tank or filler neck should be performed by a professional with the proper training to carry out this critical and potentially dangerous work. Even after cleaning and flushing, explosive fumes can remain and could explode during repair of the tank.

2 If the fuel tank is removed from the vehicle, it should not be placed in an area where sparks or open flames could ignite the fumes coming out of the tank. Be especially careful inside garages where a natural gas appliance is located because the pilot light could cause an explosion.

7 Fuel pump - removal and installation

Warning: *Gasoline is extremely flammable, so take extra precautions when you work on any part of the fuel system. Don't smoke or allow open flames or bare light bulbs near the work area, and don't work in a garage where a gas-type appliance (such as a water heater or clothes dryer) is present. Since gasoline is carcinogenic, wear latex gloves when there's a possibility of being exposed to fuel, and, if you spill any fuel on your skin, rinse it off immediately with soap and water. Mop up any spills immediately and do not store fuel-soaked rags where they could ignite. The fuel system on fuel-injected models is under constant pressure, so, if any fuel lines are to be disconnected, the fuel pressure in the system must be relieved first (see Chapter 4 for more information). When you perform any kind of work on the fuel system, wear safety glasses*

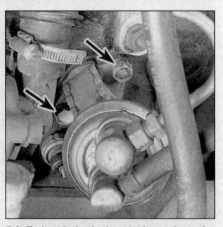

7.2 To break the fuel pump loose from the block, loosen the fuel pump mounting bolts (arrows) a couple of turns - if it still sticks to the block, have a friend turn the engine over once or twice - when the camshaft lobe actuates the fuel pump pushrod, it will break the pump loose

and have a Class B type fire extinguisher on hand.

Mechanical fuel pump (carburetor equipped vehicles)

Refer to illustration 7.2

1 Loosen the threaded fuel line fittings (at the pump) with the proper size wrench (a flare nut wrench is recommended), then retighten them until they're just snug. Don't remove the lines at this time. The outlet line is pressurized, so protect your eyes with safety goggles or wrap the fitting with a shop rag, then loosen it carefully.

2 Loosen the mounting bolts **(see illustration)** two turns. Use your hands to loosen the fuel pump. If it's stuck to the block, do not use a tool - you could damage the pump. If you cannot loosen the pump by hand, have an assistant operate the starter while you keep one hand on the pump. As the camshaft turns, the pump pushrod will operate the pump - when the pump feels loose, stop turning the engine over.

3 Disconnect the fuel lines from the pump. Be sure to use a back-up wrench on the stationary fitting to prevent the line from twisting.

4 Remove the fuel pump bolts and detach the pump and gasket. Discard the old gasket. Pull out the pushrod.

5 Inspect the pushrod for damage or excessive wear.

6 Remove all old gasket material and sealant from the engine block. If you're installing the original pump, remove all the old gasket material from the pump mating surface as well. Wipe the mating surfaces of the block and pump with a cloth saturated with lacquer thinner or acetone.

7 Install the fuel pump pushrod. Insert the bolts through the pump (to use as a guides for the new gasket) and place the gasket in position on the fuel pump mounting flange. Position the fuel pump on the block (make sure the pump arm engages the pushrod

7.13 Be sure to use a BRASS drift when loosening the lock ring on the electric fuel pump/sending unit (a spark from a steel punch or hammer could cause an explosion)

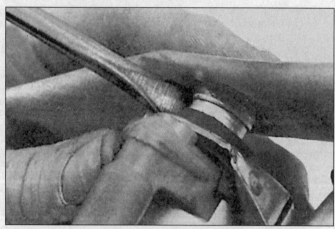

7.16 If you're replacing the pump, this step isn't necessary, but if you're going to reinstall the old pump, remove the strainer and clean it with solvent

properly). Tighten the bolts a little at a time until they're at the specified torque.

8   Attach the fuel lines to the pump. Start the threaded fitting by hand to avoid cross-threading it. Tighten the outlet nut securely. If any of the hoses are cracked, hardened or otherwise deteriorated, replace them at this time.

9   Start the engine and check for fuel leaks for two minutes.

10   Stop the engine and check the fuel line connections for leaks by running a finger under each fitting. Check for oil leaks at the fuel pump mounting gasket.

Electric fuel pump (fuel injected vehicles)

Refer to illustrations 7.13 and 7.16

In-Tank (low pressure) pump

11   Relieve the fuel system pressure (see Section 2).

12   Remove the fuel tank (see Section 5).

13   Using a brass drift only, tap the lock ring counterclockwise until it's loose **(see illustration)**.

14   Carefully pull the fuel pump/sending unit assembly from the tank.

15   Remove the old lock ring gasket and discard it.

16   If you're planning to reinstall the original fuel pump/sending unit, remove the strainer by prying it off with a screwdriver **(see illustration)**, wash it in clean solvent, then push it back onto the metal pipe on the end of the pump. If you're installing a new pump/sending unit, the assembly will include a new strainer.

17   Clean the fuel pump mounting flange and the tank mounting surface and seal ring groove.

18   Installation is the reverse of removal. Apply a thin coat of heavy grease to the new lock ring gasket to hold it in place during assembly.

Externally mounted (high pressure) pump

19   Relieve the fuel system pressure (see Section 2).

20   Raise the vehicle and place it securely on jackstands.

21   Unplug the electrical connector.

22   Disconnect the fuel line fittings from both ends of the pump. (see Section 3).

23   Remove the mounting bolts.

24   Remove the pump.

25   Installation is the reverse of removal.

---

8   Air cleaner housing - removal and installation

Four-cylinder engine

1   Detach the cable from the negative terminal of the battery.

2   Disconnect the inlet tube and idle bypass tube from the air cleaner cover.

3   Unplug the electrical connector to the throttle air bypass valve.

4   Remove the air cleaner cover and filter element.

5   Remove the three nuts holding the air cleaner tray.

6   Remove the air cleaner tray.

7   Installation is the reverse of removal.

1986 through 1989 3.0L V6 engine

8   Removal and installation of the air cleaner housing assembly on the 3.0L V6 is basically similar to the procedure outlined above for the 2.3L four, but be sure that the sealing gasket is in position between the right fender apron and the air cleaner assembly when reinstalling the air cleaner assembly.

1990 and later V6 engines

9   Detach the cable from the negative terminal of the battery.

10   Disconnect the hose from the outlet tube. On 4.0L engines, disconnect MAF sensor electrical connector.

11   Disconnect the outlet tube clamps from the air cleaner cover and the throttle body. Remove the outlet tube. Cover the throttle

body opening to prevent the entry of foreign matter.

12   Remove the air filter (see Chapter 1), then remove the two nuts securing the air cleaner assembly to the front fender apron. The nuts are located at the bottom of the air cleaner assembly, below the air filter.

13   Remove the air cleaner assembly.

14   Installation is the reverse of removal.

2.8L V6 engine

15   Detach the cable from the negative terminal of the battery.

16   Unplug the electrical connector from the air charge temperature sensor (see Chapter 6 if necessary).

17   Detach the vacuum line from the vacuum motor (on top of the snorkel).

18   Detach the air intake tube assembly.

19   Remove the single air cleaner housing cover retaining clip on the driver's side of the housing.

20   Remove the wing nut from the top of the air cleaner housing cover.

21   Remove the housing cover.

22   Remove the filter element.

23   Remove the air cleaner housing assembly from the carburetor.

24   Installation is the reverse of removal.

---

9   Throttle linkage components - removal and installation

1   Detach the cable from the negative terminal of the battery.

2   Remove the throttle linkage shield.

3   Detach the throttle cable from the throttle lever at the carburetor or throttle body.

4   Remove the throttle cable from the bracket on the intake manifold.

5   From inside the vehicle, remove the throttle cable end from the top of the accelerator pedal.

6   Collapse the retaining clips at the dash panel and push the throttle cable through the dash panel into the engine compartment.

Remove the cable assembly.

7 Installation is the reverse of removal.

10 Carburetor - removal, overhaul and installation

Refer to illustration 10.15

Warning: *Gasoline is extremely flammable, so take extra precautions when you work on any part of the fuel system. Don't smoke or allow open flames or bare light bulbs near the work area, and don't work in a garage where a gas-type appliance (such as a water heater or clothes dryer) is present. Since gasoline is carcinogenic, wear latex gloves when there's a possibility of being exposed to fuel, and, if you spill any fuel on your skin, rinse it off immediately with soap and water. Mop up any spills immediately and do not store fuel-soaked rags where they could ignite. The fuel system on fuel-injected models is under constant pressure, so, if any fuel lines are to be disconnected, the fuel pressure in the system must be relieved first (see Chapter 4 for more information). When you perform any kind of work on the fuel system, wear safety glasses and have a Class B type fire extinguisher on hand.*

10.15 Before installing the carburetor, carefully scrape away all traces of old gasket material from the manifold

Diagnosis

1 A thorough road test and check of carburetor adjustments should be done before any major carburetor service work. Specifications for some adjustments are listed on the Vehicle Emissions Control Information label found in the engine compartment.

2 Some performance complaints directed at the carburetor are actually a result of loose, out-of-adjustment or malfunctioning engine or electrical components. Others develop when vacuum hoses leak, are disconnected or are incorrectly routed. The proper approach to analyzing carburetor problems should include a routine check of the following items.

a) *Inspect all vacuum hoses and actuators for leaks and correct installation (see Chapter 6).*

b) *Tighten the intake manifold nuts and carburetor mounting nuts evenly and securely.*

c) *Perform a cylinder compression test.*

d) *Clean or replace the spark plugs as necessary.*

e) *Check the resistance of the spark plug wires.*

f) *Inspect the ignition primary wires and check the vacuum advance operation. Replace any defective parts.*

g) *Check the ignition timing according to the instructions printed on the Emissions Control Information label.*

h) *Check the fuel pump pressure.*

i) *Check the heat control valve in the air cleaner for proper operation (see Chapter 6).*

j) *Check/replace the air filter element.*

k) *Check the PCV system (see Chapter 6).*

3 Carburetor problems usually show up as flooding, hard starting, stalling, severe backfiring, poor acceleration and lack of response to idle mixture screw adjustments. A carburetor that is leaking fuel and/or covered with wet looking deposits definitely needs attention.

4 Diagnosing carburetor problems may require that the engine be started and run with the air cleaner off. While running the engine without the air cleaner, backfires are possible. This situation is likely to occur if the carburetor is malfunctioning, but just the removal of the air cleaner can lean the fuel/air mixture enough to produce an engine backfire. **Warning:** *Do not position any part of your body, especially your face, directly over the carburetor during inspection and servicing procedures.*

Removal

5 Remove the air cleaner duct (see Section 8).

6 Disconnect the throttle cable from the throttle lever (see Section 9).

7 If your vehicle is equipped with an automatic transmission, disconnect the TV rod from the throttle lever (see Chapter 7).

8 **Note:** *Label all vacuum hoses and fittings before removing them to simplify installation.* Disconnect all vacuum hoses and the fuel line from the carburetor. Use a back-up wrench on the fuel inlet fitting when removing the fuel line to avoid changing the float level.

9 Label the wires and terminals, then unplug all wire harness connectors.

10 Remove the four carburetor mounting nuts and detach the carburetor from the intake manifold. Remove the carburetor mounting gasket spacer (if so equipped).

Overhaul

11 Once it's determined that the carburetor needs adjustment or an overhaul, several options are available. If you're going to attempt to overhaul the carburetor yourself, first obtain a good quality carburetor rebuild kit (which will include all necessary gaskets, internal parts, instructions and a parts list). You'll also need some special solvent and a means of blowing out the internal passages of the carburetor with air.

12 Because the vehicles covered by this book are primarily fuel injected and because carburetor designs are constantly modified by the manufacturer in order to meet increasingly more stringent emissions regulations, it isn't feasible for us to do a step-by-step overhaul of each type. You'll receive a detailed, well illustrated set of instructions with any carburetor overhaul kit; they will apply in a more specific manner to the carburetor on your vehicle.

13 Another alternative is to obtain a new or rebuilt carburetor. They are readily available from dealers and auto parts stores. Make absolutely sure the exchange carburetor is identical to the original. A tag is usually attached to the top of the carburetor. It will aid in determining the exact type of carburetor you have. When obtaining a rebuilt carburetor or a rebuild kit, take time to make sure that the kit or carburetor matches your application exactly. Seemingly insignificant differences can make a large difference in the performance of your engine.

14 If you choose to overhaul your own carburetor, allow enough time to disassemble the carburetor carefully, soak the necessary parts in the cleaning solvent (usually for at least one-half day or according to the instructions listed on the carburetor cleaner) and reassemble it, which will usually take much longer than disassembly. When disassembling the carburetor, match each part with the illustration in the carburetor kit and lay the parts out in order on a clean work surface. Overhauls by inexperienced mechanics can result in an engine which runs poorly or not at all. To avoid this, use care and patience when disassembling the carburetor so you can reassemble it correctly.

Installation

15 Clean the gasket mating surfaces of the intake manifold and the carburetor to remove all traces of the old gasket **(see illustration)**. Place a new gasket and spacer (if so equipped) on the intake manifold. Position the carburetor on the gasket and spacer and install the four mounting nuts. To prevent distortion or damage to the carburetor body flange, tighten the nuts to the specified torque in several steps.

16 The remaining installation steps are the reverse of removal.

11 Electronic fuel injection (EFI) system - general information

The electronic fuel injection (EFI) system is a multi-point pulse time, speed density injection system. Fuel is metered into the air intake stream in accordance with engine demands through four (2.3L) or six solenoid injection valves mounted on a tuned intake manifold.

Fuel is supplied from the fuel tank by a low pressure, electric fuel pump mounted in the fuel tank. An externally mounted high pressure pump boosts fuel pressure on its way to the engine. The fuel is filtered and sent to the fuel charging manifold assembly and then to the injectors. The fuel delivery pressure is maintained by a pressure regulator connected in series with the injectors and positioned downstream from them. Excess fuel supplied by the pump but not needed by the engine is returned, through the regulator to the fuel tank by a steel fuel return line.

Air flow to the engine is controlled by a single butterfly valve mounted in a two piece, die-cast aluminum housing called a throttle body. The butterfly valve is identical in configuration to the throttle plates of a conventional carburetor and is actuated by a similar pedal and linkage arrangement.

The fuel supply manifold assembly is the component that delivers high pressure fuel from the vehicle fuel supply line to the fuel injectors. The assembly consists of two preformed tubes, or stampings, one for fuel supply and one for fuel return. The manifold is equipped with a fuel pressure relief valve on the fuel supply tube (2.3L four) or regulator mounting flange (V6).

Each fuel injector nozzle is an electro-mechanical device which meters and atomizes the fuel delivered to the engine. The injectors are mounted in the lower intake manifold and are positioned so that their tips are directing fuel just ahead of the engine intake valves. Each injector body consists of a solenoid actuated pintle and needle valve assembly.

An electrical control signal from the EEC electronic processor activates the solenoid, causing the pintle to move inward off the seat and allowing fuel to flow. The injector flow orifice is fixed and the fuel supply is constant. Therefore, fuel flow to the engine is controlled by how long the solenoid is energized.

12 Electronic Fuel Injection (EFI) system - component removal and installation (2.3L four)

Warning: *Gasoline is extremely flammable, so take extra precautions when you work on any part of the fuel system. Don't smoke or allow open flames or bare light bulbs near the work area, and don't work in a garage where a gastype appliance (such as a water heater or clothes dryer) is present. Since gasoline is carcinogenic, wear latex gloves when there's a possibility of being exposed to fuel, and, if you* spill any fuel on your skin, rinse it off immediately with soap and water. Mop up any spills immediately and do not store fuel-soaked rags where they could ignite. The fuel system on fuel-injected models is under constant pressure, so, if any fuel lines are to be disconnected, the fuel pressure in the system must be relieved first (see Chapter 4 for more information). When you perform any kind of work on the fuel system, wear safety glasses and have a Class B type fire extinguisher on hand.*

Air bypass valve

Note: *Unlike any of the other fuel injection assembly components, the air bypass valve is not actually located on the throttle body or the intake manifold - it's mounted on the air cleaner housing.*

1 Detach the cable from the negative terminal of the battery.
2 Unplug the electrical connector at the air bypass valve.
3 Remove the air cleaner cover.
4 Separate the air bypass valve and gasket from the air cleaner by removing the three retaining bolts.
5 Installation is the reverse of removal.

Throttle position sensor

6 Detach the cable from the negative terminal of the battery.
7 Unplug the wiring harness from the throttle position sensor electrical connector.
8 Remove the two bolts which attach the throttle position sensor connector to the air intake throttle body. After the connector is detached from the throttle body, note the locating dimple on the throttle body for locating the connector properly.
9 Before removing the throttle position sensor, make scribe marks on the throttle body and on the throttle position sensor to insure proper alignment during reinstallation.
10 Remove the two throttle position sensor retaining screws, the sensor and the gasket.
11 When installing the throttle position sensor, make sure that the rotary tangs on the sensor are aligned with the throttle shaft blade. Slide the rotary tangs into position over the throttle shaft blade, then rotate the throttle position sensor clockwise only to its installed position. **Caution:** *Failure to install the throttle position sensor in this manner may result in excessive idle speeds.* Tighten the sensor to the specified torque.
12 The remainder of installation is the reverse of removal.

Throttle body

Note: *If you are replacing the gasket between the upper intake manifold and the lower intake manifold (or if you are simply removing the entire fuel injection assembly for access to the lower intake manifold or gasket) it is not necessary to remove the throttle body from the upper intake manifold. Simply remove both as an assembly. The following procedure, however, must be used if you are replacing either the throttle body itself or the* gasket between the throttle body and the upper intake manifold.

13 Detach the cable from the negative terminal of the battery.
14 Remove the air intake resonator outlet tube (the duct between the resonator and the throttle body) (see Section 8).
15 Unplug the electrical connector from the throttle position sensor (see above).
16 Remove the throttle linkage shield, then detach the throttle cable and, if equipped, speed control cable (see Section 9).
17 Clearly label, then detach the air bypass and crankcase vent hoses.
18 Remove the two lower mounting nuts and the two upper mounting bolts.
19 Carefully separate the air throttle body from the upper intake manifold.
20 Remove and discard the gasket between the throttle body and the upper intake manifold.
21 Installation is the reverse of removal. Be sure that the gasket surfaces are clean and tighten the mounting bolts and nuts to the specified torque.

Upper intake manifold

22 If you are replacing the gasket between the upper and lower intake manifold or between the lower intake manifold and the cylinder head, if you are replacing/servicing an injector or if you are removing the entire fuel injection assembly to service the head, simply detach the throttle cable, cruise control cable (if equipped), vacuum lines, etc. from the throttle body (see above), but do not separate the throttle body from the upper intake manifold. Unless you are replacing the upper intake manifold itself, there is no reason to detach the throttle body.
23 Unplug the knock sensor electrical connector (located on the front side of the upper intake manifold). **Note:** *If you intend to replace the upper intake manifold with a new one, remove the knock sensor and install it in the new manifold.*
24 Detach the upper intake manifold vacuum lines from the upper intake manifold vacuum tree, disconnect the vacuum lines to the EGR valve and the fuel pressure regulator and detach the PCV hose from the fitting on the underside of the upper intake manifold. Be sure to clearly label all hoses to prevent mix-ups during reassembly.
25 Disconnect the EGR tube from the EGR valve.
26 Remove the four upper intake manifold mounting bolts and detach the upper intake manifold and gasket from the intake manifold. Be sure to discard the old gasket and use a new one for reassembly.
27 Installation is the reverse of removal. Be sure to tighten the upper intake manifold mounting bolts to the specified torque, in the sequence shown.

Fuel pressure regulator

28 Relieve the fuel pressure (see Section 2). Remove the fuel tank filler cap to relieve fuel tank pressure.

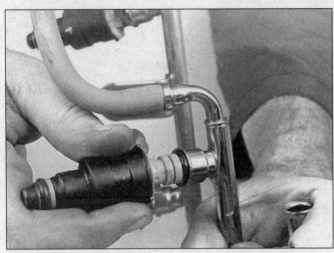

12.42 To remove an injector from the fuel supply manifold, grasp it firmly and gently pull on it while twisting it from side-to-side

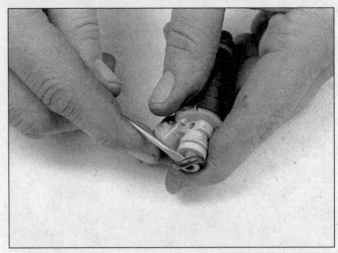

12.43 If you are planning to reinstall an old injector, inspect the two O-rings and the plastic hat which protects the pintle for signs of deterioration and replace if necessary (it's a good practice to simply replace the O-rings and hat anytime you remove the injector)

29   Detach the vacuum line from the pressure regulator.

30   Remove the three Allen retaining screws from the pressure regulator housing, then remove the regulator, gasket and O-ring. Discard the gasket and inspect the O-ring for signs of cracks or deterioration.

31   Before installing the regulator, make sure that the gasket surfaces of the fuel pressure regulator and the fuel supply manifold are clean. Then lubricate the O-ring with light oil. **Caution:** *Do not use silicone grease. It will clog the injectors.*

32   Installation is the reverse of removal. Be sure to use a new gasket and tighten the regulator Allen screws to the specified torque.

Fuel supply manifold assembly

33   Relieve the fuel pressure (see Section 2). Remove the fuel tank filler cap to relieve the fuel tank pressure.

34   Remove the air intake throttle body and the upper intake manifold (see above).

35   Unplug the four injector electrical connectors and set the harness aside.

36   Remove the two fuel supply manifold mounting bolts.

37   Carefully disengage the manifold and fuel injectors from the engine and remove the assembly.

38   If you are replacing an injector, or injectors, or injector O-rings, see below.

39   Installation is the reverse of removal. Be sure to tighten the fuel supply manifold mounting bolts to the specified torque.

Fuel injector

Refer to illustrations 12.42 and 12.43

40   Relieve the fuel pressure (see Section 2). Remove the fuel tank filler cap to relieve the fuel tank pressure.

41   Remove the fuel supply manifold assembly (see above).

42   Grasping the injector body, pull it out of the fuel supply manifold while gently rocking the injector from side-to-side **(see illustration)**.

43   Inspect the two injector O-rings and the injector plastic hat (covering the injector pintle) and washer **(see illustration)** for deterioration. Replace as necessary. **Note:** *If the hat is missing, look for it in the intake manifold.*

44   Before inserting the injector into the fuel supply manifold, be sure to lubricate the new O-rings with light grade oil and install them on the injector. **Caution:** *Do not use silicone grease. It will clog the injectors.*

45   Install the injector(s) into the fuel supply manifold using a light, twisting, pushing motion.

46   The remainder of installation is the reverse of removal. Be sure to tighten the fuel supply manifold mounting bolts to the specified torque.

---

13   Electronic fuel injection (EFI) system - component removal and installation (3.0L V6)

Warning: *Gasoline is extremely flammable, so take extra precautions when you work on any part of the fuel system. Don't smoke or allow open flames or bare light bulbs near the work area, and don't work in a garage where a gas-type appliance (such as a water heater or clothes dryer) is present. Since gasoline is carcinogenic, wear latex gloves when there's a possibility of being exposed to fuel, and, if you spill any fuel on your skin, rinse it off immediately with soap and water. Mop up any spills immediately and do not store fuel-soaked rags where they could ignite. The fuel system on fuel-injected models is under constant pressure, so, if any fuel lines are to be disconnected, the fuel pressure in the system must be relieved first (see Chapter 4 for more information). When you perform any kind of* work on the fuel system, wear safety glasses and have a Class B type fire extinguisher on hand.

Air bypass valve

Refer to illustration 13.2

1   Detach the cable from the negative terminal of the battery.

2   Unplug the wiring harness electrical connector from the air bypass valve assembly **(see illustration)**.

3   Remove the two air bypass valve retaining screws and the air bypass valve and gasket.

4   Make sure that both the throttle body and the air bypass valve gasket mating surfaces are clean. **Caution:** *Should it be necessary to scrape either surface, be extremely careful not to damage the air bypass valve or throttle body gasket surfaces or drop material into the throttle body.*

5   Installation is the reverse of removal. Be sure to tighten the air bypass retaining screws to the specified torque.

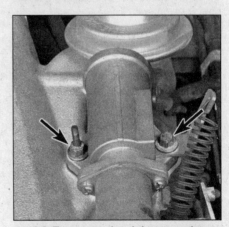

13.2 To remove the air bypass valve, simply unplug the electrical connector and remove the two mounting fasteners

13.9 Before removing the throttle position sensor, make scribe marks on the sensor housing and the throttle body to ensure proper realignment, then simply remove the two retaining screws

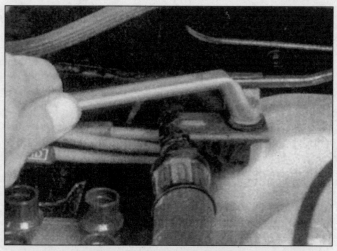

14.8 Unbolt the air conditioning hose bracket from the upper intake manifold and move it aside - DO NOT disconnect any refrigerant lines

Throttle position sensor

Refer to illustration 13.9

6   Detach the cable from the negative terminal of the battery.

7   Unplug the wiring harness electrical connector from the throttle position sensor.

8   Make scribe marks on the air throttle body and on the throttle position sensor to indicate the proper alignment during installation.

9   Remove the two throttle position sensor retaining screws and remove the throttle position sensor **(see illustration)**.

10   When installing the throttle position sensor, make sure that the rotary tangs on the sensor are aligned with the throttle shaft blade and that the red seal is inside the connector housing. Slide the rotary tangs into position over the throttle shaft blade, then rotate the throttle position sensor clockwise only to its installed position. **Caution:** *Failure to install the throttle position sensor in this manner may result in excessive idle speeds.* Tighten the sensor to the specified torque.

11   The remainder of installation is the reverse of removal.

Fuel metering and air intake/throttle body components

12   Relieve the fuel pressure (see Section 2). Remove the fuel tank filler cap to relieve fuel tank pressure.

13   Detach the cable from the negative terminal of the battery.

14   Remove the engine air cleaner outlet tube between the air cleaner and the air throttle body by loosening the two clamps.

15   Unplug the injector wiring harness electrical connectors from the throttle position sensor, air bypass valve and air charge temperature sensor.

16   Disconnect the push connect fuel supply and return lines (see Section 3).

17   Clearly label, then detach, the vacuum hoses from the vacuum fittings on the intake manifold.

18   Detach and remove the throttle cable and, if equipped, cruise control cable (see Section 9) from the throttle linkage.

19   If your vehicle is equipped with an automatic transmission, remove the transmission (TV) linkage from the throttle lever (see Chapter 7B).

20   Remove the four retaining bolts and two stud bolts and lift the air intake throttle body assembly from the guide pins on the lower intake assembly. Remove the gasket from the lower intake assembly and discard it.

21   Clean and inspect the mounting surfaces of the air intake throttle body assembly and the lower intake manifold assembly. They must be clean and flat. Also clean and oil the threads in the manifold stud bores.

22   Install a new gasket and, using the guide pins as locators, install the air intake throttle body assembly on the lower intake manifold and tighten the mounting bolts and stud bolts, a little at a time, to the specified torque.

23   The remainder of the installation is the reverse of removal.

Fuel pressure regulator

24   Relieve the fuel pressure (see Section 2). Remove the fuel tank filler cap to relieve the fuel tank pressure.

25   Detach the cable from the negative terminal of the battery.

26   Detach the vacuum line from the pressure regulator.

27   Remove the three Allen retaining screws from the regulator housing.

28   Remove the pressure regulator assembly, gasket and O-ring. Discard the gasket and inspect the O-ring for signs of cracks or deterioration.

29   Before installing the fuel pressure regulator, make sure that the gasket surfaces of the regulator and the fuel rail assembly are

clean. Lubricate the new O-ring with clean engine oil and install it on the regulator.

30   Install the regulator and a new gasket on the fuel rail assembly and tighten the regulator retaining screws to the specified torque.

31   The remainder of installation is the reverse of removal.

Fuel injector manifold assembly

32   Relieve the fuel pressure.

33   Remove the air intake throttle body (see above).

34   Carefully unplug the wiring harness electrical connectors from the injectors.

35   Detach the vacuum line from the fuel pressure regulator.

36   Remove the four fuel injector manifold retaining bolts (two on each side).

37   Carefully disengage the fuel rail assembly from the fuel injectors by lifting and gently rocking the rail.

38   Remove the injectors by lifting while gently rocking from side to side.

39   Place the removed components in a clean container to avoid dirt or other contamination. The injector(s) and fuel rail must be handled with extreme care to prevent damage to sealing areas and sensitive fuel metering orifices.

14   Electronic fuel injection - component removal and installation (4.0L V6 engine)

Warning: *Gasoline is extremely flammable, so take extra precautions when you work on any part of the fuel system. Don't smoke or allow open flames or bare light bulbs near the work area, and don't work in a garage where a gas-type appliance (such as a water heater or clothes dryer) is present. Since gasoline is carcinogenic, wear latex gloves when there's*

14.9a Remove the upper intake manifold nuts with a socket and extension

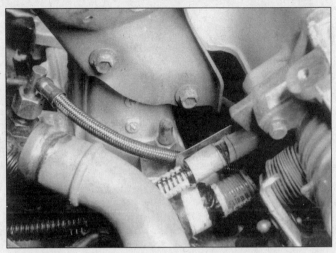

14.9b Note the cap over the front stud - remove it before unscrewing the nut

a possibility of being exposed to fuel, and, if you spill any fuel on your skin, rinse it off immediately with soap and water. Mop up any spills immediately and do not store fuel-soaked rags where they could ignite. The fuel system on fuel-injected models is under constant pressure, so, if any fuel lines are to be disconnected, the fuel pressure in the system must be relieved first (see Chapter 4 for more information). When you perform any kind of work on the fuel system, wear safety glasses and have a Class B type fire extinguisher on hand.

Upper intake manifold and throttle body

Refer to illustrations 14.8, 14.9a, 14.9b and 14.9c

Removal

1   Label and disconnect the electrical connectors and vacuum lines between the upper intake manifold and the engine and body.
2   Remove the snow shield from the accelerator linkage. Remove the accelerator cable bracket and disconnect the accelerator cable from the throttle body.
3   Remove the air inlet tube that connects the air cleaner to the throttle body.
4   Label and disconnect the vacuum lines from the upper intake manifold.
5   Detach the PCV valve from the right valve cover (see Chapter 6).
6   Detach the spark plug wires from the loom at the back of the upper intake manifold.
7   Disconnect the canister purge line at the throttle body fitting.
8   Remove the bolt that secures the air conditioning refrigerant line at the upper rear part of the manifold (see illustration). DO NOT disconnect any refrigerant lines!
9   Remove six nuts that secure the upper intake manifold (see illustration). Note the cap over the front stud (see illustration). Lift the upper manifold and throttle body together off the fuel rail (see illustration).

Installation

10   Thoroughly clean all old gasket material from the mating surfaces of the upper intake manifold and the fuel rail.
11   Install a new gasket on the fuel rail, then install the upper manifold and tighten the six nuts evenly to the torque listed in this Chapter's Specifications.
12   The remainder of installation is the reverse of the removal steps.

Air intake throttle body

Refer to illustration 14.17

13   Remove the snow shield from the accelerator linkage and disconnect the accelerator cable.
14   Disconnect the electrical connector from the throttle position sensor.
15   Unclamp and disconnect the air inlet duct at the throttle body.
16   Disconnect the canister purge hose from underneath the throttle body.
17   Remove the four throttle body mounting screws (see illustration). Remove the throttle body and gasket from the upper intake manifold.
18   Carefully clean all old gasket material from the mating surface of the throttle body

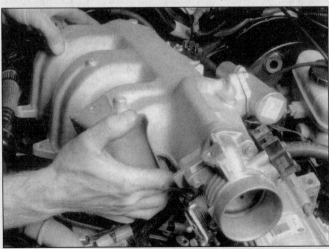

14.9c Grasp the upper intake manifold at the front and rear ends and lift it off the engine

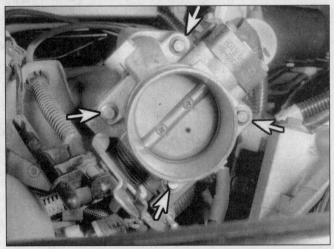

14.17 Disconnect the electrical connector and remove the throttle body mounting screws (arrows)

14.21 The Throttle Position Sensor (arrow) is mounted on the throttle body

14.26 The air bypass valve (arrow) is mounted on the upper intake manifold

and upper intake manifold. **Caution:** *Don't let the gasket material fall into the manifold.*
19   Place a new gasket on the manifold, then install the throttle body and tighten the screws to the torque listed in this Chapter's Specifications.
20   The remainder of installation is the reverse of the removal steps.

Throttle Position Sensor (TPS)

Refer to illustration 14.21
21   The throttle position sensor is mounted on the air intake throttle body **(see illustration)**.
22   Disconnect the electrical connector from the throttle position sensor.
23   Remove the throttle position sensor screws. Lift off the TPS and its gasket.
24   Install the TPS, using a new gasket, and tighten the screws securely.

Air bypass valve

Refer to illustration 14.26
25   Disconnect the air bypass valve electrical connector.
26   Remove the two mounting bolts, then take off the valve and gasket **(see illustration)**.

27   Thoroughly clean all old gasket material from the valve and it's mounting surface. **Caution:** *Don't let gasket material fall into the upper manifold.*
28   Install the valve with a new gasket and tighten its bolts to the torque listed in this Chapter's Specifications.
29   Connect the electrical connector.

Air Charge Temperature (ACT) sensor

Refer to illustration 14.30
30   Disconnect the Air Charge Temperature sensor electrical connector **(see illustration)**.
31   Unscrew the ACT sensor from the intake manifold.
32   Wrap the threads of a new sensor with thread sealing tape to prevent air leaks. Screw in a new sensor and tighten it securely.

Mass Air Flow (MAF) sensor

Refer to illustration 14.35
33   Disconnect the duct from the MAF sensor.
34   Unplug the MAF sensor electrical connector.

35   Remove four screws securing the MAF sensor to the air cleaner housing **(see illustration)**. Take off the sensor and gasket.
36   Install the sensor, using a new gasket, and tighten the screws securely.

Fuel rail and injectors

Refer to illustrations 14.43a, 14.43b. 14.43c and 14.44

Removal

37   It's a good idea to steam clean the engine before starting this procedure to prevent dirt from contaminating exposed fittings and fuel metering orifices.
38   Relieve the fuel system pressure (see Section 2).
39   Disconnect the negative cable from the battery.
40   Remove the air inlet tube the runs between the air cleaner and throttle body.
41   Remove the upper intake manifold as described in this Section.
42   Disconnect the fuel supply and return lines (see Section 3).
43   Remove six Torx stud bolts that secure the fuel rail **(see illustration)**. Carefully lift the fuel rail and injectors out with a rocking,

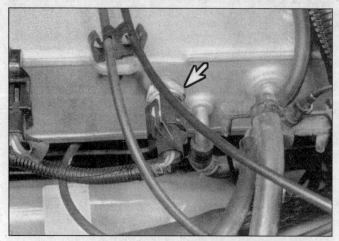

14.30 The Air Charge Temperature sensor (arrow) is screwed into the intake manifold

14.35 The Mass Air Flow sensor is mounted on the air cleaner housing and is secured by four screws (arrows)

pulling motion **(see illustrations)**.

44   The injectors may come out with the fuel rail or remain attached to the lower intake manifold. Detach the injectors as follows:

  a) Disconnect the injector electrical connector from the individual injector(s).
  b) Remove the injector retaining clip(s).
  c) Carefully pull the injector out with a rocking motion **(see illustration)**.

Installation

45   Check the injector O-rings for damage or deterioration and replace as needed.

46   Inspect the plastic "hat" that covers the end of each injector. The entire injector must be replaced if the hat is damaged or loose.

47   Lubricate the injector O-rings with a light film of engine oil. Each injector uses two O-rings. **Caution:** *Don't lubricate the injectors with silicone grease. It will clog them.*

48   Install the injectors in the fuel rail with a light rocking motion.

49   The remainder of installation is the reverse of the removal steps.

50   Turn the ignition key On and Off several times (without starting the engine) to pressurize the fuel system. Check all fuel system connections for leaks.

Fuel pressure regulator

Refer to illustration 14.57

51   Disconnect the negative cable from the battery.

52   Remove the snow shield from the throttle linkage.

53   Remove the air inlet tube that connects the air cleaner to the throttle body.

54   Relieve the fuel system pressure (see Section 2).

55   Disconnect the vacuum line from the pressure regulator.

56   Disconnect the fuel return line from the pressure regulator (see Section 3).

57   Remove the regulator mounting screws and take the regulator off **(see illustration)**.

58   Remove the O-ring and washer.

59   Check the O-ring for cracks or deterioration and replace as needed. Replace the washer whenever it's removed.

60   Check the fuel return line for kinks or worn spots and replace as needed.

61   Lubricate the O-ring with a light coat of engine oil.

62   The remainder of installation is the reverse of the removal steps, with the following additions:

14.43a Remove the Torx stud bolts that secure the fuel rail with a Torx socket

  a) Tighten the pressure regulator screws securely.
  b) To connect the fuel line, push the ends together until the shoulder on the male end touches the fitting on the female end. Install the retainer and snap its two halves together.

14.43b The fuel injectors (arrows) are mounted between the fuel rail and the lower intake manifold

14.43c Rock the fuel rail to detach the injectors, then lift it off

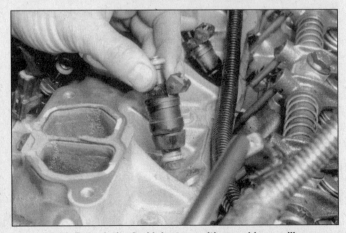

14.44 Detach the fuel injectors with a rocking, pulling motion and take them out

14.57 The fuel pressure regulator is secured by two screws (arrows)

15   Exhaust system components - replacement

Warning: *Never attempt to service any part of the exhaust system until it has cooled. Be especially careful when working around the catalytic converter. The temperature of the converter rises to a high level after only a few minutes of engine operation and, once the engine reaches its operating temperature, can remain high enough to cause burns for a long time after the engine is shut off.*

1   Detach the cable from the negative terminal of the battery.

2   Raise the vehicle and place it securely on jackstands.

Converter assembly - four-cylinder engine

3   Remove the two nuts attaching the converter pipe assembly to the muffler and the outlet pipe.

4   Remove the two nuts attaching the converter pipe assembly to the exhaust manifold.

5   Apply a soapy solution to the support slide of the converter support bracket at the support insulator.

6   Remove the converter pipe assembly from the vehicle.

Installation is the reverse of removal.

Converter assembly - V6 engines

7   Remove the two nuts attaching the converter pipe assembly to the muffler and outlet pipe. On the 2.8L engine, remove the clamp securing the Managed Thermactor Air (MTA) tube to the catalytic converter. Remove the screw holding the MTA tube bracket to the forward converter and separate the MTA tube from the converter.

8   On the 2.8L engine, disconnect the hose from the top of the MTA and check valve. It may be necessary to heat the MTA tube slightly to detach it from the nipple on the converter. Remove the MTA tube and check valve. To facilitate removal, rotate the assembly as it is being lowered.

9   Loosen and remove the exhaust manifold stud nuts. Slide the muffler and outlet pipe rearward. Remove the gasket. Move the converter assembly rearward, rotate it to clear the studs and remove it.

10   Installation is the reverse of removal.

Chapter 5
Engine electrical systems

Contents

Specifications

Drivebelt deflection .... See Chapter 1

Battery voltage
Engine off 12-volts
Engine running .... 14-to-15 volts

Ignition coil-to-distributor cap wire resistance .... 5000 ohms per foot

Ignition coil resistance
Primary resistance .... 0.3 to 1.0 ohms
Secondary resistance .... 8.0 to 11.5 K-ohms

1   General information

The engine electrical systems include all ignition, charging and starting components. Because of their engine-related functions, these components are considered separately from chassis electrical devices such as the lights, instruments, etc.

Access to the distributor, distributor cap and coil is through the engine access panel inside the vehicle. Be very careful when working on the engine electrical components. They are easily damaged if checked, connected or handled improperly. The alternator is driven by an engine drivebelt which could cause serious injury if your hands, hair or clothes become entangled in it with the engine running. Both the starter and alternator are connected directly to the battery and could arc or even cause a fire if mishandled, overloaded or shorted out.

Never leave the ignition switch on for long periods of time with the engine off. Don't disconnect the battery cables while the engine is running. Correct polarity must be maintained when connecting battery cables from another source, such as another vehicle, during jump starting. Always disconnect the negative cable first and hook it up last or the battery may be shorted by the tool being used to loosen the cable clamps.

Additional safety related information on the engine electrical systems can be found in *Safety first* near the front of this manual. It should be referred to before beginning any operation included in this Chapter.

2   Battery - removal and installation

Refer to illustration 2.2

1   Disconnect both cables from the battery terminals. **Caution:** *Always disconnect the negative cable first and hook it up last or the battery may be shorted by the tool being used to loosen the cable clamps.*

2.2 To remove the battery, detach the cables from the negative, then the positive, terminals, remove the hold down clamp nuts (arrows) and the clamp, then carefully lift out the battery

6.2 To use a calibrated ignition tester (available at most auto parts stores), simply disconnect a spark plug wire, attach the wire to the tester, clip the tester to a convenient ground (like a rocker arm cover bolt) and operate the starter - if there's enough power to fire the plug, sparks will be visible between the electrode tip and the tester body

2   Remove both battery hold-down clamp nuts and the clamp **(see illustration)**.

3   Lift out the battery. Special straps that attach to the battery posts are available - lifting and moving the battery is much easier if you use one.

4   Installation is the reverse of removal.

3   Battery - emergency jump starting

Refer to the Booster battery (jump) starting procedure at the front of this manual.

4   Battery cables - check and replacement

Note: *Additional information and illustrations on battery servicing can be found in Chapter 1.*

1   Periodically inspect the entire length of each battery cable for damage, cracked or burned insulation and corrosion. Poor battery cable connections can cause starting problems and decreased engine performance.

2   Check the cable-to-terminal connections at the ends of the cables for cracks, loose wire strands and corrosion. The presence of white, fluffy deposits under the insulation at the cable terminal connection is a sign that the cable is corroded and should be replaced. Check the terminals for distortion, missing mounting bolts and corrosion.

3   When replacing the cables, always disconnect the negative cable first and hook it up last or the battery may be shorted by the tool used to loosen the cable clamps. Even if only the positive cable is being replaced, be sure to disconnect the negative cable from the battery first.

4   Disconnect and remove the cable. Make sure the replacement cable is the same length and diameter.

5   Clean the threads of the relay or ground connection with a wire brush to remove rust and corrosion. Apply a light coat of petroleum jelly to the threads to prevent future corrosion.

6   Attach the cable to the relay or ground connection and tighten the mounting nut/bolt securely.

7   Before connecting the new cable to the battery, make sure that it reaches the battery post without having to be stretched.

8   Connect the positive cable first, followed by the negative cable.

5   Ignition system - general information

TFI-IV system (except 4.0L engine)

The ignition system is a solid state electronic design consisting of an ignition module, a coil, a distributor, the spark plug wires and the spark plugs.

All models use the Thick Film Integrated IV (TFI-IV) ignition module, which is housed in a molded thermoplastic box mounted on the base of the distributor. The TFI-IV module is controlled by the Electronic Engine Control IV (EEC-IV).

The TFI-IV/EEC-IV type distributor has neither a centrifugal nor a vacuum advance mechanism (advance is handled by the computer instead). The TFI-IV module does, however, include a "push start" mode that allows push starting of the vehicle if necessary.

A Hall Effect stator assembly consisting of a Hall sensor, a permanent magnet and a rotary vane cup, causes the ignition coil to be switched off and on by the EEC-IV and the TFI-IV models.

EDIS system (4.0L engine only)

The 4.0L V6 engine used in all models is equipped with the Electronic Distributorless Ignition System (EDIS) that is a solid-state electronic design. It consists of a crankshaft timing sensor, EDIS module, ignition coil pack, a crankshaft timing sensor (crank angle sensor), the spark plug wires and the spark plugs.

This ignition system does not have any moving parts (no distributor) and all engine timing and spark distribution is handled electronically. This system has fewer parts that require replacement and provides more accurate spark timing. During engine operation, the EDIS ignition module and the EEC-IV module calculate crankshaft angle and determine the turn-on and firing time of the ignition coil.

The crankshaft timing sensor is a variable reluctance-type consisting of a 36-tooth trigger wheel with one missing tooth that is incorporated into the crankshaft front damper. The signal generated by this sensor is called a Variable Reluctance Sensor (VRS) signal and it provides the base timing and engine RPM information to the EDIS ignition modules. The main function of the EDIS module is to synchronize the ignition coils so they are turned on and off in the proper sequence for accurate spark control. No ignition timing adjustment is possible on the EDIS system.

6   Ignition system - check

Refer to illustration 6.2

Warning: *Because of the very high secondary (spark plug) voltage generated by the ignition system, extreme care should be taken when this check is done.*

TFI-IV system

Calibrated ignition tester method

1   If the engine turns over but won't start, disconnect the spark plug lead from any spark plug and attach it to a calibrated ignition tester (available at most auto parts stores).

2   Connect the clip on the tester to a bolt or metal bracket on the engine **(see illustration)**, crank the engine and watch the end of the tester to see if bright blue, well-defined sparks occur.

3   If sparks occur, sufficient voltage is reaching the plug to fire it. Repeat the check at the remaining plug wires to verify that the distributor cap and rotor are OK. However, the plugs themselves may be fouled, so remove and check them as described in Chapter 1 or install new ones.

4   If no sparks or intermittent sparks occur, remove the distributor cap and check the cap and rotor as described in Chapter 1. If moisture is present, dry out the cap and rotor, then reinstall the cap and repeat the spark test.

5   If there's still no spark, detach the secondary coil wire from the distributor cap and hook it up to the tester (reattach the plug wire to the spark plug), then repeat the spark check.

6   If no sparks occur, check the primary (small) wire connections at the coil to make sure they're clean and tight. Refer to Section 7 and check the ignition coil supply voltage circuit. Make any necessary repairs, then repeat the check.

7   If sparks now occur, the distributor cap, rotor, plug wire(s) or spark plug(s) (or all of them) may be defective.

8   If there's still no spark, the coil-to-cap wire may be bad. Check the resistance with an ohmmeter and compare it to the Specifications. If a known good wire doesn't make any difference in the test results, the ignition coil, module or other internal components may be defective.

Alternative method

Note: *If you're unable to obtain a calibrated ignition tester, the following method will allow you to determine if the ignition system has spark, but it won't tell you if there's enough voltage produced to actually initiate combustion in the cylinders.*

9   Remove the wire from one of the spark plugs. Using an insulated tool, hold the wire about 1/4-inch from a good ground and have an assistant crank the engine.

10   If bright blue, well-defined sparks occur, sufficient voltage is reaching the plug to fire it. However, the plug(s) may be fouled, so remove and check them as described in Chapter 1 or install new ones.

11   If there's no spark, check the remaining wires in the same manner. A few sparks followed by no spark is the same condition as no spark at all.

12   If no sparks occur, remove the distribu-

tor cap and check the cap and rotor as described in Chapter 1. If moisture is present, dry out the cap and rotor, then reinstall the cap and repeat the spark test.

13   If there's still no spark, disconnect the secondary coil wire from the distributor cap, hold it about 1/4-inch from a good engine ground and crank the engine again.

14   If no sparks occur, check the primary (small) wire connections at the coil to make sure they're clean and tight. Refer to Section 7 and check the ignition coil supply voltage circuit. Make any necessary repairs, then repeat the check again.

15   If sparks now occur, the distributor cap, rotor, plug wire(s) or spark plug(s) (or all of them) may be defective.

16   If there's still no spark, the coil-to-cap wire may be bad. Check the resistance with an ohmmeter and compare it to the Specifications. If a known good wire doesn't make any difference in the test results, the ignition coil, module or other internal components may be defective.

EDIS ignition system

Calibrated ignition tester method

17   If the engine turns over but won't start, disconnect the spark plug load from any spark plug and attach it to a calibrated ignition tester (available at most auto parts stores).

18   Connect the clip on the tester to a bolt or metal bracket on the engine **(see illustration 6.2).** crank the engine and watch the end of the tester to see if bright blue, well-defined sparks occur.

19   If sparks occur, sufficient voltage is reaching the spark plug to fire it (repeat the check at the remaining plug wires to verify that all the ignition coils and wires are functioning). However, the plugs themselves may be fouled, so remove and check them as described in Chapter 1 or install new ones.

20   If no sparks or intermittent sparks occur, there's a problem in the ignition system. If there are no sparks or intermittent sparks from one or two wires only, the problem is likely a bad spark-plug wire or coil pack. Check for a bad spark plug wire by swapping wires. Refer further testing to a dealer or qualified electrical specialist.

21   If no sparks occur at any wires, check the electrical connections, particularly at the module and the coil packs. If the connections are OK, the problem is likely a bad module.

Alternative method

Note: *If you're unable to obtain a calibrated ignition tester, the following method will allow you to determine if the ignition system has spark, but it won't tell you it there's enough voltage produced to actually initiate combustion in the cylinders.*

22   Remove the wire from one of the spark plugs. Using an insulated tool, hold the wire about 1/2-inch from a good ground and have an assistant crank the engine.

23   If bright blue, well-defined sparks occur, sufficient voltage is reaching the plug to fire it. However, the plug(s) may be fouled, so remove and check them as described in Chapter 1 or install new ones.

24   If there's no spark, check the remaining wires in the same manner. A few sparks followed by no spark is the same condition as no spark at all. If there's spark at some wires but not at others, bad spark plug wires may be the problem. Swap the wires and re-test to see if the wires are the problem.

25   If there's still no spark or intermittent sparks, there's a problem in the ignition system. Refer further testing to a electrical specialist.

---

7   Ignition coil and circuits - check and coil replacement

Check (TFI-IV distributor-type ignition systems only)

Note: *Checking the ignition coils on distributorless ignition systems (EDIS) is beyond the scope of the home mechanic. Take the vehicle to a dealer service department for testing.*

Ignition coil primary circuit

1   Unplug the wiring harness connector from the ignition module. Inspect it for dirt, corrosion and damage (refer to Section 10 for a detailed illustration of the connector terminals), then plug it back in.

2   Attach a 12-volt DC test light between the coil TACH terminal and a good engine ground.

3   Crank the engine.

4   If the light flashes, or comes on but doesn't flash, refer to Step 7.

5   If the light stays off or is very dim, refer to Step 16.

6   Remove the test light.

Ignition coil primary resistance

7   Turn the ignition switch to Off.

8   Unplug the ignition coil wire harness connector. Inspect it for dirt, corrosion and damage.

9   Measure the resistance between the primary terminals of the ignition coil.

10   If the resistance is within the specified limits, proceed to Step 12.

11   If the resistance is less or more than specified, replace the ignition coil (Step 49).

Ignition coil secondary resistance

12   Measure the resistance from the negative primary terminal to the secondary terminal of the ignition coil.

13   If the resistance is within the specified limits, proceed to Step 25.

14   If the resistance is less or more than the specified resistance, replace the ignition coil (Step 49).

15   Reconnect the ignition coil wires.

Primary circuit continuity

16 Unplug the wiring harness connector from the ignition module. Inspect it for dirt, corrosion and damage.
17 Attach the negative lead of a VOM to the distributor base.
18 Measure battery voltage and record it for future reference.
19 Attach the positive lead of the VOM to a small straight pin inserted into connector terminal 2 **(see illustration 7.29). Caution:** *Do not allow the straight pin to ground against anything.*
20 Turn the ignition switch to the Run position and measure the terminal 2 voltage.
21 If the voltage is 90 percent of battery voltage, proceed to the wiring harness check (Step 25).
22 If the voltage is less than 90 percent of battery voltage, proceed to Step 35.
23 Turn the ignition switch to the Off position.
24 Remove the straight pin.

Wiring harness

Refer to illustration 7.29

25 Unplug the wiring harness connector from the ignition module. Inspect it for dirt, corrosion and damage.
26 Disconnect the wire at the S terminal of the starter relay.
27 Attach the negative lead of a VOM to the distributor base.
28 Measure battery voltage and record it for future reference.
29 Using the accompanying table **(see illustration)**, measure the connector terminal voltage by attaching the positive lead of the VOM to a small straight pin inserted into the connector terminals, one at a time, with the ignition switch in the indicated positions.
30 If the voltage is 90 percent of battery voltage at all three terminals, refer to the EEC-IV check in Section 10.
31 If the voltage is less than 90 percent of battery voltage, inspect the wiring harness

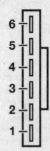

Connector Terminal	Wire/Circuit	Ignition Switch Test Position
Number 2	To ignition coil (–) terminal	Run
Number 3	Run circuit	Run and Start
Number 4	Start circuit	Start

36074-5-7.29 HAYNES

7.29 Check the wire harness connector voltage with the switch in each of the above positions

and the connectors.
32 Turn the ignition switch to the Off position.
33 Remove the straight pin.
34 Reconnect the wire to the S terminal of the starter relay.

Ignition coil primary voltage

35 Attach the negative lead of a VOM to the distributor base.
36 Measure battery voltage and record it.
37 Turn the ignition switch to the Run position.
38 Measure the voltage at the negative terminal of the ignition coil.
39 If the voltage is 90 percent of battery voltage, inspect the wiring harness between the ignition module and the coil negative terminal.
40 If the voltage is less than 90 percent of battery voltage, inspect the wiring harness between the ignition module and the coil negative terminal, then proceed to Step 42.
41 Turn the ignition switch to the Off position.

Ignition coil supply voltage

42 Unplug the ignition coil wire harness.
43 Attach the negative lead of a VOM to the distributor base.
44 Measure battery voltage.
45 Turn the ignition switch to the Run position.
46 Measure the voltage at the positive ter-

minal of the ignition coil.
47 If the voltage is 90 percent of battery voltage, inspect the ignition coil connector and terminals for dirt, corrosion and damage. If both the connector and terminals are clean, replace the ignition coil (Step 49).
48 If the voltage is less than 90 percent of battery voltage, inspect and repair the circuit between the ignition coil and the ignition switch (refer to the wiring diagrams at the end of the book). Check the ignition switch for damage and wear (refer to Chapter 12).

Ignition coil replacement (distributor type)

Refer to illustration 7.52

49 Detach the cable from the negative terminal of the battery.
50 Detach the wires from the primary terminals on the coil (some coils have a single connector for the primary wires).
51 Unplug the coil secondary lead.
52 Remove both bracket bolts **(see illustration)** and detach the coil.
53 Installation is the reverse of removal.

Ignition coil pack replacement (EDIS ignition)

Refer to illustrations 7.55, 7.56a, 7.56b, 7.58a and 7.58b

54 Disconnect the negative cable from the battery.
55 Using pieces of numbered tape, mark

7.52 A typical coil assembly (2.8L V6 shown) - to remove it, unplug the coil primary leads (terminal obscured by the high tension lead in photo) and the boot on the high tension terminal, then remove the four mounting bolts (arrows)

7.55 Squeeze the locking tabs to release the spark plug wires

7.56a Disconnect the electrical connector (arrow) from the coil pack . . .

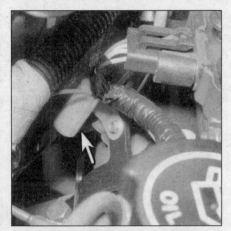

7.56b . . . and from the condenser mounted on it (arrows)

7.58a Remove the two upper bolts (arrow) . . .

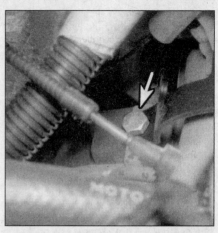

7.58b . . . and one lower bolt (arrow)

the spark plug wires to the coil terminals (if no numbers are present on the spark plug wires and the coil terminals). Squeeze the locking tab of the wire retainer by hand and remove the spark plug wires from the ignition coil pack with a twisting and pulling motion **(see illustration)**. DO NOT just pull on the wires to disconnect them.

56   Disconnect the engine wiring harness from the ignition coil pack and disconnect the condenser wire **(see illustrations)**.

57   If necessary, remove the bolt holding the air conditioning hose to the plenum. Move the air conditioning tube enough to provide working access to the upper coil pack bolts. **Note:** *It may be possible to remove the upper coil pack bolts with a socket and universal joint adapter instead of removing the air conditioning hose bolt.*

58   Remove the three bolts securing the ignition coil pack to the mounting bracket on the engine **(see illustrations)**.

59   Installation is the reverse of the removal procedure with the following additions:

a)   **Note:** *Whenever a spark plug wire is removed from either the spark plug or the ignition coil pack, the boot should be coated with silicone dielectric compound.*

b)   *Insert each spark plug wire into the proper terminal of the ignition coil pack. Push the wire into the terminal and make sure the boots are fully seated and both locking tabs are engaged properly.*

8   Distributor (TFI-IV ignition) - removal and installation

Refer to illustrations 8.3, 8.4a and 8.4b

Removal

1   Detach the cable from the negative terminal of the battery.

2   Detach the coil secondary lead from the coil and the wires from the plugs, then remove the distributor cap and wires as an assembly (refer to Chapter 1).

3   Unplug the module electrical connector **(see illustration)**.

4   Make a mark on the edge of the distributor base **(see illustration)** directly below the rotor tip and in line with it (if the rotor on your engine has more than one tip, use the center one for reference). Mark the distributor base and the engine block to ensure that the distributor is installed correctly **(see illustration)**.

5   Remove the distributor hold down bolt and clamp, then pull the distributor straight up to remove it. Be careful not to disturb the intermediate driveshaft. **Note:** *If the crankshaft is turned while the distributor is removed, or if a new distributor is required, the alignment marks will be useless.*

Installation (crankshaft not turned after distributor removal)

6   Insert the distributor into the engine in exactly the same relationship to the block that it was in when removed.

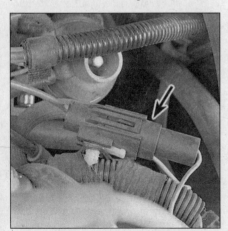

8.3 Before removing (or timing) the distributor, always unplug this connector (arrow) to detach the single wire lead from the ignition module

8.4a Before removing the distributor, remove the spark plug cap (see Chapter 1), mark the direction in which the rotor is pointing by scribing or painting an alignment mark on the edge of the distributor base (arrow) immediately below the rotor tip . . .

8.4b . . . and paint or scribe an alignment mark on the distributor base and the block (arrow) to ensure proper reinstallation, then remove the distributor hold down bolt and clamp (arrow) and lift out the distributor assembly

7   To mesh the helical gears on the camshaft and the distributor it may be necessary to turn the rotor slightly. If the distributor doesn't seat completely, the hex shaped recess in the lower end of the distributor shaft is not mating properly with the oil pump shaft. Recheck the alignment marks between the distributor base and the block to verify that the distributor is in the same position it was in before removal. Also check the rotor to see if it is aligned with the mark you made on the edge of the distributor base. Then, using a socket and a breaker bar on the crankshaft pulley bolt, turn the crankshaft in the normal direction of rotation. Because the gear on the distributor shaft is engaged with the gear on the camshaft, their relationship to one another will not change as long as the distributor is not lifted from the engine. When the hex shaped recess in the end of the distributor shaft and the oil pump shaft are aligned, the distributor will drop down over the pump shaft and the distributor housing will seat against the block.

Installation (crankshaft turned after distributor removal)

8   Position the number one piston at TDC on the compression stroke (see Chapter 2).
9   Position the rotor so it is pointing to the number one spark plug tower on the distributor cap.
10   Lower the distributor into place.
11   To mesh the helical gears on the camshaft and the distributor, it may be necessary to turn the rotor slightly.
12   If the distributor doesn't seat completely against the block, the lower end of the distributor shaft is not mating properly with the oil pump shaft. Recheck the alignment marks between the distributor housing and the block to verify that the distributor is in the same position it was in before removal. Then, using a socket and a breaker bar on the crankshaft pulley bolt, turn the crankshaft in the normal direction of rotation. Because the gear on the distributor shaft is engaged with the gear on the camshaft, their relationship to one another will not change as long as the distributor is not lifted from the engine. When the hex shaped recess in the end of the distributor shaft and the oil pump shaft are aligned, the distributor will drop down over the pump shaft and the distributor housing will seat against the block.

Final installation

13   Place the hold down clamp in position and loosely install the bolt.
14   Install the distributor cap and tighten the cap screws securely.
15   Plug in the module electrical connector.
16   Reattach the spark plug wires to the plugs.
17   Connect the cable to the negative terminal of the battery.
18   Check the ignition timing (refer to Section 9) and tighten the distributor hold down bolt securely.

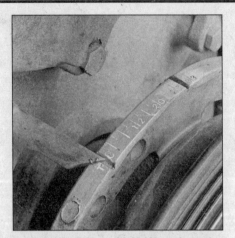

9.6a On the 2.8L and 3.0L V6 engines, the timing marks are located on the crankshaft vibration damper (2.8L V6 shown) - verify that the 10 degrees mark (arrow) - or whatever mark is specified by the VECI label on your vehicle - is aligned with the stationary pointer on the front of the water pump housing

9   Ignition timing (TFI-IV only) - check and adjustment

Refer to illustrations 9.6a and 9.6b
Note: *Check the Vehicle Emission Control Information label on your vehicle to see if a different procedure and/or ignition timing specification is given. If so, follow the information on the VECI label, as it contains information specific to your vehicle.*

1   Apply the parking brake and block the wheels. Place the transmission in Park (automatic) or Neutral (manual). Turn off all accessories (heater, air conditioner, etc.).
2   Start the engine and warm it up. Once it has reached operating temperature, turn it off.
3   Unplug the single wire connector located immediately above the harness connector for the module (see illustration 8.3).
4   Connect an inductive timing light and a tachometer in accordance with the manufacturer's instructions. **Caution:** *Make sure that the timing light and tach wires don't hang anywhere near the electric cooling fan or they may become entangled in the fan blades when it comes on.*
5   Start the engine.
6   Point the timing light at the timing marks on the crankshaft pulley/vibration damper or camshaft pulley (2.3L engine) and note whether the timing mark **(see illustrations)** is aligned with the timing pointer.
7   If the mark indicating the number of degrees advance called for on the VECI label isn't aligned with the stationary pointer, loosen the distributor hold down bolt just enough so that the distributor can be rotated. Turn the distributor until the desired mark is aligned with the pointer. Tighten the distributor hold down bolt securely when the timing is correct and recheck it to make sure it didn't

9.6b On the 2.3L engine, the timing marks are located on the edge of the camshaft pulley - to see the marks, remove the small access cover in the front of the timing belt cover

change when the bolt was tightened.
8   Turn off the engine.
9   Plug in the single wire connector.
10   Remove the timing light and tachometer.
11   The universal distributor used in all vehicles incorporates an octane adjustment feature. The adjustment is accomplished by replacing the standard zero degree rod located in the distributor with a three degree or six-degree retard rod. Consult your local dealer for additional information on this feature.

10   Ignition module - check and replacement

Refer to illustrations 10.16, 10.17 and 10.18
Caution: *The ignition module is a delicate and relatively expensive electronic component. The following tests must be done with the right equipment by someone who knows how to use it properly. Failure to follow the step-by-step procedures could result in damage to the module and/or other electronic devices, including the EEC-IV microprocessor itself. Moreover, all devices under computer control are protected by a Federally mandated extended warranty. Check with your dealer before attempting to diagnose them yourself.*

Check (TFI-IV only)

1   Unplug the wiring harness connector from the ignition module. Inspect it for dirt, corrosion and/or damage, then plug it back in.
2   Unplug the single wire connector located immediately above the ignition module connector (see Section 8).
3   Using a calibrated spark tester, check for spark (see Section 6).
4   If there is no spark, proceed to the distributor/TFI-IV module check (Step 7).

10.16 To remove the TFI-IV ignition module from the distributor base, remove the two screws (arrows) . . .

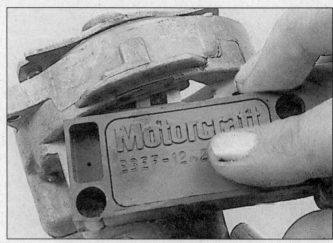

10.17 . . . then pull the module straight down to detach the spade terminals from the stator connector

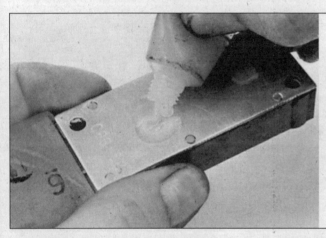

10.18 Be sure to wipe the back of the module clean and apply a film of dielectric grease (essential for cool operation of the module) - DO NOT use any other type of grease!

5   If there is spark, the problem lies either with the inferred mileage sensor (IMS) or within the EEC-IV electronic control module (ECM).
Diagnosis of these items is beyond the scope of the home mechanic. Take the vehicle to a dealer service department.
6   Remove the spark tester and reconnect the single wire connector.

TFI-IV module

Note: *You must purchase a new ignition module before performing the following check. Since the check can result in only one of two possibilities (you will need a new module, or you won't), the odds are 50/50 that you'll be buying a new module that you may not need. Electronic components can't be returned once they're purchased, so if you're unwilling to invest in a new module that you may not need until later take the vehicle to a dealer and have the module checked.*

Distributor - mounted TVI-IV module

7   Remove the distributor (see Section 8).
8   Install a new module on the distributor. Connect the body harness to the TFI-IV. Make sure the unit is grounded with a jumper lead from the distributor to the engine. Rotate

the distributor shaft by hand and check for spark with the ignition tester (see Section 6).
9   If there is spark, the old module has failed. Leave the new module on the distributor and install the distributor (see Section 8).
10   If there is no spark, the sensor has failed. Your old module is okay but you need a new or rebuilt distributor. Install the new/rebuilt distributor with the old module.

Remote-mounted TFI-IV module

11   Remove and replace the ignition module with a known good module, as discussed later in this Section.
12   Perform ignition system check (refer to Section 6 this Chapter).
13   If there is spark, the old module is faulty.
14   If there is still no spark, your old module is probably okay and the problem is elsewhere. Take the vehicle to a dealer service department for testing.

Replacement

TFI-IV ignition

Distributor - mounted TFI-IV module

15   Disconnect the negative battery cable and the wiring connector at the side of the module.
16   Remove the two module mounting

screws **(see illustration)**.
17   Pull straight down on the module to disconnect the spade connectors from the stator connector **(see illustration)**.
18   Whether you are installing the old module or a new one, wipe the back of the module clean with a soft, clean rag and apply a film of silicone dielectric grease to the back of the module **(see illustration)**.
19   Installation is the reverse of removal. Wipe a thin film of dielectric grease on the rear of the module before installation. When plugging in the module, make sure that the three terminals are inserted all the way into the stator connector.

Remote-mounted TFI-IV module

20   Remove the two screws retaining the module/heatsink assembly to the radiator support bracket.
21   Disconnect the electrical harness connector from the module.
22   Remove the two screws retaining the ignition module to the heatsink and remove the module.
23   Installation is the reverse of removal. Coat the metal baseplate of the ignition module with silicone dielectric compound approximately 1/32-inch thick prior to assembly to the heatsink.
24   Tighten the attaching screws securely.

EDIS ignition module

25   Disconnect the negative cable from the battery.
26   Locate the EDIS ignition module against the right inner fender apron.
27   Disconnect the electrical connector from the module.
28   Remove the module attaching screw and slide the module towards the front of the vehicle to release it from its mounting.
29   Installation is the reverse of removal. Insert the rivet of the module into the teardrop mounting hole and push it reward into proper position.
30   Tighten the attaching screw securely.

11.9 With the distributor shaft housing locked securely in a vise lined with several shop rags (to prevent damage to the housing), drive out the roll pin

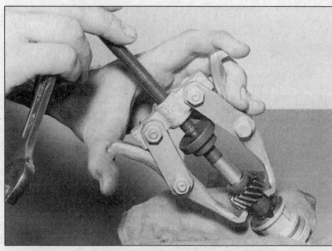

11.11 With the distributor shaft pointing up, use a small puller to separate the drive gear from the shaft

11 Distributor stator assembly (TFI-IV ignition) - check and replacement

Refer to illustrations 11.9, 11.11, 11.12, 11.13, 11.14, 11.15, 11.19, 11.28a and 11.28b

Check

1　The factory doesn't specify a check for the TFI-IV system stator (sensor). If the ignition module is good, but the system won't function normally, the stator must be replaced.

Replacement

2　Remove the distributor cap and position it out of the way with the wires attached.
3　Disconnect the TFI module from the wire harness.
4　Remove the distributor (see Section 8).
5　Remove the rotor (see Chapter 1).
6　Although not absolutely necessary, it's a good idea to remove the ignition module (see Section 10) to prevent possible damage to the module while the distributor is being disassembled.
7　Clamp the lower end of the distributor housing in a vise. Place a shop rag in the vise jaws to prevent damage to the distributor and don't overtighten the vise.
8　Before removing the drive gear, note that the roll pin is slightly offset. When the distributor is reassembled, the roll pin cannot be reinstalled through the drive gear and distributor shaft holes unless the holes are perfectly lined up. Use a bit of paint to mark the positions of these components.
9　With an assistant holding the distributor steady in the vise, use a pin punch to drive the roll pin out of the shaft **(see illustration)**.
10　Loosen the vise and reposition the distributor with the drive gear facing up.
11　Remove the drive gear with a small puller **(see illustration)**.

11.12 Inspect the distributor shaft for burrs or residue buildup in the vicinity of the hole for the drive gear roll pin - remove it with emery cloth to prevent damage to the distributor shaft bushing when removing and installing the shaft

11.13 As soon as you remove the distributor shaft, note how the washer is installed before removing it (or it can easily fall out and be lost)

11.14 To detach the octane rod from the distributor, remove the retaining screw - note the condition of the small square rubber grommet that seals the octane rod hole when you pull the rod out (it seals the interior of the distributor to prevent moisture from damaging the electronics)

12　Before removing it from the distributor, check the shaft for burrs or built up residue, particularly around the drive gear roll pin hole **(see illustration)**. If burrs or residue are evident, polish the shaft with emery cloth and wipe it clean to prevent damage to the lip seal and bushing in the distributor base.
13　After removing any burrs/residue, remove the shaft assembly by gently pulling on the plate. Note the relationship of the spacer washer to the distributor base before removing the washer **(see illustration)**.
14　Remove the octane rod retaining screw **(see illustration)**.
15　Lift the inner end of the rod off the stator retaining post **(see illustration)** and pull the octane rod from the distributor base. Note: Don't lose the grommet installed in the octane rod hole. The grommet protects the electronic components of the distributor from moisture.
16　Remove the two stator screws **(see illustration 11.15)**.

11.15 To remove the octane rod, lift the inner end of the rod off the stator assembly post (A) - to remove the stator assembly, remove both mounting screws (B) and lift the stator straight up and off the posts

17   Gently lift it straight up and remove the stator assembly from the distributor.
18   Check the shaft bushing in the distributor base for wear or signs of excessive heat buildup. If signs of wear and/or damage are evident, replace the complete distributor assembly.
19   Inspect the O-ring at the base of the distributor. If it's damaged or worn, remove it and install a new one **(see illustration)**.
20   Inspect the base casting for cracks and wear. If any damage is evident, replace the distributor assembly.
21   Place the stator assembly in position over the shaft bushing and press it down onto the distributor base until it's completely seated on the posts.
22   Install the stator screws and tighten them securely.
23   Insert the octane rod through the hole in the distributor base and push the inner end of

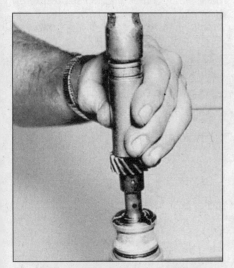

11.28a After securing the distributor assembly upside down in a vise, "eyeball" the roll pin holes in the drive gear and the shaft, then tap the drive gear onto the shaft with a deep socket and hammer

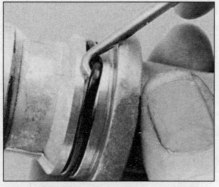

11.19 If the O-ring at the base of the distributor is worn or damaged, replace it with a new one

the rod onto the post. **Note:** *Make sure that the octane rod hole is properly sealed by the grommet.*
24   Reinstall the octane rod screw and tighten it securely.
25   Apply a light coat of engine oil to the distributor shaft and insert the shaft through the bushing.
26   Mount the distributor in the vise with the lower end up. Be sure to line the vise jaws with a few clean shop rags to protect the distributor base. Place a block of wood under the distributor shaft to support it and prevent it from falling out while the drive gear is being installed.
27   Using the paint marks you made on the drive gear and the distributor shaft housing, turn the shaft until the drive gear hole (and paint mark), the shaft hole and the paint mark on the distributor shaft housing are aligned.
28   Using a deep socket and hammer, carefully tap the drive gear back onto the distributor shaft **(see illustration)**. Make sure the hole in the drive gear and the hole in the shaft are lined up. Because the holes were drilled off center by the factory, they must be perfectly aligned or the roll pin cannot be installed **(see illustration)**.
29   Once the drive gear is seated and the holes are lined up, turn the distributor sideways in the vise and, with an assistant steadying it, drive a new roll pin into the drive gear with a pin punch. Make sure that neither end of the roll pin protrudes from the drive gear.

30   Check the distributor shaft for smooth rotation, then remove the distributor assembly from the vise.
31   Install the TFI-IV module (refer to Section 10).
32   Install the rotor (refer to Chapter 1 if necessary).
33   Install the distributor.

12 Charging system - general information and precautions

The charging system includes the alternator, either an internal or an external voltage regulator, a charge indicator, the battery, a fusible link and the wiring between all the components. The charging system supplies electrical power for the ignition system, the lights, the radio, etc. The alternator is driven by a drivebelt at the front (right end) of the engine.

The purpose of the voltage regulator is to limit the alternator's voltage to a preset value. This prevents power surges, circuit overloads, etc., during peak voltage output.

The fusible link is a short length of insulated wire integral with the engine compartment wiring harness. The link is four wire gauges smaller in diameter than the circuit it protects. Production fusible links and their identification flags are identified by the flag color. See Chapter 12 for additional information regarding fusible links.

The charging system doesn't ordinarily require periodic maintenance. However, the drivebelt, battery and wires and connections should be inspected at the intervals outlined in Chapter 1.

Be very careful when making electrical circuit connections to a vehicle equipped with an alternator and note the following:

a) *When reconnecting wires to the alternator from the battery, be sure to note the polarity.*
b) *Before using arc welding equipment to repair any part of the vehicle, disconnect the wires from the alternator and the battery terminals.*
c) *Never start the engine with a battery charger connected.*
d) *Always disconnect both battery leads before using a battery charger.*

11.28b If the drive gear and shaft roll pin hole are misaligned, the roll pin cannot be driven through the drive gear and shaft holes - this drive gear must now be pulled off the shaft and realigned

13 Charging system - check

Refer to illustration 13.9

1 If a malfunction occurs in the charging circuit, don't automatically assume that the alternator is causing the problem. First check the following items:

  a) *The battery cables where they connect to the battery. Make sure the connections are clean and tight (refer to Chapter 1).*
  b) *Check the external alternator wiring harness and the connectors at the alternator and voltage regulator. They must be in good condition, clean and tight.*
  c) *Check the drivebelt condition and tension (refer to Chapter 1).*
  d) *Make sure the alternator mounting and adjustment bolts are tight.*
  e) *Check the fusible link located between the starter relay (refer to Section 19) and the alternator. If it's burned, determine the cause, repair the circuit and replace the link (refer to Chapter 12).*
  f) *Run the engine and check the alternator for abnormal noise.*

2 Using a voltmeter, check the battery voltage with the engine off. It should be approximately 12-volts.

3 Start the engine and check the battery voltage again. It should now be approximately 14-to-15 volts.

4 If the voltage reading is less or more than the specified charging voltage, replace the voltage regulator (refer to Section 15).

5 Locate the regulator terminal (screw F) on the back of the alternator.

6 With the ignition switch in the Off position, touch the F screw with the voltmeter positive lead while touching the alternator housing with the negative lead.

7 The meter should indicate battery voltage if the system is operating normally.

8 If less than battery voltage is indicated, proceed to the next Step to determine the cause of the current drain.

13.9 To check the voltage at the regulator electrical connector, tough the positive lead of a VOM to the I terminal, then the S terminal while touching the negative lead to the rear of the alternator housing

36004-5-13.9 HAYNES

9 Disconnect the wire harness from the regulator and connect the voltmeter positive lead to terminal I in the plug **(see illustration)**. No voltage should be indicated.

10 If voltage is indicated, check the I lead from the ignition switch to identify and eliminate the voltage source.

11 If no voltage is indicated, contact the voltmeter positive lead to terminal S. No voltage should be indicated.

12 If voltage is indicated, disconnect the wire harness from the alternator. Again, connect the positive voltmeter lead to terminal S in the wiring harness connector.

13 If voltage is still indicated, repair the S lead to the alternator plug to eliminate the voltage source.

14 If no voltage is indicated, the problem is with the rectifier. Replace the alternator (refer to Section 14).

14 Alternator - removal and installation

Refer to illustrations 14.4a and 14.4b

1 Detach the cable from the negative terminal of the battery.

2 Loosen the lug nuts of the right front wheel. Raise the front of the vehicle and support it securely on jackstands. Remove the wheel.

3 Unplug the electrical connectors from the alternator and the voltage regulator.

4 Loosen the alternator adjustment and pivot bolts and detach the drivebelt **(see illustrations)**. Access to the pivot bolt can be gained through the right fenderwell.

5 Remove the adjustment and pivot bolts and separate the alternator from the engine.

6 Installation is the reverse of removal.

7 After the alternator is installed, adjust the drivebelt tension (see Chapter 1).

15 Voltage regulator/alternator brushes - replacement

Note: *Starting in 1992, there are two different types of alternators installed on the 3.0L V6 engine. On alternators with an external type fan, the alternator brushes can be serviced. There are no service procedures possible on alternators with an internal type fan. If any portion of the alternator is faulty, this type of alternator must be replaced.*

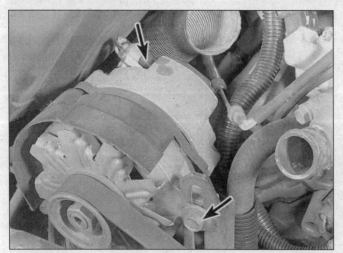

14.4a To remove the alternator, remove the adjustment bolt (arrow), remove the drivebelt, unplug both the voltage regulator (arrow) and alternator (not shown) wire harness connectors . . .

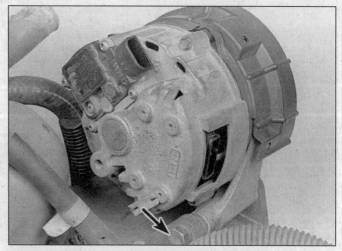

14.4b . . . then remove the alternator pivot through bolt (arrow) and carefully lift the alternator from the vehicle (2.8L V6 shown - other engines similar)

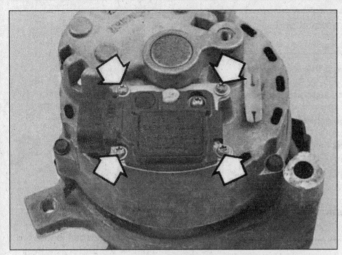

15.3 To detach the voltage regulator/brush holder assembly, remove the four screws (arrows)

15.5 To remove the brushes from the voltage regulator/brush holder assembly, detach the rubber plugs from the two brush lead wire screws and remove both screws (arrows)

Integral regulator

Refer to illustrations 15.3, 15.5 and 15.9

1   Remove the alternator (see Section 14).
2   Set the alternator on a clean workbench.
3   Remove the four voltage regulator mounting screws **(see illustration)**.
4   Detach the voltage regulator.
5   Detach the rubber plugs and remove the brush lead retaining screws and nuts to separate the brush leads from the holder **(see illustration)**. Note that the screws have Torx heads and require a special screwdriver.
6   After noting the relationship of the brushes to the brush holder assembly, remove both brushes. Don't lose the springs.
7   If you're installing a new voltage regulator, insert the old brushes into the brush holder of the new regulator. If you're installing new brushes, insert them into the brush holder of the old regulator. Make sure the springs are properly compressed and the brushes are properly inserted into the recesses in the brush holder.

15.9 Before installing the voltage regulator/brush holder assembly, insert a paper clip to hold the brushes in place during installation - after installation, pull out the paper clip

8   Install the brush lead retaining screws and nuts.
9   Insert a short section of wire, such as a paper clip, through the hole in the voltage regulator **(see illustration)** to hold the brushes in the retracted position during regulator installation.
10   Carefully install the regulator. Make sure the brushes don't hang up on the rotor.
11   Install the voltage regulator screws and tighten them securely.
12   Remove the wire or paper clip.
13   Install the alternator (see Section 14).

Remote regulator

14   Remove the negative cable from the battery terminal, then disconnect the wire harness from the regulator. Use a screwdriver to unplug the harness connector - do not pull on the wires.
15   Remove the mounting bolts and detach the regulator.
16   Installation is the reverse of removal.

16   Starting system - general information

The function of the starting system is to crank the engine to start it. The system is composed of the starter motor, starter relay, battery, switch and connecting wires.

Turning the ignition key to the Start position actuates the starter relay through the starter control circuit. The starter relay then connects the battery to the starter. The battery supplies the electrical energy to the starter motor, which does the actual work of cranking the engine.

Vehicles equipped with an automatic transmission have a Neutral start switch in the starter control circuit, which prevents operation of the starter unless the shift lever is in Neutral or Park. The circuit on vehicles with a manual transmission prevents operation of the starter motor unless the clutch

pedal is depressed.

Never operate the starter motor for more than 15 seconds at a time without pausing to allow it to cool for at least two minutes. Excessive cranking can cause overheating, which can seriously damage the starter.

17   Starter motor and circuit - in-vehicle check

Note: *Before diagnosing starter problems, make sure the battery is fully charged.*
1   If the starter motor doesn't turn at all when the switch is operated, make sure the shift lever is in Neutral or Park.
2   Make sure the battery is charged and that all cables at the battery and starter solenoid terminals are secure.
3   If the starter motor spins but the engine doesn't turn over, then the drive assembly in the starter motor is slipping and the starter motor must be replaced.
4   If, when the switch is actuated, the starter motor doesn't operate at all but the starter solenoid operates (clicks), then the problem lies with either the battery, the starter solenoid contacts or the starter motor connections.
5   If the starter solenoid doesn't click when the ignition switch is actuated, either the starter solenoid is defective, the starter relay is bad, the ignition switch is faulty, the neutral start switch (automatic) or starter/clutch interlock switch (manual) is bad, or there is a problem in the wiring between the components. Check the starter circuit.
6   To check the starter circuit, remove the push-on connector from the relay (see Section 19) - this is the signal wire from the ignition switch. Make sure that the connection is clean and secure. If the connections are good, check the operation of the relay with a jumper wire. To do this, place the transmission in Park (automatic transmission) or Neutral (manual transmission). Remove the push-

on connector from the relay. Connect a jumper wire between the battery positive terminal and the exposed terminal on the relay. If the starter motor now operates, the starter relay is okay. The problem is in the ignition switch, neutral start switch, starter/clutch interlock switch or in the wiring between these components (look for open or loose connections).

7   If the starter motor still doesn't operate, bridge the two large terminals on the relay with a screwdriver. If the starter now works, replace the relay. If it doesn't operate, check for voltage to the relay (it should be available on one of the large terminals). If voltage isn't present, check the fuses, fusible links and the cable to the relay.

8   If voltage is present, check for voltage to the starter motor while the ignition key is turned to Start. If voltage is present, replace the starter assembly. If voltage is not present and the relay checked out OK, trace the wiring between the relay and the starter for an open circuit condition.

9   If the starter motor cranks the engine at an abnormally slow speed, first make sure the battery is fully charged and all terminal connections are clean and tight. Also check the connections at the starter solenoid and battery ground. Eyelet terminals should not be easily rotated by hand. If the engine is partially seized, or has the wrong viscosity oil in it, it will crank slowly.

18 Starter motor - removal and installation

1   Detach the cable from the negative terminal of the battery.
2   Raise the vehicle and support it securely on jackstands.
3   Disconnect the large cable from the terminal on the starter motor.
4   Remove the starter motor mounting bolts. Remove the starter.
5   Installation is the reverse of removal.

19 Starter relay - removal and installation

Refer to illustration 19.2
1   Detach the cable from the negative terminal of the battery.
2   Label the wires and the terminals then disconnect the Neutral safety switch wire (automatics only), the battery cable, the fusible link and the starter cable from the relay terminals **(see illustration)**.
3   Remove the mounting bolts and detach the relay.
4   Installation is the reverse of removal.

19.2 The starter relay (arrow) is located behind and below the battery (battery removed for clarity) - to replace it, clearly label then detach the electrical leads and remove the two relay bracket mounting bolts (not visible in this photo)

Chapter 6
Emissions control systems

Contents

Specifications

Torque specifications   Ft-lbs

Air pump pulley bolts .... 10 to 12.5

Air pump mounting bolt 25

1   General information

Refer to illustration 1.7

To prevent pollution of the atmosphere from incompletely burned and evaporating gases, and to maintain good driveability and fuel economy, a number of emission control systems are incorporated. They include the:

Electronic Engine Control system
Exhaust Gas Recirculation (EGR) system
Managed air thermactor system
Fuel evaporative emission control system
Positive Crankcase Ventilation (PCV) system
Inlet air temperature control system
Catalytic converter

All of these systems are linked, directly or indirectly, to the EEC-IV system.

The Sections in this Chapter include general descriptions, checking procedures within the scope of the home mechanic and component replacement procedures (when possible) for each of the systems listed above.

Before assuming that an emissions control system is malfunctioning, check the fuel and ignition systems carefully. The diagnosis of some emission control devices requires specialized tools, equipment and training. If checking and servicing become too difficult or if a procedure is beyond your ability, consult a dealer service department.

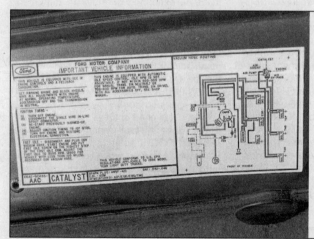

1.7 The Vehicle Emission Control Information (VECI) label located on the underside of the hood contains essential information such as spark plug type, the ignition timing procedure, fast idle speed adjustment, etc. - always defer to the information contained on the VECI label as the "last word" on the devices installed on your vehicle

This doesn't mean, however, that emission control systems are particularly difficult to maintain and repair. You can quickly and easily perform many checks and do most (if not all) of the regular maintenance at home with common tune-up and hand tools. **Note**: *The most frequent cause of emissions problems is simply a loose or broken vacuum hose or wire, so always check the hose and wiring connections first.*

Pay close attention to any special precautions outlined in this Chapter. It should be noted that the illustrations of the various systems may not exactly match the system

installed on your vehicle because of changes made by the manufacturer during production or from year-to-year.

A Vehicle Emissions Control Information label is located in the engine compartment **(see illustration)**. This label contains important emissions specifications and adjustment information, as well as a vacuum hose schematic with emissions components identified. When servicing the engine or emissions systems, the VECI label in your particular vehicle should always be checked for up-to-date information.

2.1 Digital multimeters can be used for testing many types of circuits. Because of their high impedance, they are much more accurate than analog meters for measuring voltages in computer circuits

2　On Board Diagnosis (OBD) system and trouble codes

Note: *1995 and earlier vehicles are equipped with OBD-I self diagnosis system, while 1996 and later vehicles are equipped with OBD-II self diagnosis system. Both systems require the use of a Scan tool to access trouble codes. However, many of the information sensor checks and replacement procedures do apply to both systems. Because OBD-I and OBD-II systems require a special SCAN tool to access the trouble codes, have the vehicle diagnosed by a dealer service department or other qualified automotive repair facility if the proper SCAN tool is not available. The codes indicated in the text are designed and mandated by the EPA for all 1995 and earlier OBD-I and 1996 and later OBD-II vehicles produced by automobile manufacturers. These generic trouble codes do not include the manufacturer's specific trouble codes. Consult a dealer service department for additional information. Refer to the troubleshooting tips in the beginning of this manual to gain some insight to the most likely causes to a problem.*

Diagnostic tool information

Refer to illustrations 2.1, 2.2 and 2.4

1　A digital multimeter is necessary for checking fuel injection and emission related components **(see illustration)**. A digital volt-ohmmeter is preferred over the older style analog multimeter for several reasons. The analog multimeter cannot display the volts-ohms or amps measurement in hundredths and thousandths increments. When working with electronic circuits which are often very low voltage, this accurate reading is most important. Another good reason for the digital multimeter is the high impedance circuit. The digital multimeter is equipped with a high resistance internal circuitry (10 million ohms).

2.2 Scanners like the Actron OBD II Diagnostic Tester, Actron Scantool and the AutoXray XP240 are powerful diagnostic aids - programmed with comprehensive diagnostic information, they can tell you just about anything you want to know about your engine management system

Because a voltmeter is hooked up in parallel with the circuit when testing, it is vital that none of the voltage being measured should be allowed to travel the parallel path set up by the meter itself. This dilemma does not show itself when measuring larger amounts of voltage (9 to 12 volt circuits) but if you are measuring a low voltage circuit such as the oxygen sensor signal voltage, a fraction of a volt may be a significant amount when diagnosing a problem. Obtaining the diagnostic trouble codes is one exception where using an analog voltmeter is necessary.

2　Hand-held scanners are the most powerful and versatile tools for analyzing engine management systems used on later model vehicles **(see illustration)**. Early model scanners handle codes and some diagnostics for many OBD-I systems. Each brand scan tool must be examined carefully to match the year, make and model of the vehicle you are working on. Often interchangeable cartridges are available to access the particular manufacturer (Ford, GM, Chrysler, etc.). Some manufacturers will specify by continent (Asia, Europe, USA, etc.).

3　With the arrival of the Federally mandated emission control system (OBD-II), a specially designed scanner has been developed. Several tool manufacturers have released OBD-II scan tools for the home mechanic. Ask the parts salesman at a local auto parts store for additional information concerning dates and costs.

4　Another type of code reader is available at parts stores **(see illustration)**. These tools simplify the procedure for extracting codes from the engine management computer on OBD-I systems by simply "plugging in" to the diagnostic connector on the vehicle wiring harness. **Note:** *Some diagnostic connectors are located under the dash, kick panel or glove box while others are located in the engine compartment.*

OBD system general description

5　All 1995 and earlier vehicles are equipped with the Electronic Engine Control (EEC) IV (OBD-I) system. All 1996 and later engines and powertrain combinations described in this manual are equipped with the EEC V (OBD-II) system. Both systems consist of an onboard computer, known as the powertrain Control Module (PCM), and information sensors, which monitor various functions of the engine and send data to the PCM. Based on the data and the information programmed into the computer's memory, the PCM generates output signals to control various engine functions via control relays, solenoids and other output actuators.

6　The PCM is the "brain" of the Electronic Engine Control (EEC) system. It receives data from a number of sensors and other electronic components (switches, relays, etc.). Based on the information it receives, the PCM generates output signals to control various relays, solenoids and other actuators. The PCM is specifically calibrated to optimize the emissions, fuel economy and driveability of the vehicle.

2.4 trouble code tools simplify the task of extracting the trouble codes on OBD-I systems

7   Because of a Federally mandated warranty which covers the EEC-IV and EEC-V system components and because any owner-induced damage to the PCM, the sensors and/or the control devices may void the warranty, it isn't a good idea to attempt diagnosis or replacement of the PCM at home while the vehicle is under warranty. Take the vehicle to a dealer service department if the PCM or a system component malfunctions.

Information sensors

8   **Heated Oxygen sensors (HO2S)** - The HO2S generates a voltage signal that varies with the difference between the oxygen content of the exhaust and the oxygen in the surrounding air.

9   **Crankshaft Position (CKP) sensor** -. The CKP sensor provides information on crankshaft position and the engine speed signal to the PCM.

10   **Camshaft Position (CMP) sensor** - The CMP sensor produces a signal which the PCM uses to identify number 1 cylinder and to time the sequential fuel injection.

11   **Engine Coolant Temperature (ECT) sensor** - The ECT monitors engine coolant temperature and sends the PCM a voltage signal that affects PCM control of the fuel mixture, ignition timing, and EGR operation.

12   **Intake Air Temperature (IAT) sensor** - The IAT provides the PCM with intake air temperature information. The PCM uses this information to control fuel flow, ignition timing, and EGR system operation.

13   **Throttle Position Sensor (TPS)** - The TPS senses throttle movement and position, then transmits a voltage signal to the PCM. This signal enables the PCM to determine when the throttle is closed, in a cruise position, or wide open.

14   **Mass airflow (MAF) sensor** - The MAF sensor measures the molecular mass of the intake airflow entering the engine. The MAF sensor, along with the IAT sensor, provide mass airflow and air temperature information for the most precise fuel metering.

15   **Vehicle speed sensor (VSS)** - The vehicle speed sensor provides information to the PCM to indicate vehicle speed.

16   **Knock sensor (KS)** - The knock sensor is a piezoresistive microphone that detects the sound of engine detonation, or "pinging". The PCM uses the input signal from the knock sensor to recognize detonation and retard spark advance to avoid engine damage. The knock sensor is used only on the 4.0L SOHC V6 engine.

17   **EGR backpressure sensor** - The EGR backpressure sensor is used to monitor the rate and flow of exhaust gas recirculation into the intake system.

18   **Fuel tank pressure sensor** - The fuel tank pressure sensor used on 1996 and later OBD-II vehicles is part of the evaporative emission control system and is used to monitor vapor pressure in the fuel tank. The PCM uses this information to turn on and off the purge valves and solenoids of the evaporative emission system.

19   **Power Steering Pressure (PSP) switch** - The PSP sensor is used to increase transmission hydraulic line pressure during low-speed vehicle maneuvers.

20   **Brake On-Off (BOO) switch** - This switch is used when the driver applies the brakes, to disengage the cruise control and change fuel metering and spark advance during deceleration.

21   **Transmission sensors** - In addition to the vehicle speed sensor, the PCM receives input signals on 1995 and later vehicles from the following sensors inside the transmission or connected to it: (a) the turbine shaft speed sensor, (b) the transmission fluid temperature sensor, and (c) the transmission range sensor.

22   **A/C clutch control switch** When battery voltage is applied to the air conditioning compressor solenoid, a signal is sent to the PCM, which interprets the signal as an added load created by the compressor and increases engine idle speed accordingly to compensate.

Output actuators

23   **PCM power relay** - The main PCM power relay is activated by the ignition switch and supplies battery power to the PCM and the Electronic Engine Control (EEC) system when the switch is in the Start or Run position. Refer to Chapter 12 or your owner's manual for more information on relay location.

24   **Fuel pump relay** - The fuel pump relay is activated by the PCM with the ignition switch in the Start or Run position. When the ignition switch is turned on, the relay is activated to supply initial line pressure to the fuel system. For more information on fuel pump check and replacement, refer to Chapter 4.

25   **Fuel injectors** - The PCM opens the fuel injectors individually in firing order sequence. The PCM also controls the time the injector is open, called the "pulse width." The pulse width of the injector (measured in milliseconds) determines the amount of fuel delivered. For more information on the fuel delivery system and the fuel injectors, including injector replacement, refer to Chapter 4.

26   **Ignition Control Module (ICM)** - The ICM triggers the ignition coils and determines proper spark advance based on inputs from the PCM. An externally mounted ICM is used on all 1995 and earlier vehicles. All other models use an ignition module which is incorporated into the PCM. Refer to Chapter 5 for more information on the Ignition Control Module.

27   **Idle Air Control (IAC) valve** - The IAC valve controls the amount of air to bypass the throttle plate when the throttle valve is closed or at idle position. The IAC valve opening and the resulting airflow is controlled by the PCM. Refer to Chapter 4 for more information on the IAC valve.

28   **EGR vacuum solenoid** - The EGR vacuum solenoid is controlled by the PCM to regulate the opening of the vacuum-operated EGR valve.

29   **Canister purge valve** - The evaporative emission canister purge valve is a solenoid valve, operated by the PCM to purge the fuel vapor canister and route fuel vapor to the intake manifold for combustion.

30   **Canister vent solenoid** - The evaporative emission canister vent solenoid is operated by the PCM during the OBD-II evaporative emission monitor (see Section 3) and during an emission test of the evaporative system.

Obtaining OBD-I system codes

Refer to illustration 2.35

31   The diagnostic codes for the EEC-IV systems are arranged in such a way that a series of tests must be completed in order to extract ALL the codes from the system. If one portion of the test is performed without the others, there may be a chance the trouble code that will pinpoint a problem in your particular vehicle will remain stored in the PCM without detection. The tests start first with a Key On, Engine Off (KOEO) test followed by a computed timing test then finally a Engine Running (ER) test. Here is a brief overview of the code extracting procedures of the EEC-IV system followed by the actual test:

Quick Test - Key On Engine Off (KOEO)

32   The following tests are all included with the key on, engine off:

Self test codes - These codes are accessed on the test connector by using a jumper wire and an analog voltmeter or the factory diagnostic tool called the Star tester. These codes are also called *Hard Codes*.

Separator pulse codes - After the initial Hard Codes, the system will flash a code 111 and then will flash a series of Soft (or Continuous Memory) Codes.

Continuous Memory Codes - These codes indicate a fault that may or may not be present at the time of testing. These codes usually indicate an intermittent failure. Continuous Memory codes are stored in the system and they will flash after the normal Hard Codes. These codes are three digit codes (1994 and 1995). These codes can indicate chronic or intermittent problems. Also called *Soft Codes*.

Fast codes - These codes are transmitted 100 times faster than normal codes and can only be read by a Star Tester from the manufacturer or an equivalent SCAN tool.

Engine running codes (KOER) or (ER)

33   **Running tests** - These tests make it possible for the PCM to pick-up a diagnostic trouble code that cannot be set while the engine is in KOEO. These problems usually occur during driving conditions. Some codes are detected by cold or warm running conditions, some are detected at low rpm or high rpm and some are detected at closed throttle or WOT.

I.D. Pulse codes - These codes indicate the type of engine (4, 6 or 8 cylinder) or the correct module and Self Test mode access.

Computed engine timing test - This engine running test determines base timing for the engine and starts the process of allowing the engine to store running codes.

Wiggle test - This engine running test checks the wiring system to the sensors and output actuators as the engine performs.

Cylinder balance test - This engine running test determines injector balance as well as cylinder compression balance. **Note:** *This test should be performed by a dealer service department.*

Beginning the test

34 Position the parking brake ON, Shift lever in PARK, block the drive wheels and turn off all electrical loads (air conditioning, radio, heater fan blower etc.). Make sure the engine is warmed to normal operating temperature (if possible).

35 Perform the KOEO tests:

a) *Turn the ignition key off for at least 10 seconds*

b) *Locate the Diagnostic Test connector inside the engine compartment. Install the voltmeter leads onto the battery and pin number 4 (STO) of the test connector* **(see illustration)**. *Install a jumper wire from the test terminal to pin number 2 of the Diagnostic Test terminal (STI).*

c) *Turn the ignition key ON (engine not running) and observe the needle sweeps on the voltmeter. For example on code 123, the voltmeter will sweep once and then pause. There will be a two second pause between digits and then there will be two distinct sweeps of the needle to indicate the second digit of the code number. There will be another pause and then three distinct sweeps of the meter. It is possible to also observe the CHECK ENGINE light flash the codes. Additional codes will be separated by a four second pause and then the indicated sweeps on the voltmeter. Be aware that the code sequence may continue into the continuous memory codes (read further).* **Note:** *These models will flash the CHECK ENGINE light on the dash in place of the voltmeter.*

36 Interpreting the continuous memory codes:

a) *After the KOEO codes are reported, there will be a short pause and any stored Continuous Memory codes will appear in order. Remember that the "Separator" code is 111. The computer will not enter the Continuous Memory mode without flashing the separator pulse code. The Continuous Memory codes are read the same as the initial codes or "Hard Codes". Record these codes onto a piece of paper and continue the test.*

37 Perform the Engine Running (ER) tests.

a) *Remove the jumper wires from the Diagnostic Test connector to start the test.*

b) *Run engine until it reaches normal operating temperature.*

c) *Turn the engine OFF for at least 10 seconds.*

d) *Install the jumper wire onto Diagnostic Test connector* **(see illustration 2.35)** *and start the engine.*

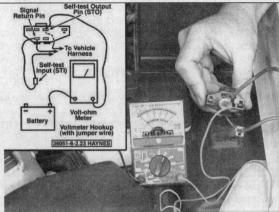

2.35 To read stored trouble codes on early models, connect a voltmeter to the diagnostic test connector as shown, then connect a jumper wire between the self-test input and pin number 2 on the larger connector. Turn the ignition key ON (engine not running) and watch the voltmeter needle or CHECK ENGINE light

e) *Observe that the voltmeter or CHECK ENGINE light will flash the engine identification code. This code indicates 1/2 the number of cylinders of the engine. For example, 4 flashes represent an 8 cylinder engine, or 3 flashes represent a six cylinder engine.*

f) *Within 1 to 2 seconds of the I.D. code, turn the steering wheel at least 1/2 turn and release. This will store any power steering pressure switch trouble codes.*

g) *Depress the brake pedal and release.* **Note:** *Perform the steering wheel and brake pedal procedure in succession immediately (1 to 2 seconds) after the I.D. codes are flashed.*

h) *Observe all the codes and record them on a piece of paper. Be sure to count the sweeps or flashes very carefully as you jot them down.*

38 On some models the PCM will request a Dynamic Response check. This test quickly checks the operation of the TPS, MAF or MAP sensors in action. This will be indicated by a code 1 or a single sweep of the voltmeter needle (one flash on CHECK ENGINE light). This test will require the operator to simply full throttle ("goose") the accelerator pedal for one second. DO NOT throttle the accelerator pedal unless it is requested.

39 The next part of this test makes sure the system can advance the timing. This is called the Computed Timing test. After the last ER code has been displayed, the PCM will advance the ignition timing a fixed amount and hold it there for approximately 2 minutes. Use a timing light to check the amount of advance. The computed timing should equal the base timing plus 20 BTDC. The total advance should equal 27 to 33 degrees advance. If the timing is out of specification, have the system checked at a dealer service department. **Note:** *Remember to remove the SPOUT from the connector as described in the ignition timing procedure in Chapter 5. This will remove the computer from the loop and give base timing.*

40 Finally perform the Wiggle Test. This test can be used to recreate a possible intermittent fault in the harness wiring system.

a) *Use a jumper wire to ground the STI lead on the Diagnostic Test connector* **(see illustration 2.35)**.

b) *Turn the ignition key ON (engine not running).*

c) *Now deactivate the self test mode (remove the jumper wire) and then immediately reactivate the self-test mode. Now the system has entered Continuous Monitor Test Mode.*

d) *Carefully wiggle, tap or remove any suspect wiring to a sensor or output actuator. If a problem exists, a trouble code will be stored that indicates a problem with the circuit that governs the particular component. Record the codes that are indicated.*

e) *Next, enter Engine Running Continuous Monitor Test Mode to check for wiring problems only when the engine is running. Start first by deactivating the Diagnostic Test connector and turning the ignition key OFF. Now start the engine and allow it to idle.*

f) *Use a jumper wire to ground the STI lead on the Diagnostic Test connector* **(see illustration 2.35)**. *Wait ten seconds and then deactivate the test mode and reactivate it again (install jumper wire). This will enter Engine Running Continuous Monitor Test Mode.*

g) *Carefully wiggle any suspect wiring to a sensor or output actuator. If a problem exists, a trouble code will be stored that indicates a problem with the circuit that governs the particular component. Record the codes that are indicated.*

41 If necessary, perform the Cylinder Balance Test. This test must be performed by a dealer service department.

Clearing OBD-I system codes

42 To clear the codes from the PCM memory, start the KOEO self test diagnostic procedure **(see illustration 2.35)** and install the jumper wire into the Diagnostic Test connector. When the codes start to display themselves on the voltmeter or CHECK ENGINE light, remove the jumper wire from the Diagnostic Test connector. This will erase any stored codes within the system. **Caution:** *Do not disconnect the battery from the vehicle to clear the codes. This will erase stored operating parameters from the KAM (Keep Alive Memory) and cause the engine to run rough for a period of time while the computer relearns the information.*

OBD-I Trouble Codes

Code	Test Condition*	Probable Cause
11	O,C,R	Pass (separator code)
12	R	RPM not within Self-test upper limit
13	R	RPM not within Self-test lower limit
14	C	Profile Ignition Pick-up circuit fault
15	O	Read Only Memory test failed
15	C	Keep Alive Memory test failed
16	R	RPM too low to perform Oxygen Sensor/fuel test
18	C	Loss of TACH input to PCM; SPOUT circuit grounded
18	R	SPOUT circuit open
19	O	Failure in EEC reference voltage
21	O,R	Coolant Temperature Sensor out of range
22	O,C	Manifold Absolute/Baro Pressure Sensor out of range
23	O,R	Throttle Position Sensor out of range
24	O,R	Intake Air Temperature sensor out of range
26	O,R	Mass Air Flow Sensor out of range
29	C	No input from Vehicle Speed Sensor
31	O,C,R	EGR Valve position Sensor out of range (low)
32	O,C,R	EGR valve not seated; closed voltage low
33	C,R	EGR valve not opening; Insufficient flow detected
34	O,C,R	EGR Valve Pressure Transducer/position Sensor sonic voltage above closed limit
35	O,C,R	EGR Valve Pressure Transducer/position Sensor voltage out of range (high)
41	R	Heated Oxygen Sensor circuit indicates system lean, right side
41	C	No Heated Oxygen Sensor switch detected, right side
42	R	Heated Oxygen Sensor circuit indicates system rich, right side
44	R	Secondary Air system inoperative, right side
45	R	Secondary Air upstream during Self-test
45	C	Distributorless Ignition system (DIS) coil pack circuit failure
46	R	Secondary Air not by-passed during Self-test
51	O,C	Coolant Temperature sensor circuit open
53	O,C	Throttle Position sensor out of range (high)
54	O,C	Intake Air Temperature sensor circuit open
56	O,C	Mass Air Flow sensor out of range (high)
61	O,C	Coolant Temperature sensor circuit grounded
63	O,C	Throttle Position sensor circuit out of range (low)
64	O,C	Intake Air Temperature sensor circuit grounded
66	C	Mass Air Flow sensor circuit out-of-range (low)
67	O	Neutral Drive Switch circuit open
72	R	Insufficient Mass Air Flow change during Dynamic Response Test
73	R	Insufficient Throttle Position output during Dynamic Response Test
74	R	Brake On/Off switch failure
75	R	Brake On/Off circuit failure
77	R	Wide Open Throttle not sensed during Self-test
79	O	Air conditioning on during self-test
81	O	Secondary Air Injection Diverter (AIRD) circuit failure
82	O	Secondary Air Injection Bypass (AIRB) circuit failure
84	O	EGR Vacuum Regulator circuit failure
85	O	Canister Purge circuit failure
87	O,C	Primary Fuel Pump circuit failure
91	R	Heated oxygen sensor indicates system lean, left side
91	C	No heated oxygen sensor switching indicated, left side
92	R	Heated oxygen sensor indicates system rich, left side
94	R	Torque Converter Clutch (TCC) solenoid circuit failure
95	O,C	Fuel Pump circuit open, PCM to motor
96	O,C	Fuel Pump circuit open, Battery to PCM
98	R	Hard Fault present
111	O,C,R	Pass
112	O,R	Intake Air Temperature sensor circuit indicates circuit grounded/above 245 degrees F
113	O,R	Intake Air Temperature sensor circuit indicates open circuit/below -40 degrees F
114	O,R	Intake Air Temperature sensor out of self-test range
116	O,R	Coolant Temperature sensor out of self-test range
117	O,C	Coolant Temperature circuit below minimum voltage/indicates above 245 degrees F
118	O,C	Coolant Temperature sensor circuit above maximum voltage/ indicates below -40 degrees F
121	O,C,R	Throttle Position sensor out of self-test range
122	O,C	Throttle Position sensor below minimum voltage
123	O,C	Throttle Position sensor above maximum voltage
124	C	Throttle Position Sensor voltage higher than expected

OBD-I Trouble Codes (continued)

Code	Test Condition*	Probable Cause
125	C	Throttle Position Sensor voltage lower than expected
126	O,C,R	MAP/BARO sensor higher than expected
128	C	MAP sensor vacuum hose damaged or disconnected
129	R	Insufficient Manifold Absolute Pressure/Mass Air Flow change during Dynamic Response Check
136	R	Heated oxygen sensor indicates lean condition, left side
137	R	Heated oxygen sensor indicates rich condition, left side
139	C	No heated oxygen sensor switching detected, left side
144	C	No heated oxygen sensor switching detected, right side
157	R,C	Mass Air Flow Sensor below minimum voltage
158	O,C,R	Mass Air Flow Sensor above maximum voltage
159	O,R	Mass Air Flow Sensor out of self-test range
167	R	Insufficient Throttle Position Sensor change during Dynamic Response Check
171	C	Heated oxygen sensor unable to switch, right side
172	R,C	Heated oxygen sensor indicates lean condition, right side
173	R,C	Heated oxygen sensor indicates rich condition, right side
174	C	Heated oxygen sensor switching slow, right side
175	C	Heated oxygen sensor unable to switch, left side
176	C	Heated oxygen sensor indicates lean condition, left side
177	C	Heated oxygen sensor indicates rich condition, left side
178	C	Heated oxygen sensor switching slow, left side
179	C	Adaptive Fuel lean limit reached at part throttle, system rich, right side
181	C	Adaptive Fuel rich limit reached at part throttle, right side
182	C	Adaptive Fuel lean limit reached at idle, right side
183	C	Adaptive Fuel rich limit reached at idle, right side
184	C	Mass Air Flow higher than expected
185	C	Mass Air Flow lower than expected
186	C	Injector Pulse-width higher than expected
187	C	Injector Pulse-width lower than expected
188	C	Adaptive Fuel lean limit reached, left side
189	C	Adaptive Fuel rich limit reached, left side
211	C	Profile Ignition Pick-up circuit fault
212	C	Ignition module circuit failure/SPOUT circuit grounded
213	R	SPOUT circuit open
214	C	Cylinder identification (CID) circuit failure
215	C	PCM detected coil 1 primary circuit failure
216	C	PCM detected coil 2 primary circuit failure
217	C	PCM detected coil 3 primary circuit failure
218	C	Loss of ignition diagnostic monitor (IDM) signal - left side (dual plug EI)
222	C	Loss of ignition diagnostic monitor (IDM) signal - right side (dual plug)
223	C	Loss of dual plug Inhibit (DPI) control (dual plug)
224	C	PCM detected coil 1,2,3 or 4 primary circuit failure (dual plug EI)
225	C	Knock sensor not detected during dynamic response test KOER
226	O	Ignition Diagnostic Module (IDM) signal not received (EI)
232	C	PCM detected coil 1,2,3 or 4 primary circuit failure (EI)
311	R	Secondary Air System inoperative, right side
312	R	Secondary Air not by-passed
313	R	Secondary Air inoperative, left side
327	O,C,R	EGR Valve Pressure Transducer/position Sensor circuit below minimum voltage
328	O,C,R	EGR valve position sensor voltage below closed limit
332	C,R	EGR valve opening not detected
334	O,C,R	EGR valve position sensor voltage above closed limit
335	O	EGR Sensor voltage out-of-range
336	R	EGR circuit higher than expected
337	O,C,R	EGR Valve Pressure Transducer/position Sensor circuit above maximum voltage
341	O	Octane adjust service pin open
411	R	Unable to control RPM during Low RPM Self-test
412	R	Unable to control RPM during High PRM Self-test
452	C	No input from Vehicle Speed Sensor
511	O	Read Only Memory test failed - replace PCM
512	C	Keep Alive Memory test failed
513	O,	Internal voltage failure in PCM
519	O	Power steering pressure switch (PSP) circuit open (1993 and 1994 only)
521	R	Power steering pressure switch (PSP) circuit did not change states (1993 and 1994 only)
522	O	Manual Lever Position (MLP) sensor circuit open/vehicle in gear
528	O	Clutch pedal position (CPP) circuit failure
536	C,R	Brake ON/OFF (BOO) circuit failure/not activated during the KOER

OBD-I Trouble Codes (continued)

Code	Test Condition*	Probable Cause
538	R	Insufficient change in RPM/operator error in Dynamic Response Check
538	R	Invalid cylinder balance test due to throttle movement during test (1995 and later models only)
538	R	Invalid cylinder balance test due to CID circuit failure (1995 and later models)
539	O	Air conditioning on during Self-test
542	O,C	Fuel Pump circuit open; PCM to motor
543	O,C	Fuel Pump circuit open; Battery to PCM
551	O	Idle Air Control (IAC) circuit failure KOEO
552	O	Air Management 1 circuit failure
552	O	Secondary Air Injection Bypass (AIRB) circuit failure
553	O	Secondary Air Injection Diverter (AIRB) circuit failure
556	O,C	Primary Fuel Pump circuit failure
558	O	EGR Vacuum Regulator circuit failure
565	O	Canister Purge circuit failure
566	O	3-4 shift solenoid circuit failure KOEO (A4LD transmission)
569	O	Auxiliary Canister Purge (AUX-CANP) circuit failure
617	C	1-2 shift error
618	C	2-3 shift error
619	C	3-4 shift error
621	O,C	Shift Solenoid 1 (SS 1) circuit failure KOEO
622	O	Shift Solenoid 2 (SS2) circuit failure KOEO
624	O,C	Electronic Pressure Control (EPC) circuit failure
625	O,C	Electronic Pressure Control (EPC) driver open in PCM
626	O	Coast Clutch Solenoid (CCS) circuit failure KOEO
628	C	Excessive converter clutch slippage
629	O,C	Torque Converter Clutch (TCC) solenoid circuit failure
631	O	Transmission Control Indicator Lamp (TCIL) circuit failure KOEO
632	R	Transmission Control Switch (TCS) circuit did not change states during KOER
633	O	4X4L switch closed during KOEO
634	O,C,R	Manual Lever Position (MLP) sensor voltage higher or lower than expected
636	O,R	Transmission Fluid Temp (TFT) higher or lower than expected
637	O,C	Transmission Fluid Temp (TFT) sensor circuit above maximum voltage/ -40°F (-40°C) indicated / circuit open
638	O,C	Transmission Fluid Temp (TFT) sensor circuit below minimum voltage/ 290°F (143°C) indicated / circuit shorted
639	C,R	Insufficient input from Transmission Speed Sensor (TSS)
641	O	Shift Solenoid 3 (SS3) circuit failure KOEO
643	O	Coast Clutch Solenoid (CCS) circuit failure KOEO
652	O	Torque Converter Clutch (TCC) solenoid circuit failure
654	O	Transmission Range (TR) sensor not indicating PARK during KOEO
655	O	Transmission Range (TR) sensor not indicating NEUTRAL during KOEO
656	C	Torque Converter Clutch continuous slip error
657	C	Transmission fluid temperature (TFT) overheating
667	C	Transmission range (TR) circuit voltage above minimum voltage
668	C	Transmission range (TR) circuit voltage above maximum voltage
691	C	4X4 LOW switch open or short circuit present (1995 and later models)
692	C	Transmission state does not match calculated ratio (1995 and later models)
998	O	Hard fault present

* O = Key On, Engine Off; C = Continuous Memory; R = Engine Running

Obtaining OBD-II system codes

Refer to illustration 2.44

43   On OBD-II systems, the PCM will illuminate the Malfunction Indicator Light on the dash if it recognizes a component fault for two consecutive drive cycles. It will continue to set the light until the PCM does not detect any malfunction for three or more consecutive drive cycles. Because the OBD-II system requires a SCAN tool to reset the light, if the tool is not available for diagnostics, have the system checked by a dealer service department or other qualified repair facility.

44   The diagnostic codes for the EEC-V (OBD-II) systems can be extracted from the PCM using a special SCAN tool that is programmed to interface with this new system by plugging into the DLC (see illustration). If the tool is not available, have the vehicle checked at a dealer service department.

Clearing OBD-II system codes

45   To clear the codes from the PCM memory, install the OBD-II SCAN tool, scroll the menu for the function that describes "CLEARING CODES" and follow the prescribed method for that particular SCAN tool. If necessary, have the codes cleared by a dealer service department or other qualified repair facility. **Caution:** *Do not disconnect the battery from the vehicle to clear the codes. This will erase stored operating parameters from the memory and cause the engine to run rough for a period of time while the computer relearns the information.*

OBD-II Trouble Codes

Code	Probable cause
P0102	Mass Airflow (MAF) sensor circuit low input
P0103	Mass Airflow (MAF) sensor circuit high input
P0112	Intake Air Temperature (IAT) sensor circuit low input
P0113	Intake Air Temperature (IAT) sensor circuit high input
P0117	Electronic Coolant Temperature (ECT) sensor circuit low input
P0118	Electronic Coolant Temperature (ECT) sensor circuit high input
P0121	In range Throttle Position Sensor (TPS) fault
P0122	Throttle Position Sensor (TPS) circuit low input
P0123	Throttle Position Sensor (TPS) circuit high input
P0131	Upstream heated O2 sensor circuit low voltage (Bank 1)
P0133	Upstream heated O2 sensor circuit slow response (Bank 1)
P0135	Upstream heated O2 sensor heater circuit fault (Bank 1)
P0136	Downstream heated O2 sensor fault (Bank 1)
P0141	Downstream heated O2 sensor heater circuit fault (Bank 1)
P0151	Upstream heated O2 sensor circuit low voltage (Bank 2)
P0153	Upstream heated O2 sensor circuit slow response (Bank 2)
P0155	Upstream heated O2 sensor heater circuit fault (Bank 2)
P0156	Downstream heated O2 sensor fault (Bank 2)
P0161	Downstream heated O2 sensor heater circuit fault (Bank 2)
P0171	System Adaptive fuel too lean (Bank 1)
P0172	System Adaptive fuel too rich (Bank 1)
P0174	System Adaptive fuel too lean (Bank 2)
P0172	System Adaptive fuel too rich (Bank 2)
P0191	Injector Pressure sensor system performance
P0192	Injector Pressure sensor circuit low input
P0193	Injector Pressure sensor circuit high input
P0301	Cylinder no. 1 misfire detected
P0302	Cylinder no. 2 misfire detected
P0303	Cylinder no. 3 misfire detected
P0304	Cylinder no. 4 misfire detected
P0305	Cylinder no. 5 misfire detected
P0306	Cylinder no. 6 misfire detected
P0307	Cylinder no. 7 misfire detected
P0308	Cylinder no. 8 misfire detected
P0325	Knock sensor circuit fault
P0326	Knock sensor circuit performance
P0351	Ignition coil no. 1 primary circuit fault
P0352	Ignition coil no. 2 primary circuit fault
P0353	Ignition coil no. 3 primary circuit fault
P0354	Ignition coil no. 4 primary circuit fault
P0355	Ignition coil no. 5 primary circuit fault
P0356	Ignition coil no. 6 primary circuit fault
P0357	Ignition coil no. 7 primary circuit fault
P0358	Ignition coil no. 8 primary circuit fault
P0400	EGR flow fault
P0401	EGR insufficient flow detected
P0402	EGR excessive flow detected
P0420	Catalyst system efficiency below threshold (Bank 1)
P0421	Catalyst system efficiency below threshold (Bank 1)
P0430	Catalyst system efficiency below threshold (Bank 2)
P0431	Catalyst system efficiency below threshold (Bank 2)
P0442	EVAP small leak detected
P0443	EVAP VMV circuit fault
P0452	EVAP fuel tank pressure sensor low input
P0453	EVAP fuel tank pressure sensor high input
P0500	VSS fault
P0505	IAC valve system fault
P0603	PCM Keep Alive Memory test error
P0605	PCM Read Only Memory test error

Component replacement

Note: *Because of the Federally-mandated extended warranty which covers the ECA, the information sensors and the devices it controls, there's no point in replacing any of the following components yourself unless the* warranty has expired. However, once the warranty has expired, you may wish to perform some of the following component replacement procedures yourself after having the problem diagnosed by a dealer service department or repair shop.

Air Charge Temperature (ACT) sensor

46   Detach the cable from the negative terminal of the battery.

47   Locate the ACT sensor in the intake manifold **(see illustration)**.

2.47 The Air Charge Temperature (ACT) sensor is located either on the air cleaner housing as shown (2.8L V6), the rear of the intake manifold (2.3L four) or the throttle body (3.0L V6)

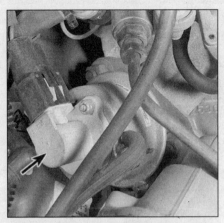

2.53 The EGR valve position (EVP) sensor is located on top of the EGR valve - to replace it, simply unplug the connector and remove the mounting bolts

2.58a To replace the engine coolant temperature (ECT) sensor (center of the intake manifold on 2.3L four, adjacent to water outlet on front end of 3.0L V6), unplug the electrical connector and unscrew the sensor (be sure to coat the threads of the new sensor with Teflon tape to prevent coolant leakage)

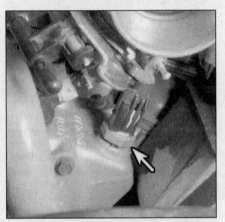

2.58b The coolant temperature sensor (arrow) fro the 4.0L V6 engine is located on the front of the intake manifold, beneath the air intake throttle body

2.64 The Manifold Absolute Pressure (MAP) sensor is located on the firewall (2.8L V6 shown) - to replace it, simply label, then detach, the vacuum hoses and remove the sensor bracket mounting screws

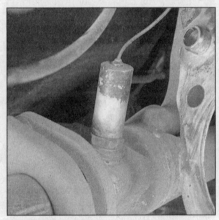

2.71 The exhaust gas oxygen (EGO) sensor is screwed into the exhaust manifold - to replace it, raise the vehicle, unplug the connector at the pigtail lead and unscrew the sensor (be sure to coat the threads of the new sensor with anti-seize compound to prevent it from welding itself to the manifold)

48   Unplug the electrical connector from the sensor.
49   Remove the sensor with a wrench.
50   Wrap the threads of the new sensor with Teflon tape to prevent air leaks.
51   Installation is the reverse of removal.

EGR Valve Position (EVP) sensor

52   Detach the cable from the negative terminal of the battery.
53   Locate the EVP sensor on the EGR valve (see illustration).
54   Unplug the electrical connector from the sensor.
55   Remove the three mounting bolts and detach the sensor.
56   Installation is the reverse of removal.

Engine Coolant Temperature (ECT) sensor

57   Detach the cable from the negative terminal of the battery.
58   Locate the ECT sensor on the thermostat housing (see illustration).

59   Unplug the electrical connector from the sensor.
60   Remove the sensor with a wrench.
61   Wrap the threads of the new sensor with Teflon tape to prevent coolant leakage.
62   Installation is the reverse of removal.

Manifold Absolute Pressure (MAP) sensor

63   Detach the cable from the negative terminal of the battery.
64   Locate the MAP sensor on the firewall (see illustration).
65   Unplug the electrical connector from the sensor.
66   Detach the vacuum line from the sensor.
67   Remove the two mounting bolts and detach the sensor.
68   Installation is the reverse of removal.

Exhaust Gas Oxygen (EGO) sensor

69   Detach the cable from the negative terminal of the battery.
70   Raise the vehicle and support it securely

on jackstands.
71   Locate the EGO sensor on the exhaust manifold (see illustration).
72   Unplug the electrical connector from the sensor.
73   Remove the sensor with a wrench.
74   Coat the threads of the new sensor with anti-seize compound to prevent the threads from welding themselves to the manifold.
75   Installation is the reverse of removal.

Throttle Position Sensor (TPS) switch

76   Don't attempt to replace the TPS switch! Specialized calibration equipment is necessary to adjust the switch once it's installed, making replacement beyond the scope of the home mechanic.

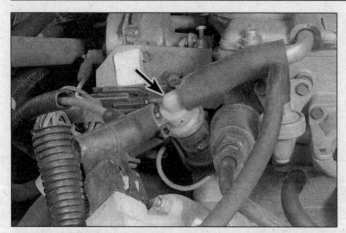

2.78 To replace the canister purge solenoid valve (arrow), unplug the electrical connector and detach the vacuum lines from the T-fitting

3.13 To remove the EGR valve, unplug the connector to the valve position sensor, disconnect the threaded fitting between the EGR pipe and the valve and remove the valve mounting bolts

Canister Purge Solenoid

77   Detach the cable from the negative terminal of the battery.

78   Locate the canister purge solenoid on the left side of the engine compartment, next to the left wheel well **(see illustration)**.

79   Unplug the electrical connector from the solenoid.

80   Label the vacuum hoses and ports, then detach the hoses.

81   Remove the solenoid.

82   Installation is the reverse of removal.

Vacuum control solenoid

83   Detach the cable from the negative terminal of the battery.

84   Locate the vacuum control solenoid on the left strut tower.

85   Unplug the electrical connector from the solenoid.

86   Label the vacuum hoses and ports, then detach the hoses.

87   Remove the solenoid/bracket screws and detach the solenoid.

88   Installation is the reverse of removal.

Mass Air Flow (MAF) sensor

Note: *The MAF sensor is located between the air cleaner housing and the air intake duct. Refer to Chapter 4 for illustrations.*

89   Disconnect the air intake duct from the MAF sensor.

90   Unplug the MAF sensor electrical connector.

91   Remove four screws securing the MAF sensor to the air cleaner housing. Take off the sensor and gasket.

92   Install the sensor, using a new gasket, and tighten the screws securely. The remainder of installation is the reverse of removal.

---

3   Exhaust Gas Recirculation (EGR) system

Refer to illustration 3.13

General description

1   The EGR system is designed to reintro-duce small amounts of exhaust gas into the combustion cycle, thus reducing the generation of nitrous oxide emissions. The amount of exhaust gas reintroduced and the timing of the cycle is controlled by various factors such as engine speed, altitude, manifold vacuum, exhaust system backpressure, coolant temperature and throttle angle.

2   All EGR valves are vacuum actuated and the vacuum diagram for your particular vehicle is shown on the Emissions Control Information label in the engine compartment.

3   The electronic EGR valve used in EEC-IV systems controls EGR flow with an EGR valve position (EVP) sensor attached to the top of the valve. The valve is operated by a vacuum signal from the dual EGR solenoid valves or the electronic vacuum regulator which actuates the valve diaphragm. As supply vacuum overcomes the spring load, the diaphragm is actuated, lifting the pintle off the seat and allowing exhaust gas to recirculate. The amount of flow is proportional to the pintle position. The EVP sensor sends an electrical signal indicating its position to the ECA.

Checking

Note: *Aside from the following check and maintenance steps, the electronic EGR valve cannot be diagnosed or serviced by the home mechanic. Additional checks must be done by a dealer service department.*

4   Make sure the vacuum hoses are in good condition and hooked up correctly.

5   Clean the inlet and outlet ports with a wire brush or scraper. Do not sandblast the valve or clean it with gasoline or solvents as they will damage the valve.

6   To perform a leakage test, connect a vacuum pump to the EGR valve.

7   Apply 5-to-6 in-Hg of vacuum to the valve.

8   Trap the vacuum - it should not drop more than 1 in-Hg in 30 seconds.

9   If the specified conditions are not met, the EGR valve, O-ring or EVP must be replaced.

Component replacement

10   Detach the cable from the negative terminal of the battery.

11   Unplug the electrical connector from the EGR valve position sensor.

12   Unscrew the threaded fitting that attaches the EGR pipe to the EGR valve.

13   Remove the two mounting bolts **(see illustration)** and detach the valve.

14   Remove the old gasket.

15   If you're replacing the EGR valve but not the position sensor, remove the sensor from the old valve and install it on the new valve.

16   Installation is the reverse of removal.

---

4   Managed air thermactor system (2.8L V6)

Refer to illustrations 4.47 and 4.48

General description

1   The thermactor (air injection) exhaust emission control system reduces carbon monoxide and hydrocarbon content in the exhaust gases by injecting fresh air into the hot exhaust gases leaving the exhaust ports. When fresh air is mixed with hot exhaust gases, oxidation is increased, reducing the concentration of hydrocarbons and carbon monoxide and converting them into harmless carbon dioxide and water.

2   Vehicles with a 2.8L V6 utilize a "managed air" thermactor system, which diverts thermactor air either upstream to the exhaust manifold check valve or downstream to the rear section check valve and dual bed catalyst. An extended idle air bypass system in carburetor equipped vehicles also vents thermactor air to the atmosphere during extended idling.

3   An air control valve is used to direct the air upstream or downstream. An air bypass valve is used to dump air to the atmosphere. On the 2.8L V6, the two valves are combined into a single air bypass/control valve.

Checking

Extended idle air bypass system

4 The normally closed Idle Tracking Switch (ITS) opens when the throttle returns to idle, signaling the EEC-IV module to de-energize the normally closed solenoid. When this occurs, vacuum is removed from the normally closed bypass valve and causes the bypass valve to dump secondary thermactor air to the atmosphere.

5 With the engine warmed up and the transmission in Neutral, momentarily increase engine speed, then allow it to return to idle. If, after 2-1/2 minutes, the thermactor bypass valve dumps secondary air through the vents, the system is okay. If it doesn't, check the routing and condition of the hoses. If the hoses are okay, check the bypass valve function. If the bypass valve is okay, check battery voltage to the ITS and continuity through the (normally closed) ITS while manually cycling the switch. If the ITS is okay, verify that the solenoid functions properly (that it actually opens and closes). If the solenoid is okay, check the vacuum signal to the solenoid. If the vacuum signal is okay, the problem is with the ECA. Further checking of the system must be performed by a dealer service department.

Air supply pump

6 Check and adjust the drivebelt tension (refer to Chapter 1).

7 Disconnect the air supply hose at the air bypass valve inlet.

8 The pump is operating satisfactorily if air flow is felt at the pump outlet with the engine running at idle, increasing as the engine speed is increased.

9 If the air pump doesn't pass the above tests, replace it with a new or rebuilt unit.

Air bypass valve

10 With the engine running at idle, disconnect the hose from the valve outlet.

11 Remove the vacuum hose from the port and remove or bypass any restrictors or delay valves in the vacuum hose.

12 Verify that vacuum is present in the vacuum hose by putting your finger over the end.

13 Reconnect the vacuum hose to the port.

14 With the engine running at 1500 rpm, the air pump supply air should be felt or heard at the air bypass valve outlet.

15 With the engine running at 1500 rpm, disconnect the vacuum hose. Air at the valve outlet should be decreased or shut off and air pump supply air should be felt or heard at the silencer ports.

16 Reconnect all hoses.

17 If the normally closed air bypass valve doesn't successfully pass the above tests, check the air pump (refer to Steps 4 and 5).

18 If the air pump is operating satisfactorily, replace the air bypass valve with a new one.

Air supply control valve

19 With the engine running at 1500 rpm, disconnect the hose at the air supply control valve inlet and verify that air is flowing through the hose.

20 Reconnect the hose to the valve inlet.

21 Disconnect the hoses at the vacuum port and at outlets A and B.

22 With the engine running at 1500 rpm, air flow should be felt at outlet B with little or no air flow at outlet A.

23 With the engine running at 1500 rpm, connect a line from any manifold vacuum fitting to the vacuum port.

24 Air flow should be present at outlet A with little or no air flow at outlet B.

25 Reconnect all hoses.

26 If all conditions above are not met, replace the air control valve with a new one.

Combination air bypass/air control valve

27 Disconnect the hoses from outlets A and B.

28 Disconnect the vacuum hose at port D and plug the hose.

29 With the engine running at 1500 rpm, verify that air flows from the bypass vents.

30 Unplug and reconnect the vacuum hose at port D, then disconnect and plug the hose attached to port S.

31 Verify that vacuum is present in the hose to port D by momentarily disconnecting it.

32 Reconnect the vacuum hose to port D.

33 With the engine running at 1500 rpm, verify that air is flowing out of outlet B with no air flow present at outlet A.

34 Attach a length of hose to port S.

35 With the engine running at 1500 rpm, apply vacuum to the hose and verify that air is flowing out of outlet A.

36 Reconnect all hoses. Be sure to unplug the hose to Port S before reconnecting it.

37 If all conditions above are not met, replace the combination valve with a new one.

Check valve

38 Disconnect the hoses from both ends of the check valve.

39 Blow through both ends of the check valve, verifying that air flows in one direction only.

40 If air flows in both directions or not at all, replace the check valve with a new one.

41 When reconnecting the valve, make sure it is installed in the proper direction.

Thermactor system noise test

42 The thermactor system is not completely noiseless. Under normal conditions, noise rises in pitch as the engine speed increases. To determine if noise is the fault of the air injection system, detach the drivebelt (after verifying that the belt tension is correct) and operate the engine. If the noise disappears, proceed with the following diagnosis.
Caution: *The pump must accumulate 500 miles (vehicle miles) before the following check is valid.*

43 If the belt noise is excessive:

a) *Check for a loose belt and tighten as necessary (refer to Chapter 1).*
b) *Check for a seized pump and replace it if necessary.*
c) *Check for a loose pulley. Tighten the mounting bolts as required.*
d) *Check for loose, broken or missing mounting brackets or bolts. Tighten or replace as necessary.*

44 If there is excessive mechanical noise:

a) *Check for an overtightened mounting bolt.*
b) *Check for an overtightened drivebelt (refer to Chapter 1).*
c) *Check for excessive flash on the air pump adjusting arm boss and remove as necessary.*
d) *Check for a distorted adjusting arm and, if necessary, replace the arm.*

45 If there is excessive thermactor system noise (whirring or hissing sounds):

a) *Check for a leak in the hoses (use a soap and water solution to find the leaks) and replace the hose(s) as necessary.*
b) *Check for a loose, pinched or kinked hose and reassemble, straighten or replace the hose and/or clamps as required.*
c) *Check for a hose touching other engine parts and adjust or reroute the hose to prevent further contact.*
d) *Check for an inoperative bypass valve (refer to Step 10) and replace if necessary.*
e) *Check for an inoperative check valve (refer to Step 38) and replace if necessary.*
f) *Check for loose pump or pulley mounting fasteners and tighten as necessary.*
g) *Check for a restricted or bent pump outlet fitting. Inspect the fitting and remove any casting flash blocking the air passageway. Replace bent fittings.*
h) *Check for air dumping through the bypass valve (only at idle). On many vehicles, the thermactor system has been designed to dump air at idle to prevent overheating the catalytic converter. This condition is normal. Determine that the noise persists at higher speeds before proceeding.*
i) *Check for air dumping through the bypass valve (the decel and idle dump). On many vehicles, the thermactor air is dumped into the air cleaner or the remote silencer. Make sure that the hoses are connected properly and not cracked.*

46 If there is excessive pump noise, make sure the pump has had sufficient break-in time (at least 500 miles). Check for a worn or damaged pump and replace as necessary.

Component replacement

47 To replace the air bypass valve, air supply control valve, check valve, combination air bypass/air control valve or the silencer, label and disconnect the hoses leading to them, **(see illustration)**, replace the faulty

4.47 When replacing a combination bypass/control valve assembly, be sure to label the hoses correctly to ensure that they are properly reinstalled

1 *Vacuum line to control solenoid*
2 *Combination air bypass and air control valve hose*
3 *Vacuum line to control solenoid*
4 *Hose and check valve for catalytic converter (check valve not visible in photo)*
5 *Hose and check valve for exhaust manifolds*

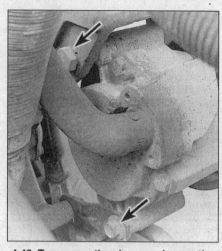

4.48 To remove the air pump, loosen the hose clamp and detach the hose, remove the lower bolt (arrow) and remove the belt, then remove the upper bolt (arrow) and remove the pump

component and reattach the hoses to the proper ports. Make sure the hoses are in good condition. If not, replace them with new ones.

48 To replace the air supply pump, first loosen the appropriate engine drivebelts (refer to Chapter 1), then remove the faulty pump from the mounting bracket **(see illustration)**. Label all wires and hoses as they're removed to facilitate installation of the new unit.

49 After the new pump is installed, adjust the drivebelts to the specified tension (refer to Chapter 1).

5   Fuel evaporative emissions control system

Refer to illustrations 5.4 and 5.16

General description

1 This system is designed to prevent hydrocarbons from being released into the atmosphere by trapping and storing fuel vapor from the fuel tank, the carburetor or the fuel injection system.

2 The serviceable parts of the system include a charcoal filled canister and the connecting lines between the fuel tank, fuel tank filler cap and the carburetor or fuel injection system.

3 Vapor trapped in the gas tank is vented through a valve in the top of the tank. The vapor leaves the valve through a single line and is routed to a charcoal canister located between the left front wheel well and the front bumper, where it's stored until the next time the engine is started.

4 The canister outlet is connected to an electrically actuated canister purge solenoid **(see illustration)** that is, in turn, connected to the air cleaner housing. The canister purge solenoid valve is normally closed. When the engine is started, the solenoid is energized by a signal from the ECA and allows intake vacuum to open the line between the canister and the air cleaner housing, which draws vapor stored in the canister through the air cleaner and into the engine where it's burned.

5 The thermal vent valve is a temperature actuated off/on valve in the carburetor-to-canister vent line and is closed when the engine compartment is cold. This prevents fuel tank vapor (generated when the engine heats up before the engine compartment does) from being vented through the carburetor float bowl and forces it instead into the charcoal canister. This effect can occur, for example, when sunlight strikes a vehicle that has been sitting out all night and begins to warm the fuel tank. With the thermal vent valve closed, the vapor cannot enter the carburetor float bowl vent valve, but is routed instead to the charcoal canister. As the engine compartment warms up during normal engine operation, the thermal vent valve opens. When the engine is again turned off, the thermal vent valve (now open because underhood temperature is above 120° F) allows fuel vapor generated in the carburetor float bowl to pass through the valve and be stored in the charcoal canister. As the thermal vent valve cools, it closes and the cycle begins again.

Checking
Charcoal canister

6 There are no moving parts and nothing to wear in the canister. Check for loose, missing, cracked or broken fittings and inspect the canister for cracks and other damage. If the canister is damaged, replace it (refer to Step 15).

Canister purge valve

7 Clearly label all vacuum hoses and ports, then detach the hoses from the valve.
8 Remove the valve.
9 Apply vacuum to port B **(see illustration 5.4)**. The valve should be closed (no air flows through it). If air does flow, the valve is open. Replace it with a new one.
10 After applying and maintaining 16 in-Hg vacuum to port A, apply vacuum to port B again. Air should pass through (the valve should open). If no air flows, the valve is closed. Replace it. **Caution:** *Never apply vacuum to port C. Doing so may dislodge the internal diaphragm and the valve will be permanently damaged.*

Carburetor fuel bowl solenoid vent valve

11 Remove the valve.
12 Apply 9-to-14 volts to the valve electrical connector terminals with jumper wires. The valve should close, preventing air from passing through. If the valve doesn't close, replace it.

Carburetor fuel bowl thermal vent valve

13 Remove the valve.
14 The vent should be fully closed at 90° F and below and open at 120° F and above. If it isn't, replace it.

Component replacement
Charcoal canister

15 Locate the canister on the right side of the engine compartment.

5.4 A typical canister purge solenoid (A) and carburetor fuel bowl thermal vent valve (B) on a 2.8L V6 - consult your VECI label to determine the approximate location of these components on your vehicle

5.16 To remove the charcoal canister, clearly label then detach the vacuum hoses and remove the mounting bolt (arrow)

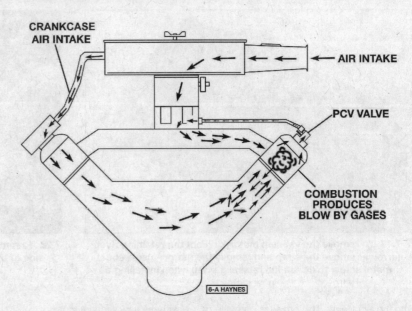

6.1 Typical Positive Crankcase Ventilation (PCV) system

16   Remove the single mounting bolt **(see illustration)** above the canister.
17   Lower the canister, detach the hose from the purge valve, or purge solenoid valve, and remove the canister.
18   Installation is the reverse of removal.

All other components

19   Label the hoses and fittings, then detach the hoses and remove the component.
20   Installation is the reverse of removal.

6   Positive Crankcase Ventilation (PCV) system

Refer to illustrations 6.1 and 6.2

General description

1   The Positive Crankcase Ventilation (PCV) system **(see illustration)** cycles crankcase vapors back through the engine where they are burned. The PCV valve regulates the amount of ventilating air and blowby gas to the intake manifold and prevents backfire from traveling into the crankcase.
2   The PCV system consists of a replaceable PCV valve **(see illustration)**, a crankcase ventilation filter (integral with the oil filler cap on some vehicles, separate on others) and the connecting hoses.
3   The air source for the crankcase ventilation system is in the air cleaner. Air passes through the PCV filter (in the rocker arm cover or the oil filler cap) and through a hose connected to the air cleaner housing. On vehicles with a PCV filter integrated into the oil filler cap, the cap is sealed at the opening to prevent the entrance of outside air. From the oil filler cap, or separate PCV filter in the rocker arm cover, the air flows into the rocker arm chamber and the crankcase, from which it circulates up into another section of the rocker arm chamber and finally enters a spring loaded regulator valve (PCV valve) that controls the amount of flow as operating conditions vary. The vapors are routed to the intake manifold through the crankcase vent hose tube and fittings. This process goes on continuously while the engine is running.

Checking

4   Checking procedures for the PCV system components are included in Chapter 1.

Component replacement

5   Component replacement involves simply installing a new valve or hose in place of the one removed during the checking procedure.

7   Inlet air temperature control system (2.8L V6)

Refer to illustrations 7.17, 7.22 and 7.23

General description

1   The inlet air temperature control system provides heated intake air during warm-up, then maintains the inlet air temperature within a 70° F to 105° F operating range by mixing warm and cool air. This allows leaner fuel/air mixture settings for the carburetor, which reduces emissions and improves driveability.
2   Two fresh air inlets - one warm and one cold - are used. The balance between the two is controlled by intake manifold vacuum. A vacuum motor, which operates a heat duct valve in the air cleaner, is controlled by an Inlet Air Solenoid (IAS) valve, located behind the vacuum reservoir on the left side of the engine compartment, which is under the control of the ECA. The ECA, in turn, monitors the temperature of the intake air with the Air Charge Temperature (ACT) sensor.
3   When the underhood temperature is cold, warm air radiating off the exhaust manifold is routed by a shroud which fits over the manifold up through a hot air inlet tube and

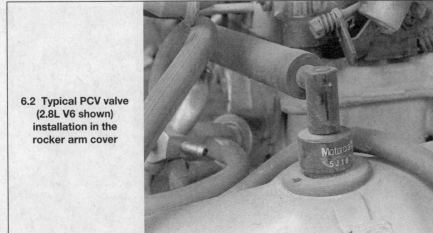

6.2 Typical PCV valve (2.8L V6 shown) installation in the rocker arm cover

7.17 To remove the vacuum motor, drill out the retaining rivet (arrow), remove the strap and remove the motor - use a sheet metal screw to attach the retaining strap when installing a new motor

7.22 To remove the vacuum reservoir, which is located on the left side of the engine compartment, remove these two screws (arrows) and lift the reservoir out of the way

into the air cleaner. This provides warm air for the carburetor, resulting in better driveability and faster warm-up. As the underhood temperature rises, a heat duct valve is gradually closed by the vacuum motor and the air cleaner draws air through a cold air duct instead. The result is a consistent intake air temperature.

Checking

Note: *Make sure that the engine is cold before beginning this test.*

4   Always check the vacuum source, the IAS valve and the integrity of all vacuum hoses between the source, the IAS valve and the vacuum motor before beginning the following test. Do not proceed until they're okay.
5   Apply the parking brake and block the wheels.
6   Detach, but do not remove, the air cleaner housing and element (see Chapter 4).
7   Turn the air cleaner housing upside down so the vacuum motor door is visible. The door should be open. If it isn't, it may be binding or sticking. Make sure that it's not rusted in an open or closed position by attempting to move it by hand. If it's rusted, it can usually be freed by cleaning and oiling the hinge. If it fails to work properly after servicing, replace it.
8   If the vacuum motor door is okay but the motor still fails to operate correctly, check carefully for a leak in the hose leading to it. Check the vacuum source to and from the bimetal sensor and the time delay valve as well. If no leak is found, replace the vacuum motor (refer to Step 16).
9   Start the engine. If the duct door has moved or moves to the "heat on" (closed to fresh air) position, go to Step 13.
10   If the door stays in the "heat off" (closed to warm air) position, place a finger over the bimetal sensor bleed. The duct door must move rapidly to the "heat on" position. If the door doesn't move to the "heat on" position, stop the engine and replace the vacuum motor (refer to Step 16). Repeat this Step

with the new vacuum motor.
11   With the engine off, cool the bimetal sensor and the Cold Weather Modulator (CWM) by spraying them with cold water (wait for 20 seconds after the liquid contacts the sensor and CWM).
12   Restart the engine. The duct door should move to the "heat on" position. If the door doesn't move or moves only partially, replace the TVS.
13   Start and run the engine briefly (less than 15 seconds). The duct door should move to the "heat on" position.
14   Shut off the engine and watch the duct door. It should stay in the "heat on" position for at least two minutes.
15   If it doesn't stay in the "heat on" position for at least two minutes, replace the IAS (refer to Step 23) and repeat this Step after cooling the CWM and bimetal again.

Component replacement

Vacuum motor

16   Detach the air cleaner housing assembly (see Chapter 4).
17   Locate the vacuum motor **(see illustration)**.

7.23 To remove the Idle Air Solenoid (IAS) valve, detach the vacuum hose and unclip the valve

18   Detach the vacuum hose and drill out the retaining strap rivet.
19   Remove the motor.
20   Installation is the reverse of removal. Use a sheet metal screw of the appropriate diameter to reattach the strap.

Inlet air solenoid (IAS) valve

21   Detach the cable from the negative terminal of the battery.
22   Remove the two mounting screws from the vacuum reservoir **(see illustration)** and remove the reservoir.
23   Detach the vacuum hose from the IAS valve **(see illustration)**.
24   Unclip the IAS solenoid valve and remove it.
25   Installation is the reverse of removal.

---

8   Catalytic converter

General description

1   The catalytic converter is designed to reduce hydrocarbon, carbon monoxide and nitrogen oxide pollutants in the exhaust. The converter "oxidizes" these components

(speeds up the heat producing chemical reaction between the exhaust gas constituents) and converts them to water and carbon dioxide.

2   The converter, which closely resembles a muffler, is located in the exhaust system immediately behind the short elbow shaped section of pipe below the exhaust manifold (you'll need to raise the vehicle to inspect or replace it).

3   **Warning:** *If large amounts of unburned gasoline enter the converter, it may overheat and cause a fire. Always observe the following precautions:*

Use only unleaded gasoline
Avoid prolonged idling
Do not run the engine with a nearly empty fuel tank
Avoid coasting with the ignition turned off

Checking

Note: *An infrared sensor is required to check the actual operation of the catalytic converter.*

Such a device is prohibitively expensive. Take the vehicle to a dealer service department or a service station for this procedure. However, there are a few things you should check whenever the vehicle is raised for any reason.

4   Check the converter for dents (maximum 3/4-inch deep) and other damage which could affect its performance.

5   Inspect the heat insulator plates above and below the catalytic converter for damage.

Component replacement

Warning: *Don't attempt to remove the catalytic converter until the complete exhaust system is cool.*

Note: *The catalytic converter cannot be unbolted from the exhaust system. It must be cut off with a torch. But a replacement converter can be bolted onto the existing system. See Chapter 4 for exploded drawings of the exhaust system.*

6   Raise the vehicle and support it securely on jackstands. Apply penetrating oil to the clamp bolts and allow it to soak in.

7   Release the hose clamp and detach the hose from the thermactor pipe inlet.

8   Remove the flange bolts from the flange between the elbow and exhaust pipe at the front and rear ends of the center section of the exhaust system. Remove the old gaskets if they are stuck to the pipes.

9   Remove the catalytic converter and center section of the exhaust as an assembly.

10   Installation is the reverse of removal. Be sure to use new exhaust pipe gaskets at the flanges at both ends.

11   It's always a good idea to inspect and, if necessary, replace the rubber exhaust pipe hangers while the vehicle is raised (refer to Chapter 4).

12   Start the engine and check carefully for exhaust leaks.

Notes

Chapter 7 Part A
Manual transmission

Contents

Specifications

Torque specifications

Ft-lbs (unless otherwise indicated)

Shift lever bolt	72 to 108 in-lbs
Driveshaft U-bolt	96 to 180 in-lbs
Clutch housing-to-engine bolt	28 to 38
Transmission insulator bolt	60 to 80
Transmission insulator stud nut	71 to 94
Transmission-to-bellhousing bolt	30 to 40
Starter-to-clutch housing bolt	15 to 20
Shift lever clamping bolt (1990 on)	46 to 68

1 General information

This portion of Chapter 7 deals with the manual transmission. Part B covers the automatic transmission. The manual transmission installed in these vehicles is a 5-speed unit manufactured by Mazda.

Due to the complexity of the transmissions and the special tools and expertise required to perform an overhaul, it is not advised that this job be undertaken by the home mechanic. Therefore, the procedures in this Chapter are limited to routine adjustments and removal and installation procedures.

Depending on the expense involved in having a faulty transmission overhauled, it may be an advantage to consider replacing the unit with either a new or rebuilt one. Your local dealer or transmission shop should be able to supply you with information concerning cost, availability and exchange policy. Regardless of how you decide to remedy a transmission problem, however, you can still save considerable expense by removing and installing the unit yourself.

2   Transmission mount - check and replacement

Refer to illustration 2.2

1   Insert a large screwdriver or pry bar into the space between the transmission extension housing and the crossmember and pry up.

2   The transmission should not spread excessively away from the insulator (**see illustration**).

3   To replace, remove the two nuts attaching the insulator to the crossmember and the two bolts attaching the insulator to the transmission.

4   Raise the transmission with a jack and remove the insulator, noting which holes are used in the crossmember for proper alignment during installation.

5   Installation is the reverse of the removal procedure.

3   Manual transmission - removal and installation

Removal

1   Disconnect the negative cable at the battery. Place the cable out of the way so it cannot accidentally come in contact with the negative terminal of the battery, as this would once again allow power into the electrical system of the vehicle.

2   Place the gearshift selector in Neutral.

3   Remove the four shift boot-to-floor bolts and lift the boot assembly up.

4   Remove the shift lever assembly (Section 6).

5   Raise the vehicle and support it securely on jackstands.

6   Remove the starter (Chapter 5).

7   Disconnect the hydraulic line from the clutch slave cylinder by removing the clip and releasing the fitting.

8   Disconnect the speedometer and electrical connections from the transmission.

9   Remove the driveshaft (Chapter 8).

2.2 Pry on the transmission mount with a bar or large screwdriver to check for excessive looseness

10   On 1992 and later models, remove the front portion of the exhaust system (Chapter 4).

11   Remove the transmission insulator-to-crossmember nuts and loosen the front insulator-to-crossmember bracket nuts and bolts.

12   Support the transmission with a jack or dolly (use chains or nylon straps to secure it to the jack).

13   Support the engine with another jack or a jackstand, using a block of wood to protect the oil pan.

14   Remove the crossmember.

15   Remove the clutch housing-to-engine bolts and carefully move the transmission to the rear until the input shaft is clear of the clutch components and the transmission can be lowered from the vehicle.

Installation

16   If the clutch slave cylinder release bearing was removed, install it over the transmission input shaft with the tabs aligned with the slots in the clutch housing. Install the bolts and tighten them to the specified torque.

17   Raise the transmission into position and move it forward, making sure the input shaft engages in the pilot bearing in the flywheel and the clutch housing fits over the pilot dowel pins on the engine.

18   Install the clutch housing-to-engine bolts. Tighten the bolts to the specified torque.

19   Install the crossmember with the bolts finger-tight.

20   Lower the transmission until the insulator studs protrude through the crossmember and install the nuts. Tighten the insulator and crossmember nuts and bolts to the specified torque and remove the jack.

21   On 1992 and later models, install the front portion of the exhaust system (Chapter 4).

22   Install the driveshaft (Chapter 8).

23   Connect the speedometer and the electrical connections to the transmission. Connect the clutch hydraulic line.

24   Install the starter (Chapter 5).

25   Install the shift lever assembly.

26   Push the shift boot down into place and install the bolts.

27   Connect the negative battery cable.

28   Bleed the clutch hydraulic system as described in Chapter 8.

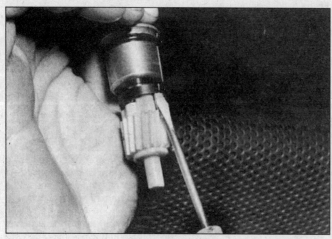

4.3 Pry the retaining clip off the pinion gear so the gear can be removed from the cable

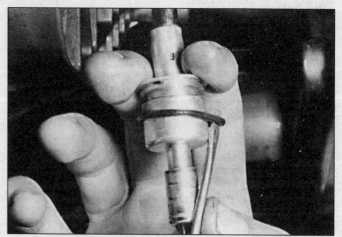

4.4 Use a small screwdriver to pry the O-ring out of its groove

4 Speedometer pinion gear and seal - removal and installation

Refer to illustrations 4.3 and 4.4

1 Remove the bolt on the speedometer cable retaining bracket and remove the bracket and bolt as an assembly.

2 Pull the pinion gear assembly straight out of the transmission extension housing.

3 Use a small screwdriver to remove the retaining clip from the pinion gear and slide the gear off the cable **(see illustration)**.

4 To replace the O-ring, use a screwdriver to remove the O-ring from the retaining groove **(see illustration)**.

5 Lubricate the new O-ring and slide it onto the pinion shaft until it seats in the retaining groove.

6 Install the new gear and secure with the retaining clip.

7 Install the pinion gear assembly and secure it with the retaining bracket and bolt.

5 Extension housing oil seal - replacement

1 Raise the vehicle and support it securely on jackstands.

2 Remove the driveshaft (Chapter 8).

Scribe marks on the driveshaft and yoke and the rear axle companion flange to assure proper positioning of the driveshaft during assembly.

3 Remove the oil seal from the end of the extension housing with a seal removing tool. If there is access, a thin blade screwdriver or chisel may also be used to remove the seal and the extension housing bushing, which is behind the seal.

4 Before installing the new seal inspect the sealing surface of the universal joint yoke for scoring. If scoring is found, replace the yoke.

5 Inspect the counterbore of the housing for burrs. Remove burrs with emery cloth or medium grit wet-and-dry sandpaper.

6 Install the new extension housing bushing and oil seal after coating the inside diameter of the seal with silicone sealant and the end of the rubber boot portion with grease.

6 Transmission shift lever assembly - removal and installation

1986 through 1989 models

1 Turn the shift knob ball counterclockwise to remove it from the lever.

2 Remove the shift boot bolts and pull the boot up the shift lever.

3 Remove the four shift lever assembly-to-transmission shift rail adapter bolts and lift the assembly from the transmission.

4 Prior to installation, lubricate the shift lever contact surfaces with multi-purpose grease.

5 Lower the shift assembly into the transmission. Coat the threads of the retaining bolts with thread sealant and install them. Tighten the bolts to the specified torque.

6 Install the shift boot and knob ball.

1990 and later models

7 Place the transmission in Neutral.

8 Remove the shift boot bolts and slide the boot up on the shift lever.

9 Remove the shift lever clamping bolt and remove the shift lever from the transmission.

10 Install the shift lever onto the transmission and tighten the bolt to the torque listed in this Chapter's Specifications.

11 Slide the rubber boot down and secure it with the bolts.

Notes

Chapter 7 Part B
Automatic transmission

Contents

Specifications

General

Transmission type
Through 1995	A4LD
1996	
3.0L V6 engine	4R44E
4.0L V6 engine	4R55E
1997	
3.0L V6 engine	4R44E
4.0L V6 engine	5R55E
Fluid type	See Chapter 1

Torque specifications

	Ft-lbs
Transmission band locknut	35 to 45
Torque converter-to-driveplate nut	30
Transmission-to-engine bolts and stud nuts	
All except 1991 and later 3.0L V6	28 to 38
1991 and later 3.0L V6	33 to 44
Crossmember through-bolt	37 to 52
Insulator nut	37 to 52
Starter bolt	15 to 20

1   General information

Due to the complexity of the automatic transmission covered in this manual and to the specialized equipment necessary to perform most service operations, this Chapter addresses only those procedures concerning routine maintenance and adjustment. Except for removal and installation of the transmission assembly, major repair operations should be undertaken by a professional mechanic with the proper facilities.

All automatic transmissions used are 4-speed types equipped with overdrive and computer-operated controls. These controls operate a piston/plate clutch in the torque converter that eliminates converter slip when applied.

The manufacturer specifies a different grade transmission fluid than other manufacturers, and this must be used when refilling or adding fluid. The fluid specification for your vehicle can be found embossed on the transmission fluid dipstick or on the certification label on the left front door post. Additional information concerning automatic transmission fluid level checking can be found in Chapter 1.

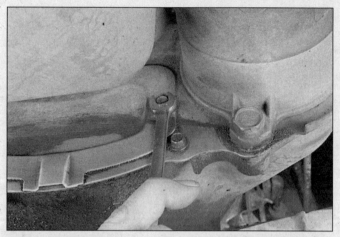

3.4 Remove the two torque converter cover bolts

3.5 Mark the torque converter stud and driveplate with white paint so they can be reinstalled in the same position

2   General diagnosis

Automatic transmission malfunctions may be caused by four general conditions: poor engine performance, improper adjustments, hydraulic malfunctions and mechanical malfunctions. Diagnosis of these problems should always begin with the easily checked items such as fluid level and condition, band adjustment and shift linkage adjustment. Perform a road test to determine if the problem has been corrected or if more diagnosis is necessary. If the problem persists after the preliminary tests and corrections are completed, additional diagnosis should be undertaken by a dealer or repair shop.

3   Automatic transmission - removal and installation

Refer to illustrations 3.4, 3.5, 3.6, 3.9, 3.19 and 3.21

Removal

1   Disconnect the negative cable at the battery. Place the cable out of the way so it cannot accidentally come in contact with the negative terminal of the battery, as this would once again allow power into the electrical system of the vehicle.
2   Raise the vehicle and support it securely on jackstands.
3   Drain the transmission fluid (Chapter 1).
4   Remove the torque converter cover **(see illustration)**.
5   Mark one of the torque converter studs and the driveplate with white paint so they can be installed in the same position **(see illustration)**.
6   Use a suitable large socket and bar on the crankshaft bolt at the front of the engine to lock the engine from turning over and then remove the four torque converter-to-driveplate nuts **(see illustration)**. Turn the crankshaft bolt for access to each nut on the torque converter. On 4-cylinder engines, make sure not to turn the engine in a counterclockwise direction (viewed from the front).
7   Remove the driveshaft (Chapter 8).
8   Disconnect the speedometer cable.
9   Disconnect the electrical connectors from the transmission **(see illustration)**.
10   Disconnect the vacuum hose from the modulator (if so equipped).
11   Remove the starter (Chapter 5).
12   Remove the transmission fluid filler tube.
13   Support the transmission with a jack and raise the transmission slightly. A safety chain should also be used around the transmission to ensure that it stays securely on the jack.
14   Unbolt the shift cable bracket from the transmission.
15   Remove the insulator-to-transmission nuts.
16   Remove the crossmember-to-frame through-bolts.
17   Lower the transmission slightly - only enough to gain access to the oil cooler fluid lines - and disconnect and plug the lines at the transmission. Be prepared to catch some spilling fluid as this is done.
18   Inside the passenger compartment, remove the engine cover and remove the top two transmission bellhousing-to-engine stud nuts. On some models the ignition coil is mounted on these studs as well.
19   Remove the remaining transmission bellhousing-to-engine bolts **(see illustration)**.
20   Disconnect the exhaust system bolts at the manifolds and lower the system to provide

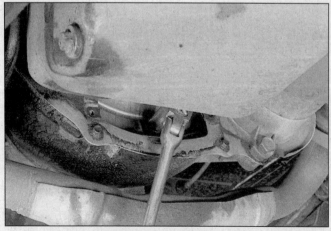

3.6 With the engine locked from turning, use a socket and extension to remove the torque converter nuts

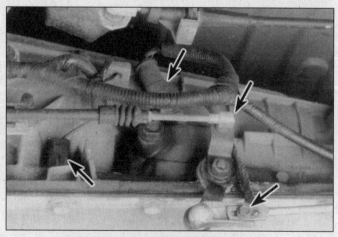

3.9 Various connections at the transmission

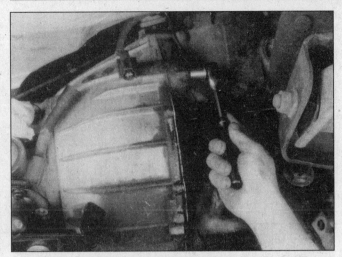

3.19 Remove the remaining transmission-to-engine bolts . . .   **3.21 . . . then pull the transmission away from the engine**

clearance for removal of the transmission.

21 Move the transmission to the rear to disengage it from the engine block dowel pins and make sure the torque converter is detached from the driveplate **(see illustration)**. Lower the transmission from the vehicle.

Installation

22 Prior to installation, make sure the torque converter hub is securely engaged in the pump. This can be checked by measuring the distance between the outer surface of the hub and the transmission flange to make sure this dimension is 7/16 to 9/16-inch.
23 Raise the transmission into position, making sure to keep it level so the torque converter does not slide forward. Connect the transmission cooler lines.
24 Turn the torque converter to line up the drive studs with the holes in the driveplate. The white paint mark on the driveplate and the stud made during Step 5 must be lined up.
25 Move the transmission carefully forward until the dowel pins are engaged and the torque converter is engaged.
26 Install the transmission housing-to-engine bolts and nuts. Tighten the bolts and nuts to the specified torque.
27 Install the torque converter-to-driveplate nuts. Tighten the nuts to the specified torque.
28 Install the transmission mount crossmember and through-bolts. Tighten the bolts and nuts to the specified torque.
29 Lower the transmission until the insulator studs extend through the crossmember and install the nuts. Tighten the nuts to the specified torque.
30 Install the fluid filler tube.
31 Install the starter.
32 Connect the vacuum modulator hose (if so equipped).
33 Connect the shift linkage and cables.
34 Plug in the transmission electrical connectors.
35 Install the torque converter cover.
36 Connect the driveshaft.

37 Connect the speedometer.
38 Adjust the shift linkage (Section 6).
39 Lower the vehicle.
40 Fill the transmission (Chapter 1), run the vehicle and check for fluid leaks.

4   Automatic transmission selector lever assembly - removal and installation

1 Under the vehicle, remove the retaining clip and disconnect the shift cable from the selector lever.
2 Inside the passenger compartment, grasp the floor carpet center retainer securely and pull up sharply to remove it from the base of the lever assembly.
3 Remove the four retaining bolts and lift the selector lever assembly from the vehicle.
4 If it is necessary to remove the selector handle and button, grasp the handle securely with the shifter in Drive and pull straight up. If the handle does not come off easily, tap from below with a rubber mallet. To install, place the handle/button in position, making sure the handle chrome ring is lined up with the selector handle groove. Strike the center of handle with a rubber mallet until it is seated.
5 To install, place the selector assembly in position and install the bolts. Tighten the bolts securely.
6 Place the carpet retainer in position and press down evenly until it snaps in place.
7 Under the vehicle, connect the shift cable to the selector lever.
8 Adjust the linkage (Section 6).

5   Neutral start switch (A4LD) - removal and installation

Note: *1996 and 1997 models do not use a neutral start switch. Instead, a digital transmission range sensor is mounted at the shift lever assembly on the transmission case.*

Because diagnosis and adjustment of this component requires special tools, it is best to consult a specialist.
1 Jack up the vehicle and support it securely on jackstands to gain access to the neutral start switch, located on the side of the transmission.
2 Disconnect the ground cable at the battery.
3 Disconnect the electrical connector from the neutral start switch.
4 Carefully unscrew the switch from the transmission.
5 Install the switch and tighten securely.
6 Install the electrical connector and lower the vehicle.
7 Connect the negative battery cable.
8 Check that the engine starts only when the selector is in the Neutral and Park positions.

6   Automatic transmission shift linkage - adjustment

1986 through 1991 models

1 Raise the vehicle and support it securely on jackstands.
2 In the passenger compartment, place the shift lever in the D (Overdrive) position.
3 Under the vehicle, loosen the adjustment screw on the shift cable end and disconnect the manual lever ballstud.
4 Place the manual lever on the transmission in the D (Overdrive) position by moving it all the way to the rear and then moving it three detents forward. Hold the shift lever against the rear stop.
5 Connect the cable end fitting to the manual shift lever and tighten the adjustment screw securely.
6 After adjustment, check for proper Park engagement; the control lever must move to the right when engaged in the Park detent.
7 Lower the vehicle and check the control lever in all detent positions with the engine

running to ensure the correct detent/transmission action, readjusting as necessary.

1992 and later models

8   Raise the vehicle and support it securely on jackstands.
9   In the passenger compartment, place the column shift lever in the OD (overdrive) position.
10   Hang a three-pound weight on the end of the column shift lever.
11   Under the vehicle, release the lock tab on the top side of the cable by pushing down on the two tangs.
12   Disconnect the shift cable from the transmission manual shift lever.
13   Place the manual shift lever on the transmission into overdrive by moving it all the way forward (counterclockwise) and then move it rearward three detent positions.
14   Connect the shift cable end fitting to the transmission manual shift lever.
15   Completely push down on the lock tab on the shift cable and make sure it fully engages, locking the cable in place.
16   Make sure the shift cable is clipped to the floor pan.
17   In the passenger compartment, adjust the shift pointer while the transmission is still in overdrive.
18   Remove the weight from the column shift lever. Check for proper engagement in all gears, including park; readjust if necessary.

7   Overdrive band (A4LD) - adjustment

Refer to illustration 7.2

1   Clean all dirt from the front (overdrive) band adjusting screw area located on the left

side of the transmission. Remove and discard the locknut.
2   Install a new locknut on the adjusting screw. Using a torque wrench, tighten the adjusting screw to 120 in-lbs **(see illustration)**.
3   If equipped with a 4.0L engine, back off the adjusting screw exactly 3-1/2 turns. On all others, back off the adjusting screw exactly two turns.
4   Hold the adjusting screw to keep it from turning and tighten the locknut.

8   Intermediate band (A4LD) - adjustment

1   Clean all dirt from the band adjusting screw area located on the left side of the transmission to the rear of the overdrive band screw (see Section 7). Remove and discard the locknut.
2   Install a new locknut on the adjusting screw.
3   Tighten the adjusting screw using torque wrench set to 120 in-lbs **(see illustration 7.2)**.
4   Back off the adjusting screw exactly two turns.
5   Hold the adjusting screw to keep it from turning and tighten the locknut.

9   Extension housing oil seal - replacement

Refer to illustrations 9.3 and 9.7

1   Raise the vehicle and support it securely on jackstands.
2   Remove the driveshaft (Chapter 8). Scribe marks on the driveshaft and yoke and the rear axle companion flange to assure

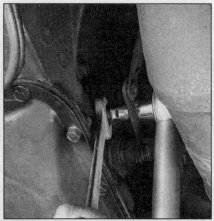

7.2 Tighten the band adjusting screw to 120 inch-lbs, then back it off the specified number of turns (see text)

proper positioning of the driveshaft during assembly.
3   Remove the oil seal from the end of the extension housing with a seal removing tool. If there is access, a thin blade screwdriver or chisel may also be used to remove the seal **(see illustration)**.
4   Before installing the new seal inspect the sealing surface of the universal joint yoke for scoring. If scoring is found, replace the yoke.
5   Inspect the counterbore of the housing for burrs. Remove any burrs with emery cloth or medium grit wet-and-dry sandpaper.
6   Coat the inside diameter of the seal with multi-purpose grease.
7   Install the new seal into the extension housing **(see illustration)**.
8   The remainder of the installation is the reverse of the removal procedures.

9.3 Be very careful when prying out the extension seal so that the sealing surface is not damaged

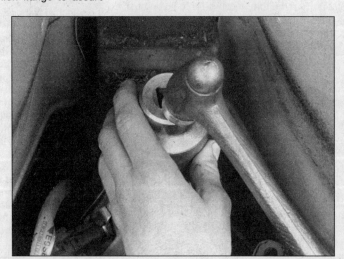

9.7 A large socket and hammer can be used to tap the new seal evenly into the bore

Chapter 8
Clutch and drivetrain

Contents

Specifications

Clutch
Slave cylinder pushrod travel 19/64-inch

Torque specifications
Ft-lbs (unless otherwise indicated)
Pressure plate-to-flywheel bolts .... 15 to 24
Clutch housing-to-engine bolts .... 28 to 38
Clutch housing-to-transmission nuts .... 30 to 40
Clutch master cylinder bolts
    1986 through 1990 .... 15 to 20
    1991 on .... 96 to 144 in-lbs
Clutch slave cylinder bolts .... 13 to 19

Driveshaft
Universal joint strap nuts .... 96 to 180 in-lbs

Rear axle
Differential cover bolts .... 15 to 20
Pinion shaft lock bolt .... 15 to 30
Brake backing plate/bearing retainer nuts (Dana model 30) .... 25 to 35
Rear lower control arm-to-rear axle housing .... 100 to 129
Rear upper control arm-to-rear axle housing .... 100 to 145

1   General information

The information in this Chapter deals with the components from the rear of the engine to the rear wheels, except for the transmission, which is dealt with in the previous Chapter. For the purposes of this Chapter, these components are grouped into three categories; clutch, driveshaft and rear axle. Separate Sections within this Chapter offer general descriptions and checking procedures for each of these three groups.

As nearly all the procedures covered in this Chapter involve working under the vehicle, make sure it is firmly supported on sturdy jackstands or on a hoist where the vehicle can be easily raised and lowered.

2   Clutch - description and check

1   All models equipped with a manual transmission feature a single dry plate, diaphragm spring-type clutch. The actuation is through a hydraulic system.
2   When the clutch pedal is depressed, hydraulic fluid (under pressure from the clutch master cylinder) flows into the slave cylinder. The slave cylinder is connected to the release bearing which is constantly in contact with the pressure plate diaphragm spring release fingers. The slave cylinder pushes the release bearing against the pressure plate release fingers, disengaging the clutch plate.
3   The hydraulic system locates the clutch pedal and provides clutch adjustment automatically, so no adjustment of the linkage or pedal is required.
4   Terminology can be a problem regarding the clutch components because common names have in some cases changed from that used by the manufacturer. For example, the driven plate is also called the clutch plate or disc, the clutch release bearing is sometimes called a throwout bearing, the slave cylinder is sometimes called the operating cylinder.
5   Other than to replace components with obvious damage, some preliminary checks should be performed to diagnose a clutch system failure.

a) The first check should be of the fluid level in the clutch master cylinder (see Fluid level checks in Chapter 1). If the fluid level is low, add fluid as necessary and re-test. If the master cylinder runs dry, or if any of the hydraulic components are serviced, bleed the hydraulic system as described in Section 8.
b) To check "clutch spin down time", run the engine at normal idle speed with the transmission in Neutral (clutch pedal up - engaged). Disengage the clutch (pedal down), wait nine seconds and shift the transmission into Reverse. No grinding noise should be heard. A grinding noise would indicate component failure in the

3.7   After removal of the transmission, this will be the view of the clutch components. Before removing the pressure plate (1), mark its relationship to the flywheel (2) so it can be reinstalled in the same position

pressure plate assembly or the clutch disc.
c) To check for complete clutch release, run the engine (with the brake on to prevent movement) and hold the clutch pedal approximately 1/2-inch from the floor mat. Shift the transmission between 1st gear and Reverse several times. If the shift is not smooth, component failure is indicated.
d) Visually inspect the clutch pedal bushing at the top of the clutch pedal to make sure there is no sticking or excessive wear.

3   Clutch components - removal, inspection and installation

Refer to illustrations 3.7, 3.12, 3.14 and 3.16
Warning: *Dust produced by clutch wear and deposited on clutch components contains asbestos, which is hazardous to your health. DO NOT blow it out with compressed air and DO NOT inhale it. DO NOT use gasoline or petroleum-based solvents to remove the dust. Brake system cleaner should be used to flush the dust into a drain pan. After the clutch components are wiped clean with a rag, dispose of the contaminated rags and cleaner in a covered container.*

Removal

1   Access to the clutch components is normally accomplished by removing the transmission, leaving the engine in the vehicle. If, of course, the engine is being removed for major overhaul, then the opportunity should always be taken to check the clutch for wear and replace worn components as necessary. The following procedures will assume that the engine will stay in place.
2   Disconnect the hydraulic line from the slave cylinder (see Section 6).
3   Referring to Chapter 7 Part A, remove

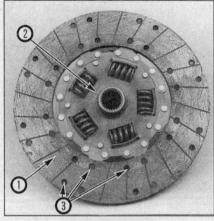

3.12   The clutch disc
1   *Lining* - this will wear down in use
2   *Markings* - "Flywheel side" or similar
3   *Rivets* - these secure the lining and will damage the flywheel or pressure plate if allowed to contact the surfaces

the transmission from the vehicle. Support the engine while the transmission is out. Preferably, an engine hoist from above should be used. However, if a jack is used underneath the engine, make sure a piece of wood is used between the jack and oil pan to spread the load. **Caution:** *The pickup for the oil pump is very close to the bottom of the oil pan. If the pan is bent or distorted in any way, engine oil starvation could occur.*
4   Remove the clutch housing-to engine bolts. Remove the housing. It may have to be gently pried off the alignment dowels with a screwdriver or pry bar.
5   The slave cylinder and release bearing can remain attached to the housing for the time being.
6   To support the clutch disc during removal, install a clutch alignment tool through the middle of the clutch.
7   Carefully inspect the flywheel and pressure plate for indexing marks. These can be an X, an O, or a white-painted letter. If these cannot be found, scribe marks yourself so that the pressure plate and the flywheel will be in the same alignment upon installation (see illustration).
8   Turning each bolt only one turn at a time, slowly loosen the pressure plate-to-flywheel bolts. Work in a diagonal pattern, again, loosening each bolt a little at a time until all spring pressure is relieved. Then grasp the pressure plate firmly and completely remove the bolts, followed by the pressure plate and clutch disc.

Inspection

9   Ordinarily, when a problem occurs in the clutch, it can be attributed to wear of the clutch driven plate assembly (clutch disc). However, all components should be inspected at this time.
10   Inspect the flywheel for cracks, heat checking, grooves or other signs of obvious defects. If the imperfections are slight, a

 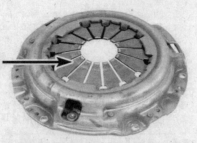 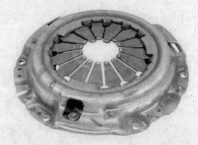

NORMAL FINGER WEAR   **EXCESSIVE FINGER WEAR**   **BROKEN OR BENT FINGERS**

EXCESSIVE WEAR

3.14 Replace the pressure plate if any of these conditions are noted

machine shop can machine the surface flat and smooth, which is highly recommended regardless of the surface appearance. Refer to Chapter 2 for the flywheel removal and installation procedure.

11 Inspect the pilot bearing as outlined in Section 4.

12 Inspect the facing on the clutch disc. There should be at least 1/16-inch of lining above the rivet heads. Check for loose rivets, distortion, cracks, broken springs and any other obvious damage **(see illustration)**. As mentioned above, ordinarily the disc is replaced as a matter of course, so if in doubt about the quality, replace it with a new one.

13 Check the release bearing for roughness or binding. If bearing operation is anything but smooth, replace it (see Section 7).

14 Check the flatness of the pressure plate with a straightedge. Look for signs of overheating, cracks, deep grooves and ridges. The inner end of the diaphragm spring fingers should not show any signs of uneven wear. Replace the pressure plate with a new one if its condition is in doubt **(see illustration)**.

Installation

15 Before installation, carefully wipe clean the flywheel and pressure plate machined surfaces. It is important that no oil or grease is on these surfaces or the facing of the clutch disc. Handle these parts only with clean hands.

16 Position the clutch disc and pressure plate against the flywheel with the clutch disc held in place with a clutch alignment tool **(see illustration)**. Make sure the disc is installed properly (most replacement discs will be marked "flywheel side" or similar. If not marked, install the disc with the damper springs toward the transmission).

17 Tighten the pressure plate-to-flywheel bolts only finger tight, working around the pressure plate.

18 Center the clutch disc by inserting the alignment tool through the splined hub and into the pilot bearing in the flywheel, if not already done. Tighten the pressure plate-to-flywheel bolts a little at a time, working in a criss-cross pattern to prevent distorting the cover. After all of the bolts are snug, tighten

them to the specified torque. Remove the alignment tool.

19 Install the transmission and all components removed previously, tightening all fasteners to the specified torque.

4 Pilot bearing - inspection and replacement

Refer to illustrations 4.5 and 4.6

1 The clutch pilot bearing is a needle roller type bearing which is pressed into the rear of the crankshaft. It is pre-greased at the factory and does not require additional lubrication. Its primary purpose is to support the front of the transmission input shaft. The pilot bearing should be inspected whenever the clutch components are removed from the engine. Due to its inaccessibility, if you are in doubt as to its condition, replace it with a new one. **Note:** *If the engine has been removed from the vehicle, disregard the following steps which do not apply.*

2 Remove the transmission (refer to Chapter 7 Part A)

3 Remove the clutch components (Section 3).

4 Inspect for any excessive wear, scoring,

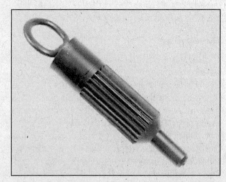

3.16 Use a clutch alignment tool to hold the clutch disc in place while installing the pressure plate and also center the disc when tightening the bolts. This particular tool, made of plastic, is available at most auto parts stores and isn't very expensive

lack of grease, dryness or obvious damage. If any of these conditions are noted, the bearing should be replaced. A flashlight will be helpful to direct light into the recess.

5 Removal can be accomplished with a slide hammer and puller attachment **(see illustration)**.

6 To install the new bearing, lightly lubricate the outside surface with lithium-based grease, then drive it into the recess with a hammer and socket **(see illustration)**.

4.5 A small slide hammer puller is handy for removing the pilot bearing

4.6 Tap the bearing into place with a bearing installer or a socket that is slightly smaller in diameter than the outside diameter of the bearing

7   Install the clutch components, transmission and all other components removed previously, tightening all fasteners properly.

5   Clutch master cylinder - removal, overhaul and installation

Note: *Before beginning this procedure, contact local parts stores or dealers concerning the purchase of a rebuild kit or a new master cylinder. Availability and cost of the necessary parts may dictate whether the cylinder is rebuilt or replaced with a new one. If it is decided to rebuild the cylinder, inspect the inside bore as described in Step 7 before purchasing parts.*

Removal

1   Disconnect the cable from the negative battery terminal.
2   Under the dashboard, disconnect the pushrod from the top of the clutch pedal. It is held in place with a retainer bushing and must be pried off with a screwdriver.
3   Disconnect the hydraulic line at the clutch slave cylinder (see Section 6) then free the hydraulic line from the clips on the underbody siderail. Have rags handy as some fluid will be lost as the line is removed. **Caution:** *Do not allow brake fluid to come into contact with paint as it will damage the finish.*
4   Remove the two bolts which secure the master cylinder to the engine firewall.
5   Slide the clutch fluid reservoir off its mounting slots and remove the cylinder, line and reservoir from the vehicle, again being careful not to spill any fluid.

Overhaul

6   It is not necessary to disconnect the fluid reservoir hose from the master cylinder unless the cylinder is being replaced.
7   Disassemble the cylinder and inspect the bore of the master cylinder for deep scratches, scoring, pitting or ridges.
The surface must be smooth to the touch. If the bore is not perfectly smooth, the master cylinder must be replaced with a new or factory rebuilt unit.
8   If you elect to rebuild the cylinder, follow the instructions included with the rebuild kit and be sure to install all of the parts that are supplied.
9   Before reassembly, lubricate the inside bore of the cylinder and the seals with plenty of fresh brake fluid (DOT 3).

Installation

10   Guide the master cylinder into its hole in the firewall, install the two mounting bolts and tighten them to the specified torque. Position the reservoir in its mounting slots.
11   Connect the hydraulic line to the slave cylinder, then secure the hydraulic line to the routing clips.
12   Inside the vehicle, connect the pushrod to the clutch pedal and install the

retainer bushing.
13   Fill the clutch master cylinder reservoir with brake fluid conforming to DOT 3 specifications and bleed the clutch system as outlined in Section 8.
14   Reconnect the negative battery cable.

6   Clutch slave cylinder - removal, overhaul and installation

Warning: *Dust produced by clutch wear and deposited on clutch components may contain asbestos, which is hazardous to your health. DO NOT blow it out with compressed air and DO NOT inhale any of it. DO NOT use gasoline or petroleum-based solvents to remove the dust. Brake system cleaner should be used to flush the dust into a drain pan. After the clutch components are wiped clean with a rag, dispose of the contaminated rags and cleaner in a covered container.*
Note: *Before beginning this procedure, contact local parts stores or dealers concerning the purchase of a rebuild kit or a new slave cylinder. Availability and cost of the necessary parts may dictate whether the cylinder is rebuilt or replaced with a new one. If it is decided to rebuild the cylinder, inspect the bore of the cylinder before purchasing parts.*

Removal

1   Disconnect the cable from the negative battery terminal.
2   Raise the vehicle and support it securely on jackstands.
3   Disconnect the hydraulic line at the slave cylinder. On 1986 and 1987 models, remove the retainer clip from the hydraulic fitting and slide the line from the slave cylinder. On 1988 models, slide the white plastic sleeve toward the slave cylinder while lightly pulling on the hydraulic line. Have a small can or some rags handy to catch the spilling fluid.
4   Remove the transmission as described in Chapter 7 Part A.
5   On 1986 and 1987 models, remove the four nuts that retain the clutch housing to the transmission. Separate the two and remove the slave cylinder from the clutch housing.
6   On 1988 and later models, remove the two slave cylinder bolts and pull the cylinder off of the transmission input shaft.
7   Separate the release bearing from the slave cylinder as described in Section 7.

Overhaul

8   Remove the piston from the slave cylinder body, drain the fluid and inspect the bore for scratches, scoring, ridges or pitting. If any imperfections are found, the slave cylinder must be replaced.
9   If it is decided to rebuild the cylinder, follow the instructions included in the rebuild kit and be sure to install all of the parts that are supplied.
10   Before reassembly, lubricate the piston, cylinder and seals with plenty of clean brake fluid (DOT 3).

Installation

11   Install the release bearing to the slave cylinder (Section 7).
12   On 1986 and 1987 models, place the slave cylinder in the clutch housing and assemble the housing to the transmission, tightening the nuts to the specified torque.
13   On 1988 and later models, slide the cylinder over the transmission input shaft and install the two bolts, tightening them to the specified torque.
14   Install the transmission.
15   Connect the hydraulic line to the slave cylinder, fill the fluid reservoir with brake fluid conforming to DOT 3 specifications and bleed the clutch system as outlined in Section 8.
16   Reconnect the negative battery cable.

7   Clutch release bearing - removal, inspection and installation

Warning: *Dust produced by clutch wear and deposited on clutch components may contain asbestos, which is hazardous to your health. DO NOT blow it out with compressed air and DO NOT inhale any of it. DO NOT use gasoline or petroleum-based solvents to remove the dust. Brake system cleaner should be used to flush the dust into a drain pan. After the clutch components are wiped clean with a rag, dispose of the contaminated rags and cleaner in a covered container.*

Removal

1   Remove the transmission following the procedure described in Chapter 7 Part A.
2   On 1986 and 1987 vehicles, dislodge the bearing from the bearing carrier by bending back the plastic tabs with a small screwdriver.
3   On 1988 and later vehicles, turn the bearing/carrier assembly until resistance is felt, then turn it slightly further which will disengage the carrier from the slave cylinder.

Inspection

4   Inspect the bearing for cracks, wear and other damage. Hold the center of the bearing and turn the outer race while applying pressure to it. If the bearing doesn't turn smoothly or if it is noisy, replace it with a new one. It is a good idea to replace the bearing whenever a clutch job is performed, to decrease the possibility of bearing failure in the future, although the manufacturer states that this is not absolutely necessary.
5   If it is decided to re-use the old release bearing, clean the external surfaces and inside diameter with a solvent-moistened rag. Do not immerse the bearing in solvent, as it is packed with grease from the factory (sealed for life) and to do so would ruin it.

Installation

6   Fill the groove in the inside diameter of the bearing with lithium based grease. Also

apply a thin coat of this grease to the entire inner diameter.

7   The remainder of the assembly procedure is the reverse of removal. Be sure to bleed the clutch hydraulic system as described in Section 8.

8 Clutch hydraulic system - bleeding

1   The hydraulic system should be bled of all air whenever any part of the system has been removed or if the fluid level has been allowed to fall so low that air has been drawn into the master cylinder. The procedure is very similar to bleeding a brake system.

2   Fill the master cylinder with new brake fluid conforming to DOT 3 specifications. **Caution:** *Do not re-use any of the fluid coming from the system during the bleeding operation or use fluid which has been inside an open container for an extended period of time.*

3   Raise the vehicle and place it securely on jackstands to gain access to the slave cylinder bleeder screw, which is located next to the hydraulic line inlet connection.

4   Remove the dust cap (if equipped) which fits over the bleed screw. Place a box end wrench over the bleed screw and then push a length of plastic hose over the screw. Place the other end of the hose into a clear container with about two inches of brake fluid. The hose end must be in the fluid at the bottom of the container.

5   Have an assistant depress the clutch pedal and hold it. Open the bleeder valve on the slave cylinder, allowing fluid to flow through the hose. Close the bleeder valve when your assistant signals that the clutch pedal is at the bottom of its travel. Once closed, have your assistant release the pedal.

6   Continue this process until all air is evacuated from the system, indicated by a full, solid stream of fluid being ejected from the bleeder valve each time and no air bubbles in the hose or container. Keep a close watch on the fluid level inside the master cylinder; if the level drops too low, air will be sucked back into the system and the process will have to be started all over again.

9 Clutch pedal - removal and installation

1   Disconnect the cable from the negative battery terminal.

2   Disconnect the clutch master cylinder pushrod from the pedal by prying the retainer bushing off with a screwdriver.

3   Disconnect and remove the starter/clutch interlock switch (see Section 10).

4   Remove the pedal pivot pin retainer clip and slide the pedal to the left to remove it. Insert a long narrow screwdriver or rod through the opposite side of the bracket to hold the brake pedal in place while the clutch

pedal is removed.

5   Wipe clean all parts; however, do not use cleaning solvents on the bushings. Replace all worn parts with new ones and lightly lubricate the bushings with engine oil.

6   Installation is the reverse of removal. Check that the starter/clutch interlock switch allows the vehicle to be started only with the clutch pedal fully depressed.

10 Starter/clutch interlock switch - removal and installation

1   Detach the wire harness connector from the interlock switch.

2   Remove the interlock switch-to-bracket screw and pry the end of the rod off of the clutch pedal pin.

3   Position the adjuster clip approximately 1-inch from the end of the rod.

4   Install the end of the rod onto the clutch pedal pin.

5   With the clutch pedal all the way up, swing the switch down into place. Install the mounting screw and tighten it securely.

6   Push the clutch pedal to the floor which automatically adjusts the switch.

7   Confirm that the engine starts only when the clutch pedal is depressed to the floor. If it doesn't, remove the adjusting clip from the rod and position it closer to the switch. Push the pedal to the floor and check the switch operation again. If the engine still doesn't crank over, or cranks over when the pedal is in the released position, replace the switch with a new one.

11 Driveshaft and universal joints - description and check

1   The driveshaft (or prop shaft as it is often called) is a tube running between the transmission and the rear end. Universal joints are located at either end of the driveshaft and permit power to be transmitted to the rear wheels at varying angles.

2   The driveshaft features a splined yoke at the front, which slips into the rear extension of the transmission. This arrangement allows the driveshaft to slide back and forth within the transmission as the vehicle is in operation. An oil seal is used to prevent leakage of fluid at this point and to keep dirt and contaminants from entering the transmission. If leakage is evident at the front of the driveshaft, replace this oil seal referring to the procedures in Chapter 7 Part A (manual transmission) or Chapter 7 Part B (automatic transmission).

3   The driveshaft assembly requires very little service. The universal joints are lubricated for life and require replacement should a fault develop. The driveshaft must be removed from the vehicle for this procedure.

4   Since the driveshaft is a balanced unit, it is important that no undercoating, mud, etc. be allowed to stay on it. When the vehicle is

raised for service it is a good idea to clean the driveshaft and inspect for any obvious damage. Also check that the small weights used to originally balance the driveshaft are in place and securely attached. Whenever the driveshaft is removed it is important that it be reinstalled in the same relative position to preserve this balance.

5   Problems with the driveshaft are usually indicated by a noise or vibration while driving the vehicle. A road test should verify if the problem is the driveshaft or another vehicle component:

a) *On an open road, free of traffic, drive the vehicle and note the engine speed (rpm - tachometer) at which the problem is most evident.*

b) *With this noted, drive the vehicle again, this time manually keeping the transmission in 1st, then 2nd, then 3rd gear ranges and running the engine up to the engine speed noted.*

c) *If the noise or vibration occurs at the same engine speed regardless of which gear the transmission is in, the driveshaft is not at fault because the speed of the driveshaft varies in each gear.*

d) *If the noise or vibration decreased or was eliminated, visually inspect the driveshaft for damage, material on the shaft which would effect balance, missing weights or damaged universal joints. Another possibility for this condition would be tires which are out of balance.*

6   To check for worn universal joints:

a) *On an open road, free of traffic, drive the vehicle slowly until the transmission is in high gear. Let off on the accelerator, allowing the vehicle to coast, then accelerate. A clunking or knocking noise will indicate worn universal joints.*

b) *Drive the vehicle at a speed of about 10 to 15 mph and then place the transmission in Neutral, allowing the vehicle to coast. Listen for abnormal driveline noises.*

c) *Raise the vehicle and support it securely on jackstands. With the transmission in Neutral, manually turn the driveshaft, observing the universal joints for excessive play.*

12 Driveshaft - removal and installation

Refer to illustration 12.3

Removal

1   Disconnect the cable from the negative terminal of the battery.

2   Block the front tires to prevent movement and raise the vehicle. Support it securely on jackstands. Place the transmission in Neutral with the parking brake off.

3   Using white paint or a hammer and punch, place marks on the driveshaft and the differential pinion flange in line with each other **(see illustration)**. This is to make sure

the driveshaft is reinstalled in the same position to preserve the balance of the system.

4   Remove the rear universal joint strap bolts and the straps. Turn the driveshaft (or tires) as necessary to bring the bolts into the most accessible position.

5   Tape the bearing caps to the spider to prevent the caps from coming off during removal.

6   Lower the rear of the driveshaft and then slide the front from the transmission.

7   To prevent loss of fluid and protect against contamination while the driveshaft is out, wrap a plastic bag over the transmission housing and hold it in place with a rubber band.

Installation

8   Remove the plastic bag on the transmission and wipe clean the area. Inspect the oil seal carefully. Procedures for replacement of this seal can be found in Chapter 7 Part B.

9   Slide the front of the driveshaft into the transmission.

10   Raise the rear of the driveshaft into position, checking to be sure that the marks are in perfect alignment. If not, turn the rear wheels to match the pinion flange and the driveshaft.

11   Remove the tape securing the bearing caps and install the straps and bolts. Tighten to the proper torque specification.

13   Universal joints - replacement

Refer to illustrations 13.2a, 13.2b, 13.4 and 13.9

Note: *A press or large vise will be required for this procedure. It may be advisable to take the driveshaft to a local dealer or machine shop where the universal joints can be replaced for you, normally at a reasonable charge.*

1   Remove the driveshaft as discussed in the previous Section.

2   Using small pliers, remove the snap-rings from the spider **(see illustrations)**.

12.3 Before removing the driveshaft, mark the relationship of the shaft to the differential pinion yoke. A screwdriver inserted through the joint will keep it from turning while the nuts are loosened

3   Supporting the driveshaft, place it in position on either an arbor press or on a workbench equipped with a vise. **Caution:** *Never clamp the driveshaft tube or yoke in the jaws of a vise.*

4   Place a piece of pipe (about 1-1/4 inch inside diameter) or a socket with the same inside diameter over one of the bearing caps. Position a socket which is of slightly smaller diameter than the cap on the opposite bearing cap **(see illustration)** and use the vise or press to force the cap out (inside the pipe or large socket), stopping just before it comes completely out of the yoke. Use the vise or large pliers to work the cap the rest of the way out.

5   Transfer the sockets to the other side and press the opposite bearing cap out in the same manner.

6   Pack the new universal joint bearings with grease. Ordinarily, specific instructions for lubrication will be included with the universal joint servicing kit and should be followed carefully.

7   Position the spider in the yoke and par-

13.2a Remove the snap-rings with a pair of pliers

tially install one bearing cap into the yoke.

8   Start the spider into the bearing cap and then partially install the other cap. Align the spider and press the bearing caps into position, being careful not to damage the dust seals.

9   Install the snap-rings. If difficulty is encountered in seating the snap rings, strike the driveshaft yoke sharply with a hammer. This will spring the yoke ears slightly and allow the snap ring groove to move into position **(see illustration)**.

10   Follow the same procedures to replace the rear universal joint, noting that only one-half of the spider needs to be pressed out, since the other two ends are held to the pinion flange with straps and bolts.

11   Install the driveshaft and all components previously removed.

14   Rear axle - description and check

1   There are three types of rear axles available on the vehicles covered by this manual - the Ford 7.5 inch ring gear, the Dana Model

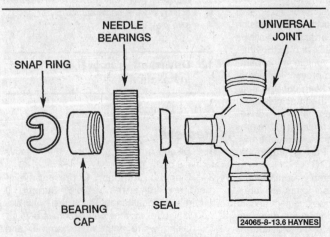

13.2b Exploded view of the universal joint components

SNAP RING

NEEDLE BEARINGS

UNIVERSAL JOINT

BEARING CAP

SEAL

24065-8-13.6 HAYNES

13.4 To press the universal joint out of the driveshaft, set it up in a vise as shown. The small socket (at left) will push the joint and bearing cap into the large socket (at right)

13.9 If the snap-ring will not seat in its groove, strike the yoke with a brass hammer. This will relieve the tension that has set up in the yoke, and slightly spring the yoke ears. This should also be done if the joint feels tight when assembled

15.3a Use a large screwdriver placed between the rear axle case and a ring gear bolt to keep the differential case from turning when removing the pinion shaft lock bolt

30, or as an option, a limited-slip version of the Ford 7.5 inch rear axle. This rear end, called the Traction-Lok Limited Slip differential, incorporates two sets of multi-disc clutches to control differential action.

2   Information regarding the rear axle can be found on the Axle Identification Tag, which is located under the differential cover-to-carrier bolt in the 2 o'clock position.

3   Many times a fault is suspected in the rear axle area when, in fact, the problem lies elsewhere. For this reason, a thorough check should be performed before assuming a rear axle problem.

4   The following noises are those commonly associated with rear axle diagnosis procedures:

a) *Road noise is often mistaken for mechanical faults. Driving the vehicle on different surfaces will allow you to determine whether the road surface is the cause of the noise. Road noise will remain the same if the vehicle is under power or coasting.*

b) *Tire noise is sometimes mistaken for*

mechanical problems. Tires which are worn or low on pressure are particularly susceptible to emitting vibrations and noises. Tire noise will remain about the same during varying driving situations, where rear axle noise will change during coasting, acceleration, etc.

c) *Engine and transmission noise can be deceiving because it will travel along the driveline. To isolate engine and transmission noises, make a note of the engine speed at which the noise is most pronounced. Stop the vehicle and place the transmission in Neutral and run the engine to the same speed. If the noise is the same, the rear axle is not at fault.*

5   Overhaul and general repair of the rear axle components is beyond the scope of the home mechanic due to the many special tools and critical measurements required for the many parts to operate properly. Thus, the procedures listed here will involve seal replacements to stop oil leakage, axle and bearing replacement and removal of the entire unit for repair or replacement.

15 Axleshaft - removal and installation

Refer to illustrations 15.3a, 15.3b, 15.4, 15.5, 15.9, and 15.10

1   Raise the rear of the vehicle, support it securely and remove the wheel and brake drum (refer to Chapter 9).

Ford 7.5 Inch rear end

2   Unscrew and remove the cover from the differential carrier and allow the oil to drain into a container.

3   Unscrew and remove the lock bolt from the differential pinion shaft. Slide the notched end of the pinion shaft out of the differential case as far as it will go **(see illustrations)**.

4   Push the outer (flanged) end of the axleshaft in and remove the C-ring from the inner end of the shaft **(see illustration)**.

5   Withdraw the axleshaft, taking care not to damage the oil seal in the end of the axle housing as the splined end of the axleshaft passes through it **(see illustration)**.

15.3b Rotate the differential case 180~ and slide the pinion shaft out of the case until the stepped part of the shaft contacts the ring gear

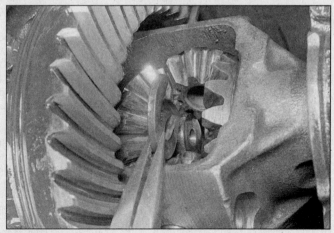

15.4 Push in on the axle flange and remove the C-lock from the inner end of the axleshaft

15.5 Pull the axle out of the housing, supporting it with one hand to prevent damage to the seal

15.9 Access to the four bearing retainer nuts is through the hole in the axle flange

15.10 A slide hammer and axle flange adapter will be necessary to remove the axle/bearing assembly from the housing

16.2 If a seal remover tool isn't available, a pry bar or even the end of the axle can be used to pop the seal out of its bore

6   Installation is the reverse of removal. Tighten the lock bolt to the specified torque.

7   Refer to Chapter 1 for the proper cover installation procedure.

8   Refill the axle with the correct quantity and grade of lubricant (Chapter 1).

Dana Model 30 rear end

9   Remove the four brake backing plate/bearing retainer nuts, working through the hole in the axle flange **(see illustration)**. These nuts must be replaced with new ones upon reassembly.

10   Attach a slide hammer and axleshaft puller adapter to the axle flange and remove the axleshaft from the housing **(see illustration)**.

11   Installation is the reverse of the removal procedure. Be sure to lubricate the bearing with gear lubricant, install new backing plate/bearing retainer nuts and tighten them to the specified torque.

16 Axleshaft oil seal - replacement

Refer to illustrations 16.2 and 16.3

1   Remove the axleshaft as described in the preceding Section.

Ford 7.5 Inch rear end

2   Pry out the old oil seal from the end of the axle housing, using a large screwdriver or the inner end of the axleshaft itself as a lever **(see illustration)**.

3   Using a large socket or section of pipe as a seal driver **(see illustration)**, tap the seal into position so that the lips are facing in and the metal face is visible from the end of the axle housing. When correctly installed, the face of the oil seal should be flush with the end of the axle housing. Lubricate the lips of the seal with gear oil.

4   Installation of the axleshaft is described in the preceding Section.

16.3 A large socket or a piece of pipe (as shown here) can be used to install the seal. The diameter of the tool should be just slightly smaller than the outside diameter of the seal

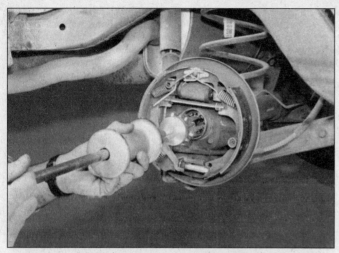

17.2 To remove the axle bearing, insert a bearing removal tool attached to a slide hammer through the center of the bearing, pull the tool up against the bearing and use the slide hammer to yank the bearing from the axle housing

17.4 A correctly-sized bearing driver must be used to drive the bearing into the housing

Dana Model 30 rear end

5 The axleshaft oil seal/bearing/retainer are pressed onto the axleshaft and require special tools and a hydraulic press to replace them. For service regarding these components, take the axleshaft to an automotive machine shop.

17 Axleshaft bearing - replacement

Refer to illustrations 17.2 and 17.4

Ford 7.5 Inch rear end

1 Remove the axleshaft (refer to Section 15) and the oil seal (refer to Section 16).
2 A bearing puller will be required or a tool which will engage behind the bearing will have to be fabricated **(see illustration)**.
3 Attach a slide hammer and pull the bearing from the axle housing.
4 Clean out the bearing recess and drive in the new bearing using a properly sized

bearing driver **(see illustration)**.
5 Discard the old oil seal and install a new one (Section 16), then install the axleshaft.

Dana Model 30 rear end

6 Remove the axleshaft as described in Section 15. The axleshaft bearing, oil seal and retainer are pressed onto the axleshaft and require special tools and a hydraulic press to replace them. For service regarding these procedures, take the axleshaft to an automotive machine shop.

18 Rear axle assembly - removal and installation

Refer to illustrations 18.5, 18.8 and 18.9

Removal

1 Loosen the rear wheel lug nuts, raise the rear of the vehicle and support it securely on jackstands positioned under the frame.

Remove the wheels.
2 Support the rear axle assembly with a floor jack placed under the differential.
3 Disconnect the two rear parking brake cables from the equalizer and push the cable casings through the crossmember (see Chapter 9).
4 Remove the driveshaft following the procedure described in Section 12.
5 Disconnect the rear brake hose from the hydraulic line at the frame **(see illustration)**. Plug the hydraulic line and the flexible hose to prevent excessive fluid loss and contamination.
6 Remove the lower mounting bolts from the shock absorbers.
7 Remove the coil springs (see Chapter 10).
8 Unbolt the lower control arms from the rear axle housing **(see illustration)**.
9 Remove the upper control arm-to-rear axle bolt and separate the control arm from the axle housing **(see illustration)**.
10 Scribe a line on the upper control arm

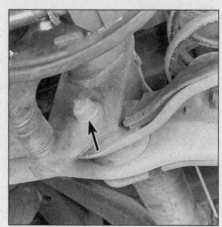

18.5 Disconnect the rear brake hose from the hydraulic line at the frame bracket. Use a backup wrench on the fitting of the hose to prevent twisting the line

18.8 Unbolt the rear lower control arms from the axle housing

18.9 Unbolt the rear upper control arm from the axle housing

bushing marking the position of the cam adjuster in the bushing.

11   Slowly lower the axle assembly from the vehicle.

Installation

12   Raise the axle into position and fit the upper control arm over the bushing, checking to be sure that the cam adjuster is still in alignment. Install the bolt and nut, tightening them snugly. Do not fully tighten the nut at this time.

13   Connect the lower control arms to the axle housing and install the bolts. Do not tighten these bolts to the specified torque at this time.

14   Install the coil springs (Chapter 10).

15   Raise the axle far enough to allow the shock absorber lower mounting bolts to be installed. Again, don't fully tighten these bolts yet.

16   Connect the brake hose to the hydraulic line on the frame and tighten the fitting securely. Make sure the hose isn't kinked when in the installed position.

17   Install the wheels.

18   Lower the vehicle and tighten the lug nuts to the specified torque. Jounce the rear of the vehicle a couple of times, allowing the suspension to come to rest at normal ride height. Slide under the vehicle and tighten the upper control arm-to-axle housing bolt nut, the lower control arm-to-axle housing mounting bolt nuts and the shock absorber lower mounting bolt nuts to the specified torque values.

19   Reconnect the parking brake cables.

20   Bleed the brakes as described in Chapter 9.

Chapter 9   Brakes

Contents

Specifications

General
Brake fluid type .... See Chapter 1

Disc brakes
Disc minimum thickness .... Cast on inside hub surface of disc
Disc runout limit .... 0.003 in
Minimum brake pad thickness .... See Chapter 1

Drum brakes
Drum maximum diameter .... Cast into the brake drum
Minimum brake lining thickness .... See Chapter 1

Torque specifications
	Ft-lbs
Master cylinder-to-booster nuts	13 to 25
Power brake booster nuts	13 to 25
Brake backing plate-to-axle housing	25 to 35
Wheel cylinder bolts	9 to 13

1   General information

Description

All models are equipped with disc type front and drum type rear brakes which are hydraulically operated and vacuum assisted.

The front brakes feature a single piston, floating caliper design. The rear drum brakes are single anchor type with a single two-piston wheel cylinder.

The front disc brakes automatically compensate for pad wear during usage. The rear drum brakes also feature automatic adjustment.

Front brake pads tend to wear at a faster rate than rear brake shoes. Consequently, it's important to inspect the brake pads frequently to make sure they haven't worn to the point where the disc itself is scored or damaged.

All models are equipped with a cable actuated parking brake which operates the rear brakes.

The hydraulic system is a dual line type with a dual master cylinder and separate systems for the front and rear brakes. In the event of brake line or seal failure, half the brake system will still operate. On 1986 models, the master cylinder also incorporates two pressure control valves that reduce the pressure to the rear brakes in order to limit rear wheel lockup during hard braking. On 1987 and later vehicles, a remote combination valve is used which performs the same function. This valve also allows full pressure to the rear brakes if the front system pressure is greatly reduced.

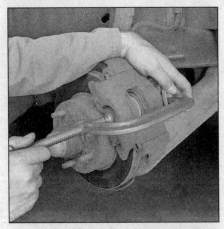

2.4a Using a large C-clamp, push the piston back into the caliper bore just enough to allow the caliper to easily slide off the rotor

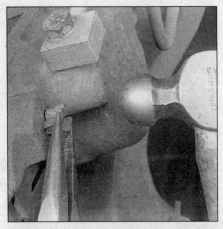

2.4b Squeeze the end of the lower caliper pin with a pair of pliers and knock the pin into the groove as far as possible, . . .

2.4c . . . then use a hammer and punch to drive the pin out completely. Proceed to remove the upper caliper pin in the same manner

2.4d Slide the caliper off the brake disc

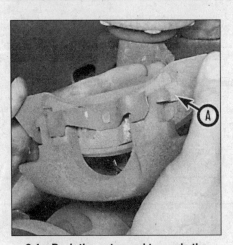

2.4e Push the outer pad towards the piston to dislodge the torque buttons (A), then slide the pad out of the caliper

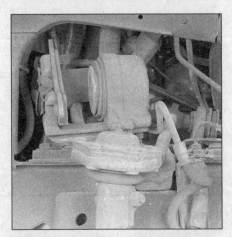

2.4f Set the caliper on top of the upper control arm in such a way that it won't fall off. DO NOT let it hang by the brake hose

Precautions

Use only DOT 3 brake fluid.

The brake pads and linings may contain asbestos fibers, which are hazardous to your

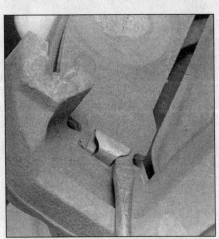

2.4g Using a screwdriver engaged behind the anti-rattle spring, pry the lower end of the inner pad out of the spindle flank

health if inhaled. When working on brake system components, carefully clean all parts with denatured alcohol or brake cleaner. Don't allow the fine dust to become airborne.

Safety should be paramount when working on brake system components. Don't use parts or fasteners that aren't in perfect condition and be sure that all clearances and torque specifications are adhered to. If you're at all unsure about a certain procedure, seek professional advice. When finished working on the brakes, test them carefully under controlled conditions before driving the vehicle in traffic. If a problem is suspected in the brake system, don't drive the vehicle until the fault is corrected.

2   Disc brake pads - replacement

Refer to illustrations 2.4a through 2.4l
Warning: *Disc brake pads must be replaced on both front wheels at the same time - never replace the pads on only one wheel. Also, the*

dust created by the brake system may contain asbestos, which is harmful to your health. Never blow it out with compressed air and don't inhale any of it. An approved filtering mask should be worn when working on the brakes. Do not, under any circumstances, use petroleum-based solvents to clean brake parts. Use brake cleaner or denatured alcohol only!

1   Remove about two-thirds of the fluid from the master cylinder reservoir.
2   Loosen the wheel lug nuts, raise the vehicle and support it securely on jackstands. Remove the front wheels.
3   Check the brake disc carefully as outlined in Section 4. If machining is necessary, follow the procedure in Section 4 to remove the disc.
4   Follow the accompanying photos, beginning with illustration 2.4a, for the actual pad replacement procedure. Be sure to stay in order and read the information in the caption under each illustration.
5   Once the new pads are in place and the caliper pins have been installed, install the

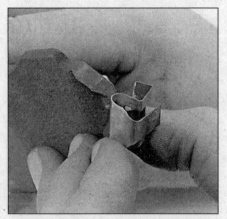

2.4h Install a new anti-rattle clip on the bottom of the inner brake pad

2.4i Place the lower end of the inner pad into its slot on the spindle then push the top of the pad into position against the brake disc

2.4j Push the piston into the cylinder bore completely to provide room for the new pads - use a block of wood and a C-clamp, but don't use excessive force or damage to the plastic piston will result

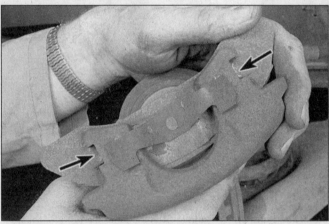

2.4k Push the outer pad into position on the caliper ears, making sure the torque buttons seat fully into the retention notches

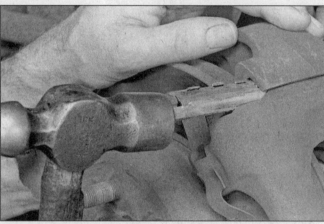

2.4l Place the caliper and outer pad assembly over the disc and inner pad, then drive the caliper pins into their grooves. When both pins have been installed, check to be sure that the pin tabs on each side of the spindle flank are exposed

wheels and lower the vehicle to the ground. **Note:** *If the brake hose was disconnected from the caliper for any reason, the brake system must be bled as described in Section 9.*

6   Fill the master cylinder reservoir with new brake fluid and slowly pump the brakes a few times to seat the pads against the rotor.

7   Check the fluid level in the master cylinder reservoir(s) one more time and then road test the vehicle carefully before driving it in traffic.

3   Caliper - removal, overhaul and installation

Refer to illustrations 3.2, 3.4, 3.5, 3.6, 3.13, 3.15, 3.16a and 3.16b

Note: *Some brake calipers on these vehicles feature cast iron caliper housings and plastic pistons (one per caliper). If an overhaul is indicated (usually because of fluid leakage) explore all options before beginning the job.*

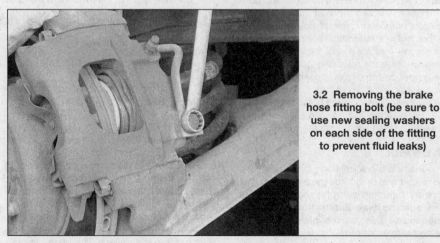

3.2 Removing the brake hose fitting bolt (be sure to use new sealing washers on each side of the fitting to prevent fluid leaks)

New and factory-rebuilt calipers are available on an exchange basis, which makes this job quite easy. If it is decided to rebuild the calipers, make sure that a rebuild kit is available before proceeding.

Removal

1   Raise the vehicle and support it securely on jackstands. Remove the wheel.

2   Disconnect the brake hose fitting from the back of the caliper by removing the bolt **(see illustration)**. Have a rag handy to catch

3.4 With the caliper padded to catch the piston, use compressed air to force the piston out of its bore. Make sure your hands or fingers are not between the piston and caliper

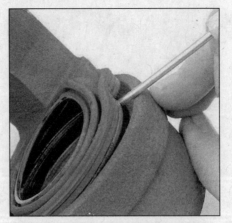

3.5 Carefully pry the dust boot out of the caliper

3.6 The piston seal should be removed with a plastic or wooden tool to avoid damage to the bore and seal groove. A pencil will do the job

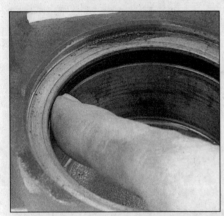

3.13 Push the new seal into the piston groove with your fingers, then check to see that it isn't twisted or kinked

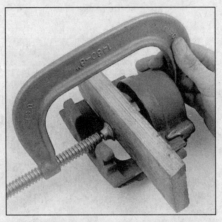

3.15 Use a C-clamp and a block of wood to bottom the piston in the caliper bore. Make sure it goes in perfectly straight, or the sides of the piston may be damaged, rendering it useless

3.16a Position the dust boot in the caliper and stretch the inner diameter of the seal over the piston. Make sure the bead on the boot seats in the piston groove

fluid spills and wrap a plastic bag around the end of the hose to prevent fluid loss and contamination. Discard the inlet fitting washers; use new ones during reassembly. **Note:** *If the caliper will not be completely removed from the vehicle - as for a pad inspection or disc removal- leave the hose connected and place the caliper on top of the upper control arm in such a way that it won't fall off. This will save the trouble of bleeding the brake system.*

3   Refer to the first few Steps in Section 2 to separate the caliper from the spindle.

Overhaul

4   Place a block of wood in the center of the caliper and then force the piston out of its bore by directing compressed air into the inlet opening **(see illustration)**. Use care when doing this, as the piston could be ejected with some force.

5   Pry the dust boot out of the caliper - be careful not to gouge the bore **(see illustration)**.

6   Using a wood or plastic tool, remove the rubber piston seal from the caliper bore **(see illustration)**. *Do not use a metal tool, as the*

bore can be easily damaged.

7   Carefully inspect the piston and the bore for score marks, nicks, cracks, corrosion and damage of any kind. If any of the above are found, the caliper should be replaced with a new or rebuilt unit.

8   Remove the bleeder valve and rubber cap.

9   Inspect the guide pins for corrosion and damage. Replace them with new ones if necessary.

10   Use clean brake fluid or denatured alcohol to clean all the parts. **Warning:** *Do not, under any circumstances, use petroleum-based solvents to clean brake parts.* Allow all parts to dry, preferably using compressed air to blow out all passages. Make sure the compressed air is filtered, as a harmful lubricant residue will be present in unfiltered systems.

11   Check the fit of the piston in the bore by sliding it into the caliper. The piston should move easily.

12   Install the rubber cap over the bleeder valve and thread the valve into the caliper. Tighten it securely.

13   Lubricate the new piston seal with silicone grease or clean brake fluid. Position the

seal in the caliper bore groove, making sure that the seal does not twist **(see illustration)**.

14   Using silicone grease or clean brake fluid, lubricate the piston and the caliper bore.

15   Push the piston into the caliper by hand as far as possible. Using a C-clamp and a block of wood, push the piston all the way to the bottom of the bore. Work slowly, keeping an eye on the side of the piston, making sure it enters the bore perfectly straight, with no resistance **(see illustration)**.

16   Install the dust boot in its bore with the open side of the casing face down. Stretch the inner part of the boot over the piston and drive the seal into place **(see illustrations)**.

17   Refer to Section 2 for the caliper installation procedure.

Installation

18   Install the inlet fitting bolt (with **new** washers) and tighten it securely.

19   Pump the brake pedal a few times to bring the pads into contact with the rotor.

20   Bleed the brakes as described in Section 9. This is not necessary if the inlet fitting was left connected to the caliper.

3.16b Using a hammer and punch, lightly tap the dust boot into its recess

4.3 Check the disc surface for deep grooves and score marks (be sure to inspect both sides)

4.4 Use a dial indicator to check disc runout - if the reading exceeds the maximum allowable runout limit, the rotor will have to be machined or replaced

21   Install the wheel and lower the vehicle. Test the brake operation carefully before placing the vehicle into normal service.

---

4   Brake disc - inspection, removal and installation

Refer to illustrations 4.3, 4.4, 4.5a and 4.5b

Inspection

1   Loosen the wheel lug nuts, raise the vehicle and support it securely on jackstands. Remove the wheel.
2   Remove the caliper and inner brake pad (refer to Section 2).
3   Visually inspect the disc surface for score marks and other damage. Light scratches and shallow grooves are normal after use and may not be detrimental to brake operation. Deep score marks - over 0.015-inch (0.38 mm) - require disc removal and refinishing by an automotive machine shop. Be sure to check both sides of the disc **(see illustration)**.
4   To check disc runout, attach a dial indicator to the brake caliper and locate the stem about 1/2-inch from the outer edge of the disc and just touching the disc surface **(see illustration)**. Set the indicator to zero and turn the disc. The indicator reading should not exceed the specified limit. If it does, the disc should be resurfaced by an automotive machine shop. **Note:** *Professionals recommend resurfacing of brake discs regardless of the dial indicator reading (to produce a smooth, flat surface that will eliminate brake pedal pulsations and other undesirable symptoms related to questionable discs). At the very least, if you elect not to have the discs resurfaced, deglaze the brake pad surface with medium-grit emery cloth (use a swirling motion to ensure a non-directional finish).*
5   The disc must never be machined to a thickness under the specified minimum allowable thickness, which is cast into the inside of the disc itself **(see illustration)**. The

4.5a The minimum thickness limit can be found cast into the inside of the disc (typical)

disc thickness can be checked with a micrometer **(see illustration)**.

Removal and installation

6   Refer to Section 2 and remove the brake caliper. **Warning:** *Don't allow the caliper to hang by the brake hose and don't disconnect the hose from the caliper.*
7   Refer to Chapter 1, "Front wheel bearing check, repack and adjustment" for the remainder of the rotor removal and installation procedure.

---

5   Rear brake shoes - replacement

Refer to illustrations 5.4, 5.5a through 5.5x, and 5.8
Warning: *The brake shoes must be replaced on both rear wheels at the same time - never replace the shoes on only one wheel. Also, brake system dust contains asbestos, which is harmful to your health. Never blow it out with compressed air and don't inhale any of it. Do not, under any circumstances, use*

4.5b Use a micrometer to measure disc thickness at several points around the rotor

petroleum-based solvents to clean brake parts. Use brake cleaner or denatured alcohol only. Whenever the brake shoes are replaced, the return and hold-down springs should also be replaced. Due to the continuos heating/cooling cycle that the springs are subjected to, they lose their tension over a period of time and may allow the shoes to drag on the drum and wear at a much faster rate than normal. When replacing the rear brake shoes, use only high quality, nationally recognized brand-name parts.
1   Remove about two-thirds of the brake fluid from the master cylinder reservoir.
2   Loosen the wheel lug nuts, block the front wheels and raise the rear of the vehicle. Support it on jackstands. Remove the rear wheels from the vehicle.
3   Remove the brake drums from the axle flanges. If the drums won't come off, make sure the parking brake is completely released. If it still won't come off, remove the rubber plug in the backing plate and back off the adjuster screw, retracting the brake shoes from the drum **(see illustration 5.8)**.

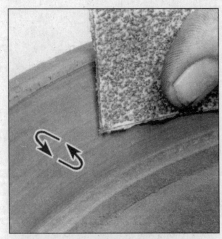

5.4 Remove glaze from the drum surface with sandpaper or emery cloth

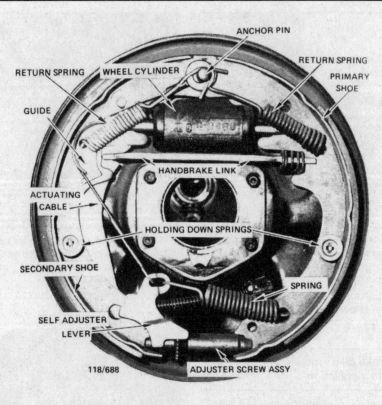

ANCHOR PIN

RETURN SPRING WHEEL CYLINDER

RETURN SPRING

PRIMARY SHOE

GUIDE

HANDBRAKE LINK

ACTUATING CABLE

HOLDING DOWN SPRINGS

SECONDARY SHOE

SPRING

SELF ADJUSTER LEVER

118/688

ADJUSTER SCREW ASSY

5.5a Rear drum brake components - right side shown

5.5b Before removing any internal drum brake components, wash them off with brake cleaner and allow them to dry - position a drain pan under the brake to catch the residue - DO NOT USE COMPRESSED AIR TO BLOW THE BRAKE DUST FROM THE PARTS!

4   Check the drum for cracks, score marks, deep grooves and signs of overheating of the shoe contact surface. If the drums have blue spots, indicating overheated areas, they should be replaced. Also, look for grease or brake fluid on the shoe contact surface. Grease and brake fluid can be removed with denatured alcohol or brake cleaner, but the brake shoes must be replaced if they are contaminated. Surface glazing, which is a glossy, highly polished finish, can be removed with medium-grit emery cloth **(see illustration)**. **Note:** *Professionals recommend*

resurfacing the drums whenever a brake job is done. Resurfacing will eliminate the possibility of out-of-round drums. If the drums are worn so much that they can't be resurfaced without exceeding the maximum allowable diameter (stamped into the drum), then new ones will be required.

5   Follow the accompanying photos **(illustrations 5.5a through 5.5x)** for the actual shoe replacement procedure. Be sure to stay in order and read the information in the caption under each illustration.

6   Once the new shoes are in place, install

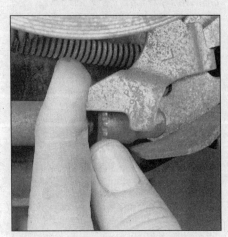

5.5c Pull outward on the adjuster lever and turn the star wheel to retract the brake shoes

5.5d Pull back on the self-adjuster cable and push the adjusting lever toward the rear, unhooking it from the secondary brake shoe

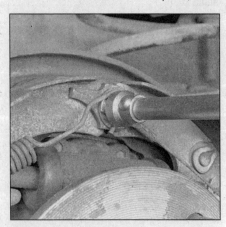

5.5e Remove the primary and secondary shoe return springs. The spring removal tool shown here can be purchased at most auto parts stores and greatly simplifies this step

5.5f Remove the self-adjuster cable and anchor pin plate from the anchor pin

5.5g Remove the secondary shoe retractor spring and cable guide from the secondary shoe

5.5h Remove the primary shoe hold down spring and pin . . .

5.5i . . . then lift the primary shoe and adjusting screw from the backing plate

5.5j Remove the parking brake link

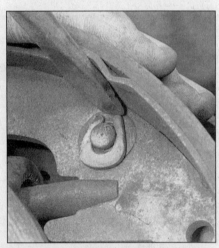

5.5k Remove the secondary shoe hold down spring and pin, lift the shoe from the backing plate. Pry the parking brake lever retaining clip off of the pivot pin . . .

5.5l . . . then separate the lever from the secondary shoe. Be careful not to lose the spring washer

5.5m Lubricate the brake shoe contact areas (arrows) with high temperature grease

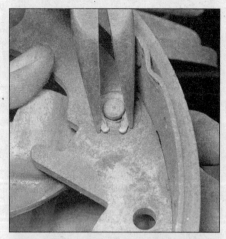

5.5n Attach the new shoe to the parking brake lever, install the spring washer and retaining clip on the pivot pin then crimp the clip closed with a pair of pliers

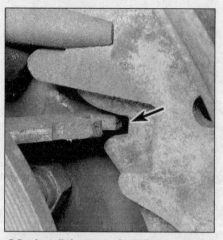

5.5o Install the secondary shoe and hold down spring to the backing plate, then position the end of the parking brake link into the notch in the shoe (arrow)

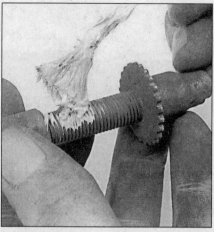

5.5p Lubricate the adjusting screw with high temperature multipurpose grease

5.5q Place the primary shoe against the backing plate then install the hold down spring. Make sure the parking brake strut and wheel cylinder pushrods engage in the brake shoe slots

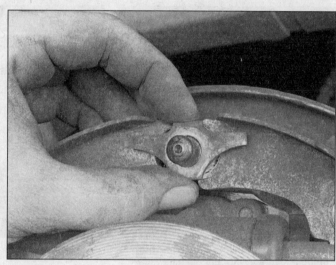

5.5r Install the anchor pin plate . . .

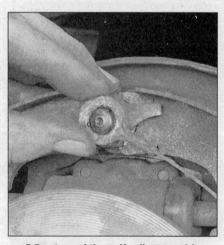

5.5s . . . and the self-adjuster cable

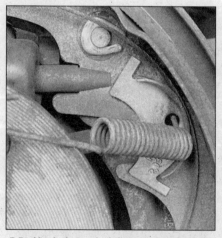

5.5t Hook the end of the secondary shoe retractor spring through the cable guide and into the hole in the shoe, then stretch the spring over the anchor pin

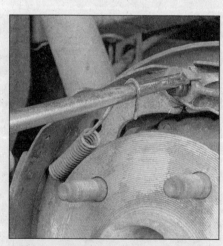

5.5u Install the primary shoe retractor spring. The tool shown here is available at most auto parts stores and makes this step much easier and safer

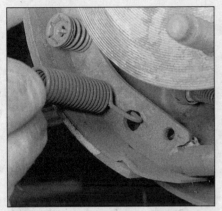

5.5v Hook the adjuster lever spring into the hole at the bottom of the primary shoe

5.5w Hook the adjuster lever spring and cable into the adjuster lever and pull the cable down and to the rear, inserting the hook on the lever into the hole in the secondary shoe

5.5x Wiggle the brake assembly to ensure that the shoes are centered on the backing plate

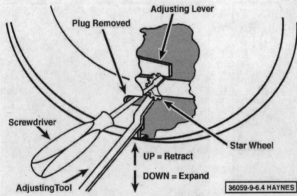

5.8 Using a screwdriver or brake adjuster tool, turn the adjuster wheel in the direction shown until the shoes drag on the brake drum. Then, insert a small screwdriver through the backing plate to move the adjusting lever away from the adjuster wheel and turn the wheel in the opposite direction until the drum turns freely

6.4 Completely loosen the inlet tube fitting then remove the two mounting bolts

the drums to the axle flanges.

7   Remove the rubber plug from the brake backing plate.

8   Insert a narrow screwdriver or brake adjusting tool through the adjustment hole and turn the star wheel until the brakes drag slightly as the drum is turned **(see illustration)**.

9   Turn the star wheel in the opposite direction until the drum turns freely. Keep the adjuster lever from contacting the star wheel or it won't turn. This can be done by pushing on it with a narrow screwdriver.

10   Repeat the adjustment on the opposite wheel.

11   Install the plugs in the backing plate access holes.

12   Install the wheels and lower the vehicle. Tighten the lug nuts to the specified torque.

13   Top up the master cylinder with brake fluid and pump the pedal several times. Lower the vehicle and make a series of quick, harsh stops in forward and reverse to allow the brakes to further adjust themselves. Confirm proper brake operation before placing the vehicle in normal traffic service.

6   Wheel cylinder - removal, overhaul and installation

Refer to illustrations 6.4 and 6.7
Note: *If an overhaul is indicated (usually*

because of fluid leakage) explore all options before beginning the job. New wheel cylinders are available, which makes this job quite easy. If it is decided to rebuild the wheel cylinder, make sure that a rebuild kit is available before proceeding.

Removal

1   Raise the rear of the vehicle and support it securely on jackstands.

2   Remove the brake shoe assembly (Section 5).

3   Carefully clean all dirt and foreign material from around the wheel cylinder.

4   Disconnect the brake fluid inlet line **(see illustration)**. Do not pull the brake line away from the wheel cylinder.

5   Remove the two bolts that retain the wheel cylinder to the brake backing plate.

6   Remove the wheel cylinder from the brake backing plate and place it on a clean workbench. Immediately plug the brake line to prevent fluid loss and contamination.

Overhaul

7   Remove the bleeder valve, seals, pistons, boots and spring assembly from the cylinder body **(see illustration)**.

6.7 Exploded view of a typical wheel cylinder

1   Pushrods
2   Dust boots
3   Pistons
4   Cups
5   Expander spring
6   Wheel cylinder

7.2 Unscrew the brake line fittings from the master cylinder - a flare nut wrench is recommended

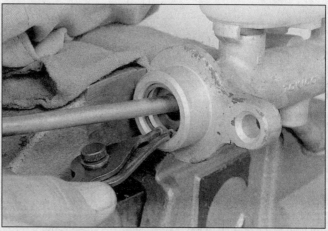

7.7 Use a Phillips head screwdriver to push the primary piston into the cylinder, then remove the snap-ring

7.8 Remove the primary piston assembly from the cylinder

7.9 Tap the master cylinder against a block of wood to eject the secondary piston assembly

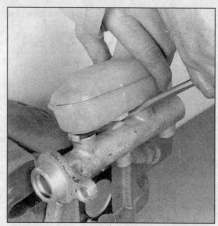

7.10 If it is necessary to remove the fluid reservoir (to replace leaking seals or a broken reservoir), gently pry it off with a screwdriver or pry bar

8   Clean the wheel cylinder with brake fluid, denatured alcohol or brake system cleaner. **Warning:** *Do not, under any circumstances, use petroleum based solvents to clean brake parts.*

9   Use compressed air to remove excess fluid from the wheel cylinder and to blow out the passages.

10   Check the cylinder bore for corrosion and scoring. Crocus cloth may be used to remove light corrosion and stains, but the cylinder must be replaced with a new one if the defects cannot be removed easily, or if the bore is scored.

11   Lubricate the new seals with brake fluid.

12   Assemble the brake cylinder, making sure the boots are properly seated.

Installation

13   Place the wheel cylinder in position against the backing plate and install the bolts, tightening them to the specified torque.

14   Connect the brake line and install the brake shoe assembly.

15   Bleed the brakes following the procedure described in Section 9.

7   Master cylinder - removal, overhaul and installation

Note: *Before deciding to overhaul the master cylinder, investigate the availability and cost of a new or factory rebuilt unit and also the availability of a rebuild kit. The master cylinders on 1990 and later models cannot be rebuilt.*

Removal

Refer to illustration 7.2

1   Place rags under the brake line fittings and prepare caps or plastic bags to cover the ends of the lines once they are disconnected. **Caution:** *Brake fluid will damage paint. Cover all body parts and be careful not to spill fluid during this procedure.*

2   Loosen the tube nuts at the ends of the brake lines where they enter the master cylinder. To prevent rounding off the flats on these nuts, a flare-nut wrench, which wraps around the nut, should be used **(see illustration)**.

3   Pull the brake lines away from the master cylinder slightly and plug the ends to pre-

vent contamination.

4   Disconnect the brake warning lamp electrical connector, remove the two master cylinder mounting nuts and remove the master cylinder from the vehicle.

5   Remove the reservoir cap, then discard any fluid remaining in the reservoir.

Overhaul

Refer to illustrations 7.7, 7.8, 7.9, 7.10, 7.14 and 7.19

6   Mount the master cylinder in a vise with the vise jaws clamping on the mounting flange.

7   Remove the primary piston snap ring by depressing the piston and extracting the ring with a pair of snap ring pliers **(see illustration)**.

8   Remove the primary piston assembly from the cylinder bore **(see illustration)**.

9   Remove the secondary piston assembly from the cylinder bore. It may be necessary to remove the master cylinder from the vise and invert it, carefully tapping it against a block of wood to expel the piston **(see illustration)**.

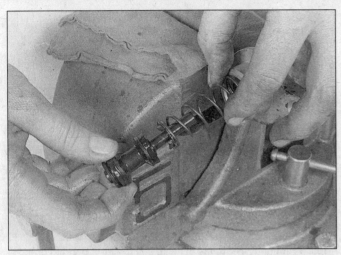

7.14 Coat the secondary piston with clean brake fluid and install it into the master cylinder - spring end first

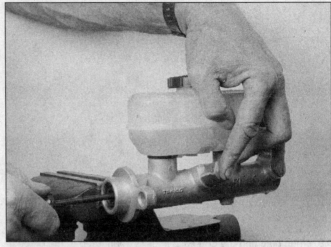

7.19 When bench bleeding the master cylinder, start with the rear brake outlet port

10   If fluid has been leaking past the reservoir grommets, pry the reservoir from the cylinder body with a screwdriver (see illustration). Remove the grommets.

11   Inspect the cylinder bore for corrosion and damage. If any corrosion or damage is found, replace the master cylinder body with a new one, as abrasives cannot be used on the bore.

12   Lubricate the new reservoir grommets with silicone lubricant and press them into the master cylinder body. Make sure they're properly seated.

13   Lay the reservoir on a hard surface and press the master cylinder body onto the reservoir, using a rocking motion.

14   Lubricate the cylinder bore and primary and secondary piston assemblies with clean brake fluid. Install the secondary piston assembly into the cylinder (see illustration).

15   Install the primary piston assembly in the cylinder bore, depress it and install the snap ring.

16   Inspect the reservoir cap and diaphragm for cracks and deformation. Replace any damaged parts with new ones and attach the diaphragm to the cover.

17   Note: Whenever the master cylinder is removed, the complete hydraulic system must be bled. The time required to bleed the system can be reduced if the master cylinder is filled with fluid and bench bled (refer to Steps 18 through 22) before the master cylinder is installed on the vehicle.

18   Clamp the master cylinder mounting flange in a vise, then fill the reservoir with brake fluid.

19   Loosen one plug at a time, starting with the rear outlet port first, and push the piston assembly into the bore to force air from the master cylinder (see illustration). To prevent air from being drawn back into the cylinder, the appropriate plug must be replaced before allowing the piston to return to its original position.

20   Stroke the piston three or four times for each outlet to ensure that all air has been expelled.

21   Since high pressure is not involved in the bench bleeding procedure, an alternative to the removal and replacement of the plugs with each stroke of the piston assembly is available. Before pushing in on the piston assembly, remove one of the plugs completely. Before releasing the piston, however, instead of replacing the plug, simply put your finger tightly over the hole to keep air from being drawn back into the master cylinder. Wait several seconds for the brake fluid to be drawn from the reservoir to the piston bore, then repeat the procedure. When you push down on the piston it will force your finger off the hole, allowing the air inside to be expelled. When only brake fluid is being ejected from the hole, replace the plug and go on to the other port.

22   Refill the master cylinder reservoirs and install the diaphragm and cover assembly.

Installation

23   Carefully install the master cylinder by reversing the removal steps, then bleed the brakes (refer to Section 9).

8   Brake hoses and lines - inspection and replacement

Refer to illustration 8.3

Inspection

1   About every six months, with the vehicle raised and supported securely on jackstands, the rubber hoses which connect the steel brake lines with the front and rear brake assemblies should be inspected for cracks, chafing of the outer cover, leaks, blisters and other damage. These are important and vulnerable parts of the brake system and inspection should be complete. A light and mirror will be helpful for a thorough check. If a hose exhibits any of the above conditions, replace it with a new one.

Flexible hose replacement

2   Clean all dirt away from the ends of the hose.

3   Disconnect the brake line from the hose fitting using a back-up wrench on the fitting (see illustration). Be careful not to bend the frame bracket or line. If necessary, soak the connections with penetrating oil.

4   Unbolt the hose bracket from the strut assembly.

5   Remove the U-clip from the female fitting at the bracket (see illustration 8.3) and remove the hose from the bracket.

6   Disconnect the hose from the caliper, discarding the copper washers on either side of the fitting block.

7   Using new copper washers, attach the new brake hose to the caliper.

8   Pass the female fitting through the frame bracket. With the least amount of twist in the hose, install the fitting in this position (use the paint stripe on the hose to help determine twist). Note: The weight of the vehicle should

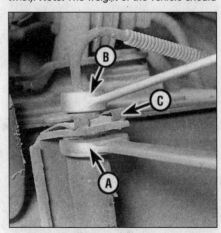

8.3 To disconnect a brake hose from the brake line fitting, place a backup wrench on the hose fitting (A) and loosen the tube nut (B) with a flare nut wrench. Remove the U-clip (C) to detach the hose from the frame bracket

9.8 When bleeding the brakes, a hose is connected to the bleeder valve at the caliper or wheel cylinder and then submerged in brake fluid. Air will be seen as bubbles in the tube and container. All air must be expelled before moving to the next wheel

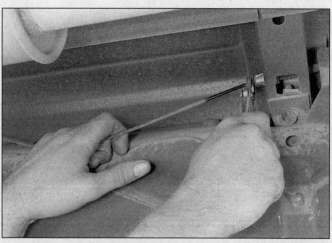

10.2 From underneath the vehicle, pull the front parking brake cable to the rear and clamp a pair of locking pliers on it, right where it exits the cable casing at the frame crossmember

be on the suspension, so the vehicle should not be raised while positioning the hose.

9   Install the U-clip in the female fitting at the frame bracket.

10   Attach the brake line to the hose fitting using a back-up wrench on the fitting.

11   Mount the brake hose bracket to the strut assembly.

12   Carefully check to make sure the suspension or steering components don't make contact with the hose. Have an assistant push on the vehicle and also turn the steering wheel from lock-to-lock during inspection.

13   Bleed the brake system as described in Section 9.

Rigid brake line replacement

14   When replacing brake lines, be sure to use the correct parts. Don't use copper tubing for any brake system components. Purchase steel brake lines from a dealer or auto parts store.

15   Prefabricated brake line, with the tube ends already flared and fittings installed, is available at auto parts stores and dealers. These lines are also bent to the proper shapes.

16   If prefabricated lines aren't available, obtain the recommended steel tubing and fittings to match the line to be replaced. Determine the correct length by measuring the old brake line (a piece of string can usually be used for this) and cut the new tubing to length, allowing about 1/2-inch extra for flaring the ends.

17   Install the fitting over the cut tubing and flare the ends of the line with a flaring tool.

18   If necessary, carefully bend the line to the proper shape. A tube bender is recommended for this. **Warning:** *Do not crimp or damage the line.*

19   When installing the new line, make sure it's securely supported in the brackets and has plenty of clearance between moving or hot components.

20   After installation, check the master

cylinder fluid level and add fluid as necessary. Bleed the brake system as outlined in the next Section and test the brakes carefully before driving the vehicle in traffic.

9   Brake hydraulic system - bleeding

Refer to illustration 9.8
Warning: *Wear eye protection when bleeding the brake system. If the fluid comes in contact with your eyes, immediately rinse them with water and seek medical attention.*

1   Bleeding the hydraulic system is necessary to remove any air that manages to find its way into the system as a result of removal and installation of a hose, line, caliper, wheel cylinder or master cylinder. Use only the specified fluid in this system or extensive damage could result. It will probably be necessary to bleed the system at all four brakes if air has entered the system due to low fluid level, or if the brake lines have been disconnected at the master cylinder.

2   If a brake line was disconnected only at one wheel, then only that caliper (or wheel cylinder) must be bled.

3   If a brake line is disconnected at a fitting located between the master cylinder and any of the brakes, that part of the system served by the disconnected line must be bled.

4   Remove any residual vacuum from the power brake booster by applying the brake several times with the engine off.

5   Remove the master cylinder reservoir cap and fill the reservoir with brake fluid. Reinstall the cover. **Note:** *Check the fluid level often during the bleeding operation and add fluid as necessary to prevent the level from falling low enough to allow air bubbles into the master cylinder.*

6   Have an assistant on hand, as well as a supply of new brake fluid, an empty clear plastic container, a length of 3/16-inch clear

plastic or vinyl tubing to fit over the bleeder screw and a wrench to open and close the bleeder screw.

7   Beginning at the right rear wheel, loosen the bleeder screw slightly, then tighten it to a point where it's snug but can still be loosened quickly and easily.

8   Place one end of the tubing over the bleeder screw and submerge the other end in brake fluid in the container **(see illustration)**.

9   Have an assistant pump the brakes a few times to get pressure in the system, then hold the pedal down.

10   While the pedal is held down, open the bleeder screw until brake fluid begins to flow. Watch for air bubbles to exit the submerged end of the tube. When the fluid flow slows after a couple of seconds, tighten the screw and have your assistant release the pedal.

11   Repeat Steps 9 and 10 until no more air is seen leaving the tube, then tighten the bleeder screw and proceed to the left rear wheel, the right front wheel and the left front wheel, in that order, and perform the same procedure. Be sure to check the fluid in the master cylinder reservoir frequently.

12   Never use old brake fluid. It contains moisture which will deteriorate the brake system components.

13   Refill the master cylinder with fluid at the end of the operation.

14   Check the operation of the brakes. The pedal should feel solid when depressed, with no sponginess. If necessary, repeat the entire process. **Warning:** *Do not operate the vehicle if you are in doubt about the effectiveness of the brake system.*

10   Parking brake cables - replacement

Refer to illustrations 10.2, 10.4 and 10.12
Note: *The parking brake system is self-adjusting.*

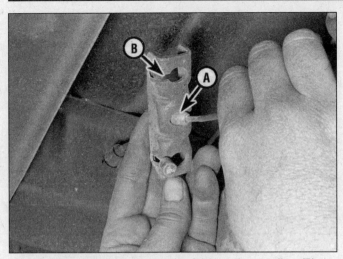

10.4 Disconnect the front cable (A) from the equalizer (B)

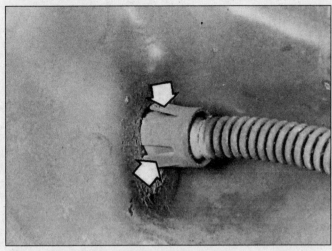

10.12 Depress the retention tangs to free the cable and casing from the brake backing plate

Front cable

1   Raise the vehicle and support it securely on jackstands. Make sure the parking brake handle is in the released position.

2   Firmly grasp the front cable near the equalizer and pull it toward the rear. Clamp a pair of locking pliers on the cable right where the cable casing fastens to the frame **(see illustration)**.

3   Remove the parking brake handle boot cover and insert a steel pin (a nail or drill bit will work) into the pawl lock-out pin hole from the inboard side (larger hole) at a slightly upward and forward angle. Push the pin in as far as it will go then maneuver it downward and to the rear to move the adjuster pawl out of the way. The pin should now pass through the outer hole.

4   Disconnect the front cable from the equalizer **(see illustration)** and remove the locking pliers.

5   Depress the tangs that retain the cable casing to the frame crossmember **(see illustration 10.2)** and push the cable and casing through the frame. Repeat this procedure to the front end of the casing at the parking brake handle reinforcement bracket.

6   Remove the cable anchor pin from the control assembly and guide the cable from the adjuster mechanism.

7   To install the new cable, reverse the removal procedure. Make sure that all of the cables are connected to the equalizer before the steel pin is removed from the parking brake handle.

Rear cable(s)

8   Perform Steps 1 and 2 of the front cable replacement procedure.

9   Unhook the rear cable from the equalizer, compress the retainer tangs and push the cable and casing through the frame crossmember.

10   Pull the cable and casing through the guide brackets near the shock absorber upper mount.

11   Remove the brake drum and brake shoe assembly following the procedure described in Section 5.

12   Lift the cable out of the slot in the parking brake lever, then depress the tangs on the cable casing retainer in the backing plate **(see illustration)**.

13   Installation is the reverse of the removal procedure.

11   Parking brake handle - removal and installation

1   Refer to Section 10, Steps 1 through 3 for the adjusting mechanism tension release procedure.

2   Remove the front cable anchor pin from the adjusting mechanism and guide the cable from the adjuster.

3   Remove the two mounting bolts and remove the parking brake handle.

4   Do not remove the lock-out pin from the hole in the handle until the handle is installed and all of the parking brake cables are connected.

5   Installation is the reverse of the removal procedure.

12   Brake pedal - removal and installation

Refer to illustration 12.2

Removal

1   Disconnect the cable from the negative battery terminal.

2   Working under the dash, unplug the connector from the brake light switch. Remove the brake booster pushrod retaining clip and slide the pushrod and switch off of the pedal pin **(see illustration)**.

3   On vehicles equipped with manual transmissions, disconnect the clutch master

cylinder pushrod from the pedal (refer to Chapter 8 if necessary).

4   Remove the right-side retainer from the pedal pivot shaft. Support the brake pedal and slide the shaft out to the left. On manual transmission vehicles, the pivot shaft is connected to the clutch pedal, so the clutch pedal must be removed as well.

5   Inspect the bushings for wear, replacing them as necessary. Lubricate the bushings and pivot shaft with a light coat of engine oil prior to installation.

Installation

6   Place the pedal into position and install the pivot shaft and retainer.

7   Attach the brake booster pushrod and brake light switch and install the retaining clip. On manual transmission vehicles install the clutch master cylinder pushrod.

8   Connect the wire harness to the brake light switch and pump the brake pedal several times to ensure proper operation. Reconnect the battery.

12.2 Remove the brake booster pushrod retaining clip (arrow) and slide the rod and brake light switch off of the pedal pin

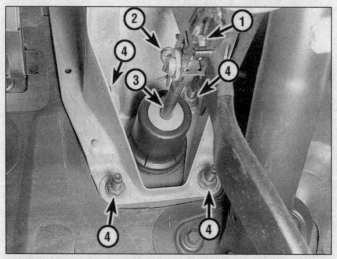

13.6 The following under-dash components must be removed to allow power brake booster removal:

1   Brake light switch
    connector
2   Pushrod retaining clip
3   Pushrod
4   Booster mounting nuts

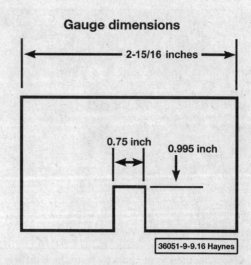

Gauge dimensions

2-15/16 inches

0.75 inch

0.995 inch

36051-9-9.16 Haynes

13.11 To adjust the booster master cylinder pushrod, fabricate a gauge block of the dimensions shown. Place it against the master cylinder mounting surface of the booster and adjust the end of the pushrod so it just touches the slot in the gauge block

13   Power brake booster - removal, installation and adjustment

Refer to illustrations 13.6 and 13.11

1   The power brake booster unit requires no special maintenance apart from periodic inspection of the vacuum hose and the case.
2   Dismantling of the brake booster requires special tools and is not ordinarily done by the home mechanic. If a problem develops, install a new or factory rebuilt unit.

Removal

3   Remove the master cylinder (refer to Section 7 in this Chapter).
4   Disconnect the vacuum hose where it attaches to the power brake booster.
5   Working in the passenger compartment under the steering column, unplug the wiring connector from the brake light switch, then remove the pushrod retaining clip and nylon washer from the brake pedal pin. Slide the pushrod off the pin.
6   Remove the nuts attaching the brake booster to the firewall **(see illustration)**.
7   Carefully detach the booster from the firewall and lift it out of the engine compartment.

Installation

8   Place the booster into position on the firewall and tighten the mounting nuts. Connect the pushrod and brake light switch to the brake pedal. Install the retaining clip in the brake pedal pin.
9   Install the master cylinder and vacuum hose. Refer to Section 7 for the master cylinder bleeding procedure.
10   Carefully check the operation of the brakes before driving the vehicle in traffic.

Adjustment

11   Some boosters feature an adjustable pushrod. They are matched to the booster at the factory and most likely will not require adjustment, but if a misadjusted pushrod is suspected, a gauge can be fabricated out of heavy gauge sheet metal using the accompanying template **(see illustration)**.
12   Some common symptoms caused by a misadjusted pushrod include dragging brakes (if the pushrod is too long) or excessive brake pedal travel accompanied by a groaning sound from the brake booster (if the pushrod is too short).
13   To check the pushrod length, unbolt the master cylinder from the booster and position it to one side. It isn't necessary to disconnect the hydraulic lines, but be careful not to bend them.
14   Place the pushrod gauge against the master cylinder mounting surface of the booster **(see illustration 13.11)**. The pushrod should just touch the inner edge of the slot in the gauge. If it doesn't, adjust it by holding the knurled portion of the pushrod with a pair of pliers and turning the end with a wrench.
15   When the adjustment is complete, reinstall the master cylinder and check for proper brake operation before driving the vehicle in traffic.

14   Brake light switch - removal and installation

Removal

1   Locate the switch near the top of the brake pedal and disconnect the wiring harness **(see illustration 13.6)**.

2   Remove the pushrod retaining clip and nylon washer from the brake pedal pin and slide the pushrod off far enough for the outer hole of the switch to clear the pin. Now pull up on the switch to remove it.

Installation

3   Position the switch so it straddles the pushrod and the slot on the inner side of the switch rests on the pedal pin. Slide the pushrod and switch back onto the pin, then install the nylon washer and retaining clip.
4   Reconnect the wiring harness.
5   Check the brake lights for proper operation.

15   Anti-lock brakes - general information

Rear anti-lock brake system (RABS) became available in 1989. If there is a fault with the RABS, a REAR ANTI-LOCK light will illuminate on the instrument panel. Due to the complex nature of RABS, the do-it-yourself mechanic should take the vehicle to a dealer service department or to a repair shop that can service RABS.

Chapter 10
Suspension and steering systems

Contents

Specifications

Torque specifications

Ft-lbs

Front suspension

Upper control arm-to-adjusting arm nuts	70 to 100
Crossmember-to-frame rail bolts	
Front	135 to 145
Center and rear	145 to 155
Lower control arm pivot bolts	
1986 and 1987	187 to 260
1988 on	100 to 140
Lower balljoint spindle nut*	80 to 113
Upper balljoint-to-spindle pinch bolt and nut	27 to 37

Rear suspension

Upper control arm-to-right frame bracket nut and bolt	
1986 through 1989	100 to 145
1990	155 to 210
1991 on	92 to 125
Upper control arm-to-left frame bracket nut and bolt	60 to 100
Upper control arm-to-axle housing nut and bolt	
1986 through 1989	100 to 145
1990 on	155 to 210
Lower control arm-to-axle housing nut and bolt	
1986 through 1989	100 to 145
1990 on	95 to 130
Lower control arm-to-frame bracket nut and bolt	
1986 through 1989	100 to 145
1990 on	95 to 130

Torque specifications

Ft-lbs

Steering system

Tie-rod end-to-spindle arm nuts*	52 to 74
Steering gear mounting bolts	65 to 90
Intermediate shaft-to-steering gear pinion shaft bolt	30 to 42
Intermediate shaft-to-steering column shaft nut and bolt	30 to 42
Steering wheel hub bolt	23 to 33

Tighten to the minimum specified torque, then align the next castellation in the nut with the cotter pin hole.

1.1 Front suspension and steering components

1	Stabilizer bar	4	Upper control arm	7	Steering gear	10	Tie-rod end
2	Shock absorber	5	Spindle	8	Tie-rod	11	Front crossmember
3	Lower control arm	6	Coil spring	9	Steering gear boot		

1   General information

Refer to illustrations 1.1 and 1.2
Warning: *Whenever any of the suspension or steering fasteners are loosened or removed they must be inspected and if necessary, replaced with new ones of the same part number or of original equipment quality and design. Torque specifications must be followed for proper reassembly and component retention. Never attempt to heat, straighten or weld any suspension or steering component.*

Instead, replace any bent or damaged part with a new one.

The front suspension is independent, allowing each wheel to compensate for road surface changes without appreciably affecting the other. Each wheel is connected to the frame by a spindle , balljoints and upper and lower control arms. Each side uses a coil spring mounted between the lower control arm and the frame. Shock absorbers are positioned in the center of the coil spring, with the upper end fastened to the chassis and the lower end bolted to the lower control

arm. Body side roll is controlled by a stabilizer bar.

The rear suspension is comprised of the axle, two coil springs, shock absorbers, two lower control arms and one upper control arm.

The steering system consists of the steering wheel, steering column, an articulated intermediate shaft, the steering gear (either power or manual), power steering pump and the tie rods, which connect the steering gear to the spindle.

1.2 Rear suspension components

1	Shock absorber	3	Lower control arm	5	Rear axle housing
2	Coil spring	4	Upper control arm		

2.3 A pair of locking pliers clamped to the stabilizer bar link will prevent it from turning while loosening the nut

2 Front stabilizer bar - removal and installation

Refer to illustrations 2.3 and 2.4

Removal

1   Raise the vehicle and support it securely on jackstands. Apply the parking brake.
2   Remove the air deflector (see Chapter 11).

2.4 The stabilizer bar is held to the frame with two brackets like this. The rubber bushings should be replaced if they are hard, cracked or otherwise deformed

3   Remove the stabilizer bar-to-lower control arm nuts and bolts, noting how the links, washers and bushings are positioned. Clamp a pair of locking pliers to the stabilizer bar link to prevent it from turning **(see illustration)**.
4   Remove the stabilizer bar bracket bolts and detach the bar from the vehicle **(see illustration)**.
5   Pull the brackets off the stabilizer bar and inspect the bushings for cracks, harden-

ing and other signs of deterioration. If the bushings are damaged, replace them with new ones.

Installation

6   Position the stabilizer bar bushings on the bar with the slits facing the front of the vehicle.
7   Push the brackets over the bushings and raise the bar up to the frame. Install the bracket bolts but don't tighten them completely at this time.
8   Install the stabilizer bar-to-lower control arm nuts and rubber bushings and tighten them securely.
9   Tighten the bracket bolts.

3 Front shock absorbers - removal and installation

Refer to illustrations 3.2 and 3.3

Removal

1   Loosen the wheel lug nuts, raise the vehicle and support it securely on jackstands. Apply the parking brake. Remove the wheel.
2   Remove the upper shock absorber stem nut. Use an open end wrench to keep the stem from turning. If the nut won't loosen because of rust, squirt some penetrating oil

3.2 If the shock absorber upper mounting nut is rusted tight, it will probably be necessary to hold the stem from turning with a pair of locking pliers

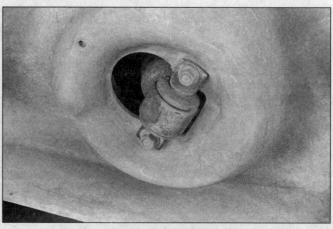

3.3 The lower end of the shock absorber is bolted to the lower control arm

on the stem threads and allow it to soak in for awhile. It may be necessary to keep the stem from turning with a pair of locking pliers, since the flats provided for a wrench are quite small **(see illustration)**.

3   Remove the two lower shock mount bolts **(see illustration)** and pull the shock absorber out through the bottom of the lower control arm. Remove the washers and the rubber grommets from the top of the shock absorber.

Installation

4   Extend the new shock absorber as far as possible. Position a new washer and rubber grommet on the stem and guide the shock up through the coil spring and into the upper mount.

5   Install the upper rubber grommet and washer and wiggle the stem back-and-forth to ensure that the grommets are centered in the mount. Tighten the stem nut securely.

6   Install the lower mounting bolts and tighten them securely.

4   Balljoints - replacement

At the time of writing, the balljoints on these vehicles were not replaceable separately. If they are in need of replacement, check your local auto parts store to see if replacement balljoints are now available. If so, remove the control arm and take it and the new balljoint to an automotive machine shop to have the old balljoint pressed out and the new one pressed in. If a replacement balljoint is not available, the entire control arm will have to be replaced.

5   Upper control arm - removal and installation

Refer to illustrations 5.3a and 5.3b

Removal

1   Loosen the wheel lug nuts, raise the

5.3a To separate the upper control arm from the spindle, support the lower control arm with a jack, remove the pinch bolt . . .

front of the vehicle and support it securely on jackstands. Apply the parking brake. Remove the wheel.

2   Support the lower control arm with a jack and raise it until the upper control arm no longer contacts the rebound bumper. The support point must be as close to the balljoint as possible to give maximum leverage on the lower control arm.

3   Remove the upper balljoint stud-to-spindle pinch bolt, then carefully pry the upper control arm from the spindle **(see illustrations)**. **Note:** *DO NOT use a "pickle fork" type balljoint separator - it may damage the balljoint seals.*

4   Remove the control arm-to-adjusting arm nuts, recording the position of any alignment shims. They must be reinstalled in the same location to maintain wheel alignment.

5   Detach the control arm from the adjusting arm.

6   To remove the adjusting arm and mounting bracket assembly, position another

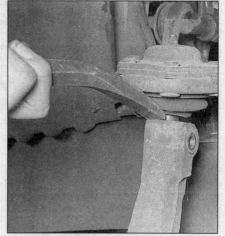

5.3b . . . then carefully pry the ball stud out of the spindle

jack under the crossmember and remove the mounting bracket-to-frame bolts and nuts. DO NOT remove the jack that is supporting the lower control arm!

Installation

7   If the adjusting arm and mounting bracket assembly was removed, position it on the frame rail and install the bolts, tightening them to the specified torque.

8   Position the control arm on the adjusting arm and install the nuts loosely. Install any alignment shims that were removed. Tighten the nuts to the specified torque.

9   Insert the balljoint stud into the spindle, install the pinch bolt and tighten the nut to the specified torque.

10   Install the wheel and lug nuts and lower the vehicle. Tighten the lug nuts to the specified torque.

11   Drive the vehicle to an alignment shop to have the front end alignment checked and, if necessary, adjusted.

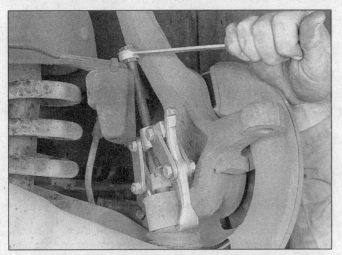

6.5 A two-jaw puller can be used to disconnect the lower balljoint from the spindle. Notice that the nut hasn't been completely removed - this is to prevent the two components from separating violently

8.4 Support the inner edge of the lower control arm with a jack and loop a length of chain through the spring and around the control arm, bolting the ends together. Remove the pivot bolts and carefully lower the jack

6   Lower control arm - removal and installation

Refer to illustration 6.5

Removal

1   Loosen the wheel lug nuts, raise the vehicle and support it securely on jackstands. Apply the parking brake. Remove the wheel.
2   Remove the shock absorber (refer to Section 3).
3   Disconnect the stabilizer bar from the lower control arm (Section 2).
4   Remove the coil spring as described in Section 8.
5   Remove the cotter pin and back off the lower control arm balljoint stud nut two turns. Using a two-jaw puller, break the balljoint loose from the spindle and remove the nut **(see illustration)**.
6   Remove the control arm from the vehicle.

Installation

7   Insert the balljoint stud into the spindle, tighten the nut to the specified torque and install a new cotter pin. If necessary, tighten the nut slightly to align a slot in the nut with the hole in the balljoint stud.
8   Install the coil spring (Section 8) and the lower control arm pivot bolts and nuts, but don't completely tighten them at this time.
9   Install the shock absorber.
10   Connect the stabilizer bar to the lower control arm.
11   Install the wheel and lug nuts, lower the vehicle and tighten the lug nuts to the specified torque.
12   With the vehicle at normal ride height, tighten the lower control arm pivot bolt nuts to the specified torque.
13   Drive the vehicle to an alignment shop to have the front end alignment checked and, if necessary, adjusted.

7   Spindle - removal and installation

Removal

1   Loosen the wheel lug nuts, raise the vehicle and support it securely on jackstands placed under the frame. Apply the parking brake. Remove the wheel.
2   Remove the brake caliper and place it on top of the upper control arm (see Chapter 9 if necessary).
3   Remove the brake disc and hub assembly (see "Wheel bearings" in Chapter 1).
4   Remove the splash shield from the spindle.
5   Separate the tie-rod end from the spindle arm (see Section 17).
6   Position a floor jack under the lower control arm and raise it slightly to take the spring pressure off the suspension stop. The jack must remain in this position throughout the entire procedure.
7   Remove the cotter pins from the lower balljoint studs and back off the nut two turns.
8   Break the balljoint loose from the spindle with a two-jaw puller **(see illustration 6.5)** or an equivalent balljoint separator. **Note:** *A pickle fork type balljoint separator may damage the balljoint seals.*
9   Remove the upper control arm-to-spindle pinch bolt and separate the control arm from the spindle **(see illustration 5.3b)**.
10   Remove the nut from the lower balljoint stud, separate the lower control arm from the spindle and remove the spindle from the vehicle.

Installation

11   Place the spindle between the upper and lower control arms and insert the balljoint studs into the knuckle, beginning with the lower balljoint. Install the bolt and nut and tighten them to the specified torque. Install a new cotter pin, tightening the nut slightly to align a slot in the nut with the hole in the

balljoint stud, if necessary.
12   Install the splash shield.
13   Connect the tie-rod end to the spindle arm and tighten the nut to the specified torque. Be sure to use a new cotter pin.
14   Install the brake disc and adjust the wheel bearings following the procedure outlined in Chapter 1.
15   Install the brake caliper.
16   Install the wheel and lug nuts. Lower the vehicle to the ground and tighten the nuts to the specified torque.

8   Front coil spring - removal and installation

Refer to illustrations 8.4, 8.8 and 8.11

Removal

1   Loosen the wheel lug nuts, raise the vehicle and support it securely on jackstands placed under the frame. Apply the parking brake. Remove the wheel.
2   Remove the shock absorber (Section 3).
3   Disconnect the stabilizer bar from the lower control arm (Section 2).
4   Position a jack under the inner edge of the lower control arm **(see illustration)**. Make sure the jack is centered from front to rear.
5   Loop a length of safety chain up through the control arm and coil spring and bolt the ends of the chain together. Make sure there's enough slack in the chain so it won't inhibit spring extension when the control arm is lowered **(see illustration 8.4)**.
6   Raise the jack slightly to relieve spring pressure from the control arm pivot bolts and remove the nuts and bolts.
7   Slowly lower the jack until the coil spring is fully extended.
8   Unbolt the safety chain and maneuver the coil spring out **(see illustration)**. Do not apply downward pressure on the lower control arm as it may damage the balljoint. If the

8.8 Once the coil spring is fully extended, it can be guided out from between the spring pocket and lower control arm

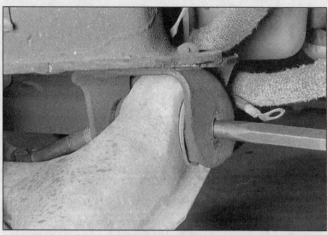

8.11 Insert a long punch through the frame bracket and into the lower control arm bushing to help align the hole

upper coil spring insulator is not on the top of the spring, reach up in the spring pocket and retrieve it.

Installation

9   Place the insulators on the top and bottom of the coil spring (the top of the spring is flat on the end, with a gripper notch near the end of the spring coil).
10   Install the top of the spring into the spring pocket and the bottom in the lower control arm. The end of the lower spring coil must be seated in the lowest recessed portion of the spring seat.
11   Place the jack under the lower control arm, install the safety chain and slowly raise the control arm into place. When the bolt holes are aligned, install the bolts with the bolt heads towards the center of the control arm. It may be necessary to insert a punch through the mounting bracket and into the control arm bushing to align the holes **(see illustration)**. Do not completely tighten the nuts at this time.
12   Install the shock absorber (Section 3).
13   Attach the stabilizer bar to the lower control arm (Section 2).

14   Install the wheel and lug nuts. Lower the vehicle and tighten the lug nuts to the specified torque.
15   Reach under the vehicle and tighten the lower control arm pivot bolt nuts to the specified torque.
16   Drive the vehicle to an alignment shop to have the front end alignment checked and, if necessary, adjusted.

9   Rear shock absorbers - removal and installation

Refer to illustration 9.2

1   Raise the rear of the vehicle and support it securely on jackstands. Block the front wheels to keep the vehicle from rolling.
2   Support the rear axle housing with a floor jack nearest the shock absorber to be removed. Remove the shock absorber lower mounting bolt and nut **(see illustration)**.
3   Remove the shock absorber upper mounting bolt and slide the shock absorber off the bolt that protrudes through the frame.
4   Installation is the reverse of the removal procedure.

10 Rear coil spring - removal and installation

Refer to illustrations 10.5 and 10.6

Removal

1   Raise the rear of the vehicle and support it securely on jackstands placed under the frame rails. Block the front wheels.
2   Place a jack under the rear axle housing and raise it slightly to take the weight off the shock absorber.
3   Unbolt the lower end of the shock absorber from the rear axle.
4   Slowly lower the jack until the springs are fully extended.
5   Remove the coil spring lower retainer nut from underneath the control arm **(see illustration)**.
6   Remove the coil spring upper retainer bolt **(see illustration)**.
7   Remove the coil spring, retainers and insulators. Inspect the insulators for cracks and general wear, replacing them if necessary.

9.2 Removing the shock absorber lower mounting bolt

10.5 Remove the nut on the underside of the lower control arm to free the bottom of the spring. This should only be done when the coil spring is fully extended.

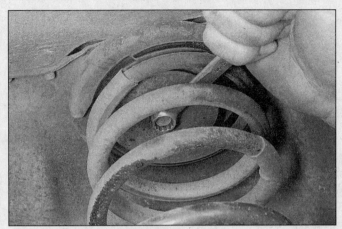

10.6 Reach through the coil spring with a wrench to remove the spring upper retainer bolt

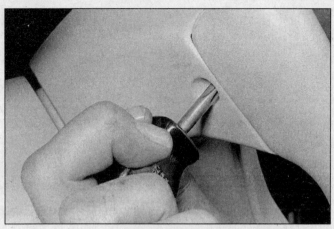

12.2 Remove the horn pad retaining screws . . .

Installation

8   Place the lower insulator on the frame and the upper insulator on top of the spring (the top of the spring is the end with the smaller diameter, tapered coils that are painted white).
9   Place the spring between the control arm and the frame. Install the upper retainer and bolt, tightening it securely.
10   Install the lower retainer and tighten the nut securely.
11   Connect the shock absorber to its lower mounting bracket, lower the vehicle and tighten the mounting bolt securely.

---

11   Rear suspension control arms - removal and installation

Upper control arm

Removal

1   Raise the rear of the vehicle and support it securely on jackstands placed beneath the frame rails. Block the front wheels.
2   Position a jack under the differential and raise it slightly.
3   Remove the shock absorber lower mounting bolts and nuts.
4   Slowly lower the jack until the springs are no longer compressed. Keep an eye on the brake hose - if the axle is lowered too far it could be damaged.
5   Remove the upper control arm-to-rear axle bolt and nut **(see illustration)**. Separate the control arm from the axle and mark the position of the cam adjuster in the axle bushing.
6   Remove the control arm-to-right side frame bracket bolt and nut.
7   Remove the control arm-to-left side frame bracket nut, washer and outer bushing.
8   Remove the control arm from the vehicle. **Note:** *The pressed-in bushing on the right side of the control arm is replaceable, but special tools are necessary to remove and install it. If the bushing is worn out, take the control arm to a dealer service department or automotive machine shop.*

Installation

9   Insert the control arm stud (with inner washer and bushing)
into the left side frame bracket. Install the outer bushing and washer then install the nut finger-tight.
10   Push the right side of the arm into the right side frame bracket and install the bolt and nut, but don't tighten it to the specified torque yet.
11   Check to be sure that the cam adjuster in the axle bushing is still in alignment with the previously applied mark.
12   Connect the arm to the rear axle and install the bolt and nut, but don't tighten it fully yet.
13   Raise the axle to normal ride height and connect the lower end of the shock absorber to the mounting brackets. Tighten the mounting bolts securely.
14   Tighten all of the upper control arm bolts to the specified torque.

Lower control arm

Removal

15   Raise the rear of the vehicle and support it securely on jackstands placed under the frame rails. Block the front wheels.
16   Place a jack under the rear axle housing and raise it slightly to take the load off the shock absorbers. Unbolt the shock absorber from the rear axle.
17   Slowly lower the jack until the coil spring is no longer compressed. Keep an eye on the brake hose - it could be damaged if the axle is lowered too far. Remove the lower spring retainer nut from the underside of the control arm **(see illustration 10.5)**.
18   Remove the lower control arm-to-axle housing bolt and nut, noting the direction in which it was installed.
19   Remove the lower control arm-to-frame bolt and nut. Remove the control arm from the vehicle. **Note:** *The pressed-in bushings at either end of the control arm are replaceable, but special tools are necessary to remove and install them. If the bushings are worn out, take the control arm to a dealer service department or automotive machine shop.*

Installation

20   Position the forward end of the control arm in the frame bracket and install the bolt and nut. The bolt head must be facing toward the inside of the vehicle. Do not fully tighten it at this time.
21   Swing the trailing end of the control arm up into the rear axle bracket and install the bolt and nut, but don't fully tighten them yet. The bolt head must be on the inboard side of the axle bracket.
22   Place the spring insulator on the lower control arm, raise the rear axle until it touches the coil spring and install the spring lower retainer. Tighten the nut securely.
23   Connect the shock absorber to the rear axle housing and install the bolt loosely.
24   Lower the vehicle to the ground and tighten the control arm mounting bolts to the specified torque. Tighten the shock absorber bolts securely.

---

12   Steering wheel - removal and installation

Refer to illustrations 12.2, 12.3, 12.4, 12.5, 12.6 and 12.8
Warning: *Some models covered by this manual are equipped with Supplemental Restraint Systems (SRS), more commonly known as airbags. Always disable the airbag system before working in the vicinity of any airbag system component to avoid the possibility of accidental deployment of the airbag(s), which could cause personal injury (see Chapter 12).*

Removal

1   Park the vehicle with the wheels pointing straight ahead. Disconnect the cable from the negative terminal of the battery. On airbag-equipped models, also disconnect the cable from the positive terminal and wait two minutes before proceeding.
2   On models without an airbag, remove the two horn pad retaining screws and pull the horn pad away from the steering wheel **(see illustration)**.

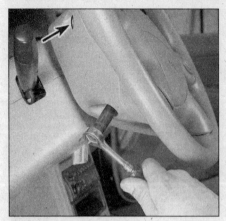

12.3 On airbag-equipped models, remove the four airbag module retaining nuts (two more, on the other side of the wheel, aren't visible in this photo)

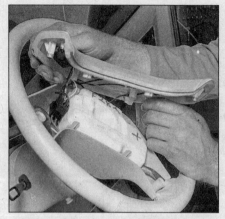

12.4 . . . then pull the horn pad off the plastic tabs and disconnect the horn electrical connector

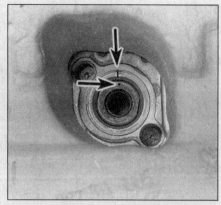

12.5 Before removing the steering wheel, check to see if any alignment marks exist (arrows) - if not, use a sharp scribe or white paint to index the steering wheel hub to the steering shaft

3   On models with an airbag, remove the four horn pad/airbag module retaining nuts **(see illustration)**.

4   Lift off the horn pad or airbag module and disconnect any wires for the horn, cruise control, airbag module, etc. **(see illustration)**. **Warning:** *When removing an airbag module, carry the module with trim side*

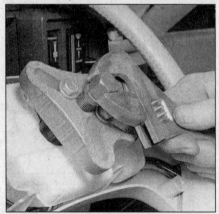

12.6 Remove the wheel from the shaft with a puller - DO NOT beat on the shaft!

(airbag opening) facing away from your body, and set it down in an isolated area with the trim side facing up).

5   Remove the steering wheel retaining bolt. Unless it's already marked, mark the relationship of the steering wheel to the steering shaft to ensure correct steering wheel alignment when the wheel is installed **(see illustration)**.

6   Use a puller to detach the steering wheel from the shaft **(see illustration)**. **Caution:** *Don't hammer on the shaft in an attempt to remove the wheel.*

7   On airbag-equipped models, tape the airbag sliding contact in place with a couple pieces of tape to prevent it from being accidentally rotated.

Installation

8   On airbag-equipped models, make sure the sliding contact is correctly aligned **(see illustration)**. **Note:** *The airbag sliding contact automatically locks when the steering wheel is removed. However, if the steering shaft has turned or the sliding contact is not in alignment, turn the sliding contact clockwise until it stops, then turn it counterclockwise 2-3/4 turns and align the arrows on the sliding con-*

tact hub and housing. **Caution:** *If the instructions on the airbag sliding contact differ from those printed here, follow the instructions on the sliding contact.*

9   Install the steering wheel on the steering shaft, aligning the mark on the steering wheel with the mark on the shaft. Install the bolt and tighten it to the torque listed in this Chapter's Specifications.

10   Plug in the electrical connectors and install the horn pad or airbag module. Tighten the airbag module retaining nuts to the torque listed in this Chapter's Specifications.

11   Connect the battery cable(s). **Note:** *If both battery cables were disconnected, connect the positive cable first, then the negative cable.*

13 Intermediate shaft - removal and installation

Refer to illustrations 13.2a and 13.2b

1   Turn the front wheels to the straight ahead position.

2   Using white paint, place alignment marks on the upper universal joint, the steer-

12.8 Before installing the steering wheel on airbag-equipped models, make sure the airbag sliding contact is centered and aligned

13.2a Place matchmarks (arrow) on the intermediate shaft-to-steering shaft joint

13.2b Also place alignment marks from the lower universal joint
to the steering gear input shaft

14.5 Location of the steering gear mounting bolts (arrows)

ing shaft, the lower universal joint and the steering gear input shaft **(see illustrations).**
3   Remove the upper and lower universal joint pinch bolts.
4   Pry the intermediate shaft out of the steering shaft universal joint with a large screwdriver, then pull the shaft from the steering gearbox.
5   Installation is the reverse of the removal procedure. Be sure to align the marks and tighten the pinch bolts to the specified torque.

14 Steering gear - removal and installation

Refer to illustration 14.5
Warning: *On models equipped with airbags, make sure the steering shaft is not turned while the steering gear or box is removed or you could damage the airbag system. To prevent the shaft from turning, turn the ignition key to the lock position before beginning work or run the seal belt through the steering wheel and clip the seal belt into place. Due to the possible damage to the airbag system, we recommend only experienced mechanics attempt this procedure.*
Note: *This procedure applies to both power and manual steering gear assemblies. When working on a vehicle equipped with a manual steering gear, simply ignore any references made to the power steering system.*

Removal
1   Raise the front of the vehicle and support it securely on jackstands. Apply the parking brake.
2   Place a drain pan under the steering gear (power steering only). Remove the power steering pressure and return lines and cap the ends to prevent excessive fluid loss and contamination.
3   Mark the relationship of the lower intermediate shaft universal joint to the steering

gear input shaft **(see illustration 13.2b).** Remove the lower intermediate shaft pinch bolt.
4   Separate the tie-rod ends from the spindle arms (see Section 17).
5   Support the steering gear and remove the steering gear-to-frame mounting bolts **(see illustration).** Lower the unit, separate the intermediate shaft from the steering gear input shaft and remove the steering gear from the vehicle.

Installation
6   Raise the steering gear into position and connect the intermediate shaft, aligning the marks.
7   Install the mounting bolts and washers and tighten them to the specified torque.
8   Connect the tie-rod ends to the spindle arms (Section 17).
9   Install the lower intermediate shaft pinch bolt and tighten it to the specified torque.
10   Connect the power steering pressure and return hoses to the steering gear and fill

the power steering pump reservoir with the recommended fluid (Chapter 1).
11   Lower the vehicle and bleed the steering system as outlined in Section 16.

15 Power steering pump - removal and installation

Refer to illustrations 15.3a, 15.3b, 15.4, 15.5, 15.8a and 15.8b

Removal
1   Disconnect the cable from the negative terminal of the battery.
2   Place a drain pan under the power steering pump. Remove the drivebelt (Chapter 1).
3   Using a special power steering pump pulley remover or a two-jaw puller, remove the pulley from the pump **(see illustrations).**
4   Remove the pressure and return hoses from the backside of the pump and allow the fluid to drain **(see illustration).** Plug the

15.3a This special tool, designed for removing power steering pulleys is available at tool stores and some auto parts stores

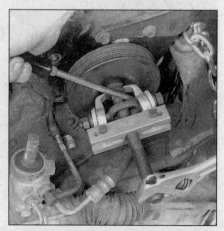

15.3b A two jaw puller can also be used to remove the pulley, but only if it is compact enough to work in the confines of the engine compartment

hoses to prevent contaminants from entering.
5   Remove the pump mounting bolts **(see illustration)** and lift the pump from the vehicle, taking care not to spill fluid on the painted surfaces.

Installation

6   Position the pump in the mounting bracket and install the bolts. Tighten the bolts securely.
7   Connect the hoses to the pump. Tighten the fittings securely.
8   Press the pulley onto the pump shaft using a special pulley installer tool **(see illustration)**. An alternative tool can be fabricated from a long bolt, nut, washer and a socket of the same diameter as the pulley hub **(see illustration)**. Push the pulley onto the shaft until the front of the hub is flush with the end of the shaft, but no further.
9   Install the drivebelt.
10   Fill the power steering reservoir with the recommended fluid and bleed the system following the procedure described in the next Section.

16   Power steering system - bleeding

1   Following any operation in which the power steering fluid lines have been disconnected, the power steering system must be bled to remove all air and obtain proper steering performance.
2   With the front wheels in the straight ahead position, check the power steering fluid level and, if low, add fluid until it reaches the Cold mark on the dipstick.
3   Start the engine and allow it to run at fast idle. Recheck the fluid level and add more if necessary to reach the Cold mark on the dipstick.
4   Bleed the system by turning the wheels from side-to-side, without hitting the stops. This will work the air out of the system. Keep

15.4 Remove the power steering pressure and return lines (arrows); use a back-up wrench on the pressure fitting - the return hose is retained by a hose clamp

the reservoir full of fluid as this is done.
5   When the air is worked out of the system, return the wheels to the straight ahead position and leave the vehicle running for several more minutes before shutting it off.
6   Road test the vehicle to be sure the steering system is functioning normally and noise free.
7   Recheck the fluid level to be sure it is up to the Hot mark on the dipstick while the engine is at normal operating temperature. Add fluid if necessary (see Chapter 1).

17   Tie-rod ends - removal and installation

Refer to illustrations 17.2a, 17.2b and 17.4

Removal

1   Loosen the wheel lug nuts. Raise the front of the vehicle, support it securely, block

15.5 Remove the three power steering pump mounting bolts and remove the pump from the mounting bracket

the rear wheels and set the parking brake. Remove the front wheel.
2   Hold the tie-rod with a pair of locking pliers and loosen the jam nut enough to mark the position of the tie-rod end in relation to the threads **(see illustrations)**.
3   Remove the cotter pin and loosen the nut on the tie-rod end stud.
4   Disconnect the tie-rod from the steering knuckle arm with a puller **(see illustration)**. Remove the nut and separate the tie-rod.
5   Unscrew the tie-rod end from the tie-rod.

Installation

6   Thread the tie-rod end on to the marked position and insert the tie-rod stud into the steering knuckle arm. Tighten the jam nut securely.
7   Install the castellated nut on the stud and tighten it to the specified torque. Install a new cotter pin.
8   Install the wheel and lug nuts. Lower the

15.8a This special tool, designed for installing the power steering pulley is available at tool stores and some auto parts stores

15.8b A long bolt with the same thread pitch as the internal threads of the power steering pump shaft, nut, washer and a socket that is the same diameter as the pulley hub can be used to install the pulley on the shaft

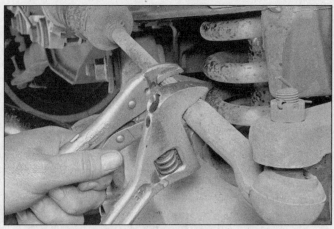

17.2a Hold the tie-rod with a pair of locking pliers and loosen the jam nut

17.2b Use white paint to mark the position of the tie-rod end on the tie-rod

17.4 Use a two-jaw puller to detach the tie-rod end from the spindle arm. Notice that the nut has been loosened, but not removed. This will prevent the components from separating violently

18.3a Remove the outer boot clamp with a pair of pliers

vehicle and tighten the lug nuts to the specified torque.

9   Have the alignment checked by a dealer service department or an alignment shop.

18 Steering gear boots - replacement

Refer to illustrations 18.3a and 18.3b

1   Loosen the lug nuts, raise the vehicle and support it securely on jackstands. Remove the wheel.

2   Refer to Section 17 and remove the tie-rod end and jam nut.

3   Remove the steering gear boot clamps and slide the boot off **(see illustrations)**.

4   Before installing the new boot, wrap the threads and serrations on the end of the steering rod with a layer of tape so the small end of the new boot isn't damaged.

5   Slide the new boot into position on the

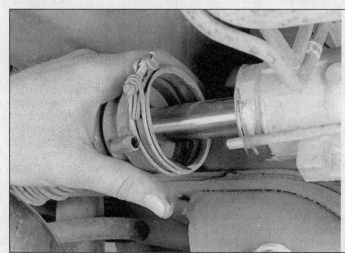

18.3b Loosen the inner boot clamp screw and slide the boot off the rack housing

steering gear until it seats in the groove in the steering rod and install new clamps.

6   Remove the tape and install the tie-rod end (Section 17).

7   Install the wheel and lug nuts. Lower the vehicle and tighten the lug nuts to the specified torque.

METRIC TIRE SIZES

P 185 / 80 R 13

TIRE TYPE
P-PASSENGER
T-TEMPORARY
C-COMMERCIAL

ASPECT RATIO
(SECTION HEIGHT)
(SECTION WIDTH)
70
75
80

RIM DIAMETER
(INCHES)
13
14
15

SECTION WIDTH
(MILLIMETERS)
185
195
205
ETC

CONSTRUCTION TYPE
R-RADIAL
B-BIAS - BELTED
D-DIAGONAL (BIAS)

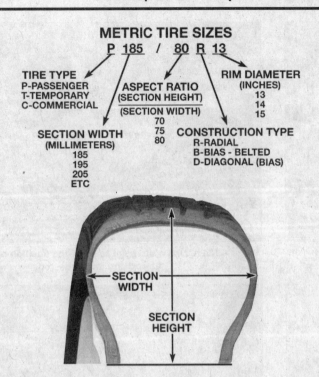

SECTION WIDTH

SECTION HEIGHT

19.1 Metric tire size code

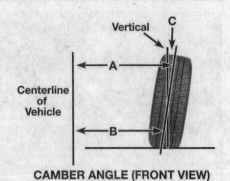

CAMBER ANGLE (FRONT VIEW)

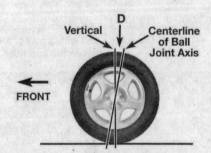

CASTER ANGLE (SIDE VIEW)

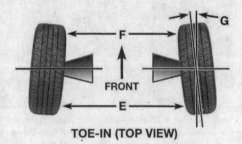

TOE-IN (TOP VIEW)

20.1 Front end alignment details

A minus B = C (degrees camber)
D = caster (measured in degrees)
E minus F = toe-in (measured in inches)
G = toe-in (expressed in degrees)

19   Wheels and tires - general information

Refer to illustration 19.1

All vehicles covered by this manual are equipped with metric-sized fiberglass or steel belted radial tires **(see illustration)**. Use of other size or type of tires may affect the ride and handling of the vehicle. Don't mix different types of tires, such as radials and bias belted, on the same vehicle as handling may be seriously affected. It's recommended that tires be replaced in pairs on the same axle, but if only one tire is being replaced, be sure it's the same size, structure and tread design as the other.

Because tire pressure has a substantial effect on handling and wear, the pressure on all tires should be checked at least once a month or before any extended trips (see Chapter 1).

Wheels must be replaced if they are bent, dented, leak air, have elongated bolt holes, are heavily rusted, out of vertical symmetry or if the lug nuts won't stay tight. Wheel repairs that use welding or peening are not recommended.

Tire and wheel balance is important in the overall handling, braking and performance of the vehicle. Unbalanced wheels can adversely affect handling and ride characteristics as well as tire life. Whenever a tire is installed on a wheel, the tire and wheel should be balanced by a shop with the proper equipment.

20   Front end alignment - general information

Refer to illustration 20.1

A front end alignment refers to the adjustments made to the front wheels so they are in proper angular relationship to the suspension and the ground. Front wheels that are out of proper alignment not only affect steering control, but also increase tire wear. The front end adjustments normally required are camber, caster and toe-in **(see illustration)**.

Getting the proper front wheel alignment is a very exacting process, one in which complicated and expensive machines are necessary to perform the job properly. Because of this, you should have a technician with the proper equipment perform these tasks. We will, however, use this space to give you a basic idea of what is involved with front end alignment so you can better understand the process and deal intelligently with the shop that does the work.

Toe-in is the turning in of the front wheels. The purpose of a toe specification is to ensure parallel rolling of the front wheels. In a vehicle with zero toe-in, the distance between the front edges of the wheels will be the same as the distance between the rear edges of the wheels. The actual amount of toe-in is normally only a fraction of an inch. Toe-in adjustment is controlled by the tie-rod end position on the inner tie-rod. Incorrect toe-in will cause the tires to wear improperly by making them scrub against the road surface.

Camber is the tilting of the front wheels from the vertical when viewed from the front of the vehicle. When the wheels tilt out at the top, the camber is said to be positive (+). When the wheels tilt in at the top the camber is negative (-). The amount of tilt is measured in degrees from the vertical and this measurement is called the camber angle. This angle affects the amount of tire tread which contacts the road and compensates for changes in the suspension geometry when the vehicle is cornering or traveling over an undulating surface.

Caster is the tilting of the front steering axis from the vertical. A tilt toward the rear is positive caster and a tilt toward the front is negative caster.

Caster is adjusted by moving shims from one end of the upper control arm mount to the other.

Chapter 11 Body

Contents

Specifications

Torque specifications

	Ft-lbs
Front door hinge bolt	13 to 20
Front seat belt D ring bolt	22 to 32
Front seat belt retractor assembly bolt	22 to 32
Seat belt anchor bolts	22 to 32
Liftgate hinge bolt	13 to 20
Dual rear door hinge bolt	13 to 20

1 General information

The vehicles covered in this manual have a separate frame and body. The body is metal with the exception of the rear liftgate, which is constructed of fiberglass.

As with other parts of the vehicle, proper maintenance of body components plays an important part in retention of the vehicle's market value. It is far less costly to handle small problems before they grow into larger ones. Information in this Chapter will tell you all you need to know to keep seals sealing, body panels aligned and general appearance up to par.

Major body components which are particularly vulnerable in accidents are remov-able. These include the hood, grille and doors. It is often cheaper and less time consuming to replace an entire panel than it is to attempt a restoration of the old one. However, this must be decided on a case-by-case basis.

2 Maintenance - body and frame

1   The condition of your vehicle's body is very important, because it is on this that the second hand value will mainly depend. It is much more difficult to repair a neglected or damaged body than it is to repair mechanical components. The hidden areas of the body, such as the fender wells, the frame, and the engine compartment, are equally important, although they obviously do not require as frequent attention as the rest of the body.

2   Once a year, or every 12,000 miles, it's a good idea to have the underside of the body and the frame steam cleaned. All traces of dirt and oil will be removed and the underside can then be inspected carefully for rust, damaged brake lines, frayed electrical wiring, damaged cables and other problems. The front suspension components should be greased after completion of this job.

3   At the same time, clean the engine and the engine compartment using either a steam cleaner or a water soluble degreaser.

4   The fender wells should be given particular attention, as undercoating can peel away and stones and dirt thrown up by the tires can cause the paint to chip and flake, allow-

ing rust to set in. If rust is found, clean down to the bare metal and apply an anti-rust paint.

5   The body should be washed as needed. Wet the vehicle thoroughly to soften the dirt, then wash it down with a soft sponge and plenty of clean soapy water. If the surplus dirt is not washed off very carefully, it will in time wear down the paint.

6   Spots of tar or asphalt coating thrown up from the road should be removed with a cloth soaked in solvent.

7   Once every six months, give the body and chrome trim a thorough waxing. If a chrome cleaner is used to remove rust from any of the vehicle's plated parts, remember that the cleaner also removes part of the chrome, so use it sparingly.

3   Maintenance - upholstery and carpets

1   Every three months remove the carpets or mats and clean the interior of the vehicle (more frequently if necessary). Vacuum the upholstery and carpets to remove loose dirt and dust.

2   If the upholstery is soiled, apply uphol-stery cleaner with a damp sponge and wipe it off with a clean, dry cloth.

4   Body repair - minor damage

See color photo sequence

Repair of minor scratches

1   If the scratch is superficial and does not penetrate to the metal of the body, repair is very simple. Lightly rub the scratched area with a fine rubbing compound to remove loose paint and built up wax. Rinse the area with clean water.

2   Apply touch-up paint to the scratch, using a small brush. Continue to apply thin layers of paint until the surface of the paint in the scratch is level with the surrounding paint. Allow the new paint at least two weeks to harden, then blend it into the surrounding paint by rubbing with a very fine rubbing compound. Finally, apply a coat of wax to the scratch area.

3   If the scratch has penetrated the paint and exposed the metal of the body, causing the metal to rust, a different repair technique is required. Remove all loose rust from the bottom of the scratch with a pocket knife, then apply rust inhibiting paint to prevent the formation of rust in the future. Using a rubber or nylon applicator, coat the scratched area with glaze-type filler. If required, the filler can be mixed with thinner to provide a very thin paste, which is ideal for filling narrow scratches. Before the glaze filler in the scratch hardens, wrap a piece of smooth cotton cloth around the tip of a finger. Dip the cloth in thinner and then quickly wipe it along

the surface of the scratch. This will ensure that the surface of the filler is slightly hollow. The scratch can now be painted over as described earlier in this section.

Repair of dents

4   When repairing dents, the first job is to pull the dent out until the affected area is as close as possible to its original shape. There is no point in trying to restore the original shape completely as the metal in the damaged area will have stretched on impact and cannot be restored to its original contours. It is better to bring the level of the dent up to a point which is about 1/8-inch below the level of the surrounding metal. In cases where the dent is very shallow, it is not worth trying to pull it out at all.

5   If the back side of the dent is accessible, it can be hammered out gently from behind using a soft-face hammer. While doing this, hold a block of wood firmly against the oppo-site side of the metal to absorb the hammer blows and prevent the metal from being stretched.

6   If the dent is in a section of the body which has double layers, or some other factor makes it inaccessible from behind, a different technique is required. Drill several small holes through the metal inside the damaged area, particularly in the deeper sections. Screw long, self tapping screws into the holes just enough for them to get a good grip in the metal. Now the dent can be pulled out by pulling on the protruding heads of the screws with locking pliers.

7   The next stage of repair is the removal of paint from the damaged area and from an inch or so of the surrounding metal. This is easily done with a wire brush or sanding disk in a drill motor, although it can be done just as effectively by hand with sandpaper. To complete the preparation for filling, score the surface of the bare metal with a screwdriver or the tang of a file or drill small holes in the affected area. This will provide a good grip for the filler material. To complete the repair, see the Section on filling and painting.

Repair of rust holes or gashes

8   Remove all paint from the affected area and from an inch or so of the surrounding metal using a sanding disk or wire brush mounted in a drill motor. If these are not available, a few sheets of sandpaper will do the job just as effectively.

9   With the paint removed, you will be able to determine the severity of the corrosion and decide whether to replace the whole panel, if possible, or repair the affected area. New body panels are not as expensive as most people think and it is often quicker to install a new panel than to repair large areas of rust.

10   Remove all trim pieces from the affected area except those which will act as a guide to the original shape of the damaged body, such as headlight shells, etc. Using metal snips or a hacksaw blade, remove all loose

metal and any other metal that is badly affected by rust. Hammer the edges of the hole inward to create a slight depression for the filler material.

11   Wire brush the affected area to remove the powdery rust from the surface of the metal. If the back of the rusted area is acces-sible, treat it with rust-inhibiting paint.

12   Before filling is done, block the hole in some way. This can be done with sheet metal riveted or screwed into place, or by stuffing the hole with wire mesh.

13   Once the hole is blocked off, the affected area can be filled and painted. See the following sub-section on filling and paint-ing.

Filling and painting

14   Many types of body fillers are available, but generally speaking, body repair kits which contain filler paste and a tube of resin hardener are best for this type of repair work. A wide, flexible plastic or nylon applicator will be necessary for imparting a smooth and contoured finish to the surface of the filler material. Mix up a small amount of filler on a clean piece of wood or cardboard (use the hardener sparingly). Follow the manufac-turer's instructions on the package, other-wise the filler will set incorrectly.

15   Using the applicator, apply the filler paste to the prepared area. Draw the applica-tor across the surface of the filler to achieve the desired contour and to level the filler sur-face. As soon as a contour that approximates the original one is achieved, stop working the paste. If you continue, the paste will begin to stick to the applicator. Continue to add thin layers of paste at 20-minute intervals until the level of the filler is just above the surrounding metal.

16   Once the filler has hardened, the excess can be removed with a body file. From then on, progressively finer grades of sandpaper should be used, starting with a 180-grit paper and finishing with a 600-grit wet-or-dry paper. Always wrap the sandpaper around a flat rub-ber or wooden block, otherwise the surface of the filler will not be completely flat. During the sanding of the filler surface, the wet-or-dry paper should be periodically rinsed in water. This will ensure that a very smooth fin-ish is produced in the final stage.

17   At this point, the repair area should be surrounded by a ring of bare metal, which in turn should be encircled by the finely feath-ered edge of good paint. Rinse the repair area with clean water until all of the dust pro-duced by the sanding operation is gone.

18   Spray the entire area with a light coat of primer. This will reveal any imperfections in the surface of the filler. Repair the imperfec-tions with fresh filler paste or glaze filler and once more smooth the surface with sandpa-per. Repeat this spray-and-repair procedure until you are satisfied that the surface of the filler and the feathered edge of the paint are perfect. Rinse the area with clean water and

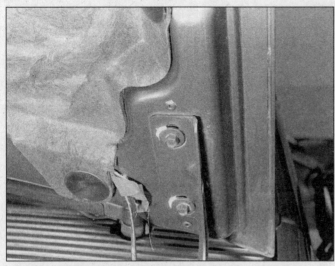

8.1 Mark around the hood bolts with white paint before removal so the hood can be reinstalled in the same position

9.3 After unbolting the latch, remove the screw and disengage the cable

allow it to dry completely.

19 The repair area is now ready for painting. Spray painting must be carried out in a warm, dry, windless and dust free atmosphere. These conditions can be created if you have access to a large indoor work area, but if you are forced to work in the open, you will have to pick the day very carefully. If you are working indoors, dousing the floor in the work area with water will help settle the dust which would otherwise be in the air. If the repair area is confined to one body panel, mask off the surrounding panels. This will help minimize the effects of a slight mismatch in paint color. Trim pieces such as chrome strips, door handles, etc., will also need to be masked off or removed. Use masking tape and several thicknesses of newspaper for the masking operations.

20 Before spraying, shake the paint can thoroughly, then spray a test area until the spray painting technique is mastered. Cover the repair area with a thick coat of primer. The thickness should be built up using several thin layers of primer rather than one thick one. Using 600-grit wet-or-dry sandpaper, rub down the surface of the primer until it is very smooth. While doing this, the work area should be thoroughly rinsed with water and the wet-or-dry sandpaper periodically rinsed as well. Allow the primer to dry before spraying additional coats.

21 Spray on the top coat, again building up the thickness by using several thin layers of paint. Begin spraying in the center of the repair area and then, using a circular motion, work out until the whole repair area and about two inches of the surrounding original paint is covered. Remove all masking material 10 to 15 minutes after spraying on the final coat of paint. Allow the new paint at least two weeks to harden, then use a very fine rubbing compound to blend the edges of the new paint into the existing paint. Finally, apply a coat of wax.

5 Body and frame repair - major damage

1 Major damage must be repaired by an auto body/frame repair shop with the necessary welding and hydraulic straightening equipment.

2 If the damage has been serious, it is vital that the frame be checked for proper alignment or the vehicle's handling characteristics may be adversely affected. Other problems, such as excessive tire wear and wear in the driveline and steering may occur.

3 Due to the fact that many of the major body components (hood, doors, etc.) are separate and replaceable units, any seriously damaged components should be replaced rather than repaired. Sometimes these components can be found in a wrecking yard that specializes in used vehicle components, often at considerable savings over the cost of new parts.

6 Maintenance - hinges and locks

Every 3000 miles or three months, the door and hood hinges should be lubricated with a few drops of oil. Lubricate the locks with graphite spray. The door striker plates should also be given a thin coat of white lithium-base grease to reduce wear and ensure free movement.

7 Windshield and fixed glass - replacement

Replacement of the windshield and fixed glass requires the use of special fast-setting adhesive/caulk materials and some specialized tools. It is recommended that

these operations should be left to a dealer or a shop specializing in glass work.

8 Hood - removal and installation

Refer to illustration 8.1

1 Mark the position of the retaining bolts with a sharp scribe or white paint **(see illustration)**.

2 Use rags or pads to protect the cowling and windshield area from the rear of the hood.

3 Disconnect the release cable and the engine compartment light from the hood.

4 With an assistant supporting its weight, remove the bolts and lift the hood from the vehicle.

5 Installation is the reverse of removal.

9 Hood release latch and cable - removal and installation

Refer to illustrations 9.3 and 9.4

Warning: *Some models covered by this manual are equipped with Supplemental Restraint Systems (SRS), more commonly known as airbags. Always disable the airbag system before working in the vicinity of any airbag system component to avoid the possibility of accidental deployment of the airbag(s), which could cause personal injury (see Chapter 12).*

Latch

1 Mark its position with white paint or by scribing with a pencil, disconnect the cable, remove the two retaining bolts and lift the latch from the vehicle.

2 Installation is the reverse of removal.

Cable

3 In the engine compartment, disconnect the cable at the latch and retaining clips **(see illustration)**.

These photos illustrate a method of repairing simple dents. They are intended to supplement *Body repair - minor damage* in this Chapter and should not be used as the sole instructions for body repair on these vehicles.

1 If you can't access the backside of the body panel to hammer out the dent, pull it out with a slide-hammer-type dent puller. In the deepest portion of the dent or along the crease line, drill or punch hole(s) at least one inch apart . . .

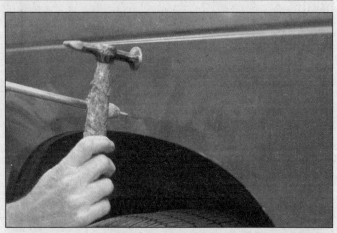

2 . . . then screw the slide-hammer into the hole and operate it. Tap with a hammer near the edge of the dent to help 'pop' the metal back to its original shape. When you're finished, the dent area should be close to its original contour and about 1/8-inch below the surface of the surrounding metal

3 Using coarse-grit sandpaper, remove the paint down to the bare metal. Hand sanding works fine, but the disc sander shown here makes the job faster. Use finer (about 320-grit) sandpaper to feather-edge the paint at least one inch around the dent area

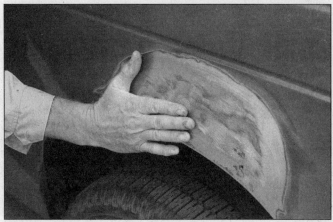

4 When the paint is removed, touch will probably be more helpful than sight for telling if the metal is straight. Hammer down the high spots or raise the low spots as necessary. Clean the repair area with wax/silicone remover

5 Following label instructions, mix up a batch of plastic filler and hardener. The ratio of filler to hardener is critical, and, if you mix it incorrectly, it will either not cure properly or cure too quickly (you won't have time to file and sand it into shape)

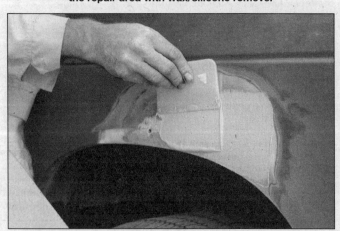

6 Working quickly so the filler doesn't harden, use a plastic applicator to press the body filler firmly into the metal, assuring it bonds completely. Work the filler until it matches the original contour and is slightly above the surrounding metal

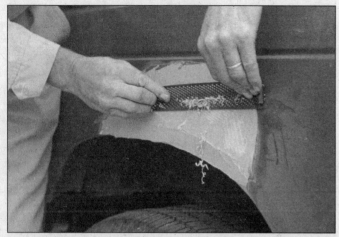

7 Let the filler harden until you can just dent it with your fingernail. Use a body file or Surform tool (shown here) to rough-shape the filler

8 Use coarse-grit sandpaper and a sanding board or block to work the filler down until it's smooth and even. Work down to finer grits of sandpaper - always using a board or block - ending up with 360 or 400 grit

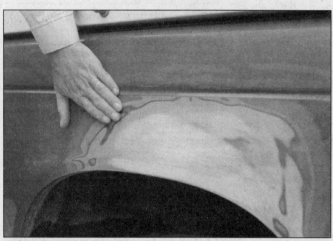

9 You shouldn't be able to feel any ridge at the transition from the filler to the bare metal or from the bare metal to the old paint. As soon as the repair is flat and uniform, remove the dust and mask off the adjacent panels or trim pieces

10 Apply several layers of primer to the area. Don't spray the primer on too heavy, so it sags or runs, and make sure each coat is dry before you spray on the next one. A professional-type spray gun is being used here, but aerosol spray primer is available inexpensively from auto parts stores

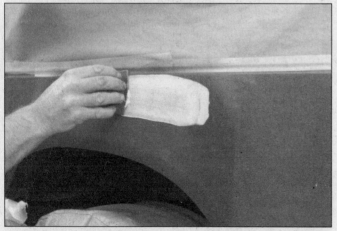

11 The primer will help reveal imperfections or scratches. Fill these with glazing compound. Follow the label instructions and sand it with 360 or 400-grit sandpaper until it's smooth. Repeat the glazing, sanding and respraying until the primer reveals a perfectly smooth surface

12 Finish sand the primer with very fine sandpaper (400 or 600-grit) to remove the primer overspray. Clean the area with water and allow it to dry. Use a tack rag to remove any dust, then apply the finish coat. Don't attempt to rub out or wax the repair area until the paint has dried completely (at least two weeks)

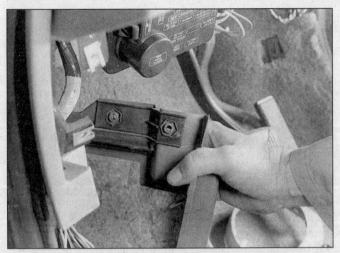

9.4 Remove the trim panel for access to the cable release handle bolts

10.2 With the fasteners removed, the cowl vent grille can be lifted from the opening

4   In the passenger compartment, remove the retaining screw, pull back the trim panel and remove the two cable release lever retaining bolts **(see illustration)**.

5   Connect a thin piece of wire or string to the end of the cable in the engine compartment and then pull the cable through into the engine compartment. Attach the wire or string to the new latch cable assembly and pull it through into the engine compartment.

6   Connect the cable to the latch and clips in the engine compartment and install the release lever and trim panel in the passenger compartment.

10 Cowl ventilator grille - removal and installation

Refer to illustration 10.2

Warning: *Some models covered by this manual are equipped with Supplemental Restraint Systems (SRS), more commonly known as airbags. Always disable the airbag system before working in the vicinity of any airbag*

system component to avoid the possibility of accidental deployment of the airbag(s), which could cause personal injury (see Chapter 12).

1   Remove the windshield wiper arms (Chapter 12).

2   Remove the fasteners and lift the grille from the vehicle **(see illustration)**.

3   Installation is the reverse of removal.

11 Engine cover - removal and installation

1   Release the four engine cover retaining latches (two on each side).

2   Rotate the engine cover down and rearward to remove it. On floorshift models it may be necessary to shift the lever out of the Park position for additional clearance (set the parking brake and block the wheels to prevent the vehicle from rolling).

3   To install, place the cover in position, making sure the rubber gaskets are aligned to make a good seal, and then secure it by pushing the latches forward.

12 Front fender - removal and installation

1   Remove the grille (Section 13) and headlight(s) (Chapter 12).

2   Remove the retaining bolts and clips which secure the fender along the inside of the engine compartment, at the door and along the bottom.

3   Carefully lift the fender away. On the passenger side, disconnect the antenna cable.

4   Installation is the reverse of removal.

13 Radiator grille - removal and installation

Refer to illustrations 13.2 and 13.4

Warning: *Some models covered by this manual are equipped with Supplemental Restraint Systems (SRS), more commonly known as airbags. Always disable the airbag system before working in the vicinity of any airbag*

13.2 Push down on the bumper cover for access to the two lower grille screws

13.4 Hold the headlight bezel out of the way while lifting the grille out

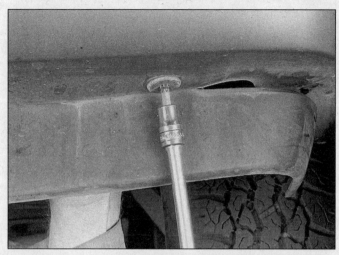

14.1 Use a Torx head tool to remove the air deflector screws

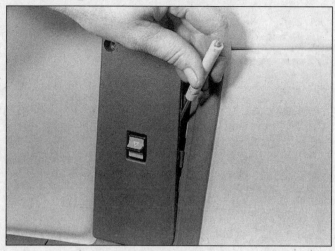

17.2 Use a screwdriver to pry out the power window switch

system component to avoid the possibility of accidental deployment of the airbag(s), which could cause personal injury (see Chapter 12).

1   Remove the six retaining screws along the top edge of the grille.

2   Press down on the bumper cover for access and remove the two lower grille screws **(see illustration)**.

3   Loosen the headlight bezel screws and pull the tops of the bezels forward to provide clearance for grille removal.

4   Hold the bezels out of the way, grasp the grille securely and lift it from the vehicle **(see illustration)**.

5   Installation is the reverse of removal.

14   Front air deflector - removal and installation

Refer to illustration 14.1

1   Remove the Torx-head screws retaining the air deflector to the bumper **(see illustra-tion)**.

2   Use a large screwdriver or pry bar to pry the retaining clips loose and disengage the air deflector from the bumper.

3   To install, place the air deflector in posi-tion, press the clips in place and install the screws.

15   Bumpers - removal and installation

Front bumper

1   Remove the retaining screws and clips and remove the bumper cover.

2   Remove the bumper end bracket assemblies and lower the bumper from the vehicle.

3   Installation is the reverse of removal.

Rear bumper

4   Remove the retaining bolts and push

pins and lower the bumper from the vehicle.

5   Installation is the reverse of removal.

16   Side window - removal and installation

1   Remove the window trim moulding.

2   Remove the window assembly retaining nuts.

3   With an assistant supporting the win-dow from the outside of the vehicle, push the assembly out and remove it from the vehicle.

4   Prior to installation, clean the old sealer from around the window assembly and open-ing.

5   Apply a 1/4-inch bead of butyl sealer around the window assembly opening.

6   Place the window assembly in place and install the retaining nuts. Tighten the nuts securely.

17   Door trim panel - removal and installation

Refer to illustrations 17.2 and 17.3

Front door

1   Remove the screws and pry the finish panel out of the door.

2   On manual window regulator models, remove the screw and lift off the crank. On power regulator models, pry off the control switch assembly and unplug it **(see illustra-tion)**.

3   Insert a screwdriver between the door panel and the door and, working around the outer circumference, disengage the clips **(see illustration)**.

4   Once the clips are disengaged, pull the door panel out at the bottom and lift straight up to remove it from the door.

5   To install, lower the top of the panel into position in the door, line up the clips and press them into place, working around the panel until they are all seated.

Sliding door

6   Remove the retaining nuts and screws and pry off the garnish moulding.

7   Remove the retaining screws, insert a screwdriver between the door panel and the door and, working around the outer circum-ference, disengage the clips.

8   Lift the trim panel off.

9   Installation is the reverse of removal.

Rear door

10   Remove the window moulding.

11   Remove the door trim retaining screws.

12   Lift the trim panel off.

13   Installation is the reverse of removal.

18   Front door - removal and installation

1   Remove the door trim panel (Section 17) and unplug any electrical connectors.

2   Carefully scribe around the hinges and bolts with a sharp awl or paint. This will locate the present door alignment and help during installation.

17.3 Use a large screwdriver to disengage the door panel retainers

19.1 Use a Phillips head screwdriver to remove the sliding door filler retaining screw (arrow)

19.2a Mark the locations of the bolts and bracket with a scribe or pencil and then remove the bolts from the upper bracket . . .

3   With an assistant supporting its weight, remove the retaining bolts and lift the door off the hinges. **Caution:** *The door is quite heavy, so support it fully during removal.*

4   To install, place the door in position and install the bolts. Align the door using the marks made on removal and tighten the bolts to the specified torque.

19   Sliding door - removal, installation and adjustment

Refer to illustrations 19.1, 19.2a and 19.2b

Removal and installation

1   Remove the body side filler assembly from the sliding door track **(see illustration)**.
2   With an assistant supporting the door weight, remove the bolts from the upper bracket and lower guide **(see illustrations)**.
3   Support the door and guide it rearward and out of the track.
4   Installation is the reverse of removal.

Adjustment

Fore-and-aft

5   Fore-and-aft adjustment of the sliding door is accomplished by loosening the three center hinge-to-door bolts and adjusting the hinge assembly as necessary to secure the proper door fit. After adjustment, tighten the bolts securely.

Front upper

6   Loosen the two body side door bracket-to-upper guide roller assembly nuts and move the roller in or out as required to obtain flush fit of the door. Tighten the nuts securely after adjustment.

Front lower

7   Adjust the lower front edge of the sliding door by loosening the adjusting nut on the lower guide assembly and then moving the door in or out until a flush fit is obtained. Tighten the adjusting nut securely.

Striker

8   The rear latch striker can be adjusted in or out to properly position the rear edge of the door for correct fit.

20   Rear door - removal and installation

Refer to illustration 20.1

Liftgate

1   With an assistant supporting its weight, disconnect the liftgate support struts after pushing off the clips with a screwdriver **(see illustration)**.
2   Remove the hinge screw and washer assemblies and lift off the liftgate.
3   Installation is the reverse of removal.
4   The liftgate can be adjusted by loosening the hinge nut and washer which runs through the interior header. These nuts are accessible after removing the overhead trim panel across the rear opening. To move the liftgate in or out, spacer plates can be used (or removed) between the hinge assembly and the body.

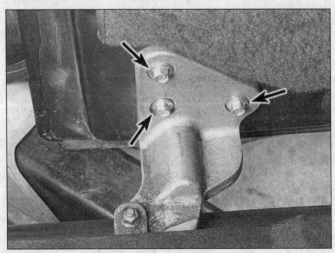

19.2b . . . and the lower guide

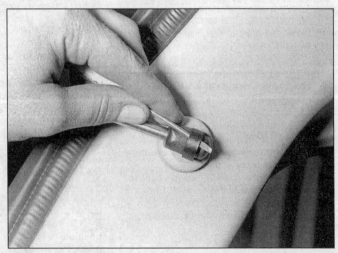

20.1 Use a screwdriver to pry off the liftgate support strut clips

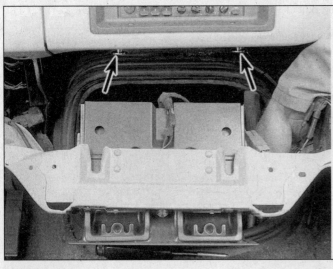

24.6 Pull the panel off the retaining pins (arrows) and lower it from the dash

Dual door

5   The procedure for these doors is essentially the same as for the front doors (see Section 18).

21   Front door window regulator - removal and installation

1   Remove the door trim panel (Section 17).
2   Remove the front door window glass (Section 22).
3   Use a suitable punch and hammer to drive out the heads of the rivets retaining the regulator to the door. Drill out the remainder of the rivets with a 1/4-inch drill bit, taking care not to enlarge the holes. On power regulators, unplug the connector. Remove the regulator from the vehicle.
4   Lift the regulator from the door.
5   To install, place the regulator in position on the door and install the new rivets. If rivets are not available, use 1/4-inch bolts with nuts to retain the regulator.
6   The remainder of installation is the reverse of removal.

22   Front door window glass - removal and installation

1   Disconnect the negative cable at the battery.
2   Remove the door trim panel and watershield.
3   With the glass in the full up position, remove the rear glass run from the door.
4   Lower the glass to provide access to the glass bracket and retaining rivets.
5   Use a suitable punch and hammer to drive out the heads of the rivets retaining the glass to the bracket. Drill out the remainder of the rivets with a 1/4-inch drill bit, taking care not to enlarge the holes.
6   Lift the glass from the bracket.
7   To install, lower the glass into position.

8   Align the glass and bracket retaining holes and install the rivets.
9   Raise the glass to the full up position.
10   Install the rear glass run and retainer.
11   Install the watershield and door trim panel.

23   Exterior mirror - removal and installation

Western-type mirror

1   Remove the four screws attaching the mirror and lift the mirror off the vehicle.
2   Installation is the reverse of removal.

Conventional mirror

3   With the window lowered, remove the door trim panel (Section 17).
4   Remove the two retaining nuts and lift the mirror from the vehicle.
5   Installation is the reverse of removal.

Power mirror

6   Disconnect the power mirror electrical connector behind the rear view mirror while the door is open.
7   Remove the two nuts attaching the outside rear view mirror to the front door.
8   Remove the rear view mirror from the front door.
9   Installation is the reverse of removal.

24   Lower instrument trim panel - removal and installation

Refer to illustration 24.6
Warning: *Some models covered by this manual are equipped with Supplemental Restraint Systems (SRS), more commonly known as airbags. Always disable the airbag system before working in the vicinity of any airbag system component to avoid the possibility of accidental deployment of the airbag(s), which could cause personal injury (see Chapter 12).*
1   Remove the two screws located in the

glove compartment.
2   Remove the screw located under the lower right of the lower panel to-cowl side bracket.
3   Pry off the fuse panel cover.
4   Remove the vertical screw located on the right side of the steering column.
5   Unplug the wiring connector.
6   Pull the lower panel sharply rearward to unsnap it from the two retaining pins and lower from the instrument panel **(see illustration)**.
7   Installation is the reverse of removal.

25   Seats - removal and installation

Front seats

1   Remove the retaining bolts, disconnect any electrical or other connectors and lift the seat from the vehicle.
2   Installation is the reverse of removal. Tighten the retaining bolts to the specified torque.

Rear seats

3   Lift the seat latch handle (one for bucket seats, two for bench seats) to the rear, which will cause the seat to move rearward.
4   Lift the latch J-rod hook end out of the locking hole(s) in the attaching plate.
5   Push the seat rearward and lift it from the vehicle.
6   To install, lower the seat into position with the front and rear retainers in the floor openings at the rear edge.
7   Engage the J-rod hook end into the slot at the front floor attaching plate and then rotate the latch handle(s) forward, moving the seat forward, engaging it with the floor attachments and continue pushing until the seat is locked in place. The handle must be rotated all the way forward so that it contacts the positive stop on the front attachment.

26   Seat belts - check and replacement

Check

1   Inspect the seat belt system for frayed belts, binding of the retractor mechanism and loose bolts. Replace any worn or damaged components with new ones.

Replacement

Front seat belts

2   Remove the D ring cover or trim panel and retractor assembly bolt.
3   Lift up on the carpet and remove the floor anchor bolt.
4   Remove the D ring bolt and remove the belt from the vehicle.
5   To install, place the belt in position and install the D ring bolt. Tighten the bolt to the specified torque.
6   Feed the belt into the slotted shoulder

strap loop, making sure not to twist it.

7   Install the D ring cover or trim panel.

8   Feed the anchor and tongue through the slot in the trim panel and install the floor anchor bolt. Tighten the bolt to the specified torque.

Rear bucket seat belts

9   Remove the plug from the buckle boot and snap open the cover for access to the retaining bolt.

10   Remove the retaining bolts and lift off the belt assemblies.

11   Installation is the reverse of removal.

Bench seat belts

12   Remove the belt assembly-to-seat structure bolts.

13   Remove the belts by pulling the belt and anchor plates through the grommet in the seat cushion.

14   Installation is the reverse of removal.

Chapter 12
Chassis electrical system

Contents

1 General information

The electrical system is a 12-volt, negative ground type. Power for the lights and all electrical accessories is supplied by a lead/acid-type battery which is charged by the alternator.

This Chapter covers repair and service procedures for the various electrical components not associated with the engine. Information on the battery, alternator, distributor and starter motor can be found in Chapter 5.

When working on electrical system components, disconnect the negative battery cable from the battery to prevent electrical shorts and/or fires.

2 Electrical troubleshooting - general information

A typical electrical circuit consists of an electrical component, any switches, relays, motors, etc. related to that component and the wiring and connectors that connect the component to both the battery and the chassis. To aid in locating a problem in any electrical circuit, wiring diagrams are included at the end of this book.

Before tackling any troublesome electrical circuit, first study the appropriate diagrams to get a complete understanding of what makes up that individual circuit. Trouble spots, for instance, can often be narrowed down by noting if other components related to that circuit are operating properly or not. If several components or circuits fail at one time, chances are the problem lies in the fuse or ground connection, as several circuits are often routed through the same fuse and ground connections.

Electrical problems often stem from simple causes, such as loose or corroded connections, a blown fuse or a melted fusible link. Always visually inspect the condition of the fuse, wires and connections in a problem circuit before troubleshooting it.

If testing instruments are going to be utilized, use the diagrams to plan ahead of time where you will make the necessary connections in order to accurately pinpoint the trouble spot.

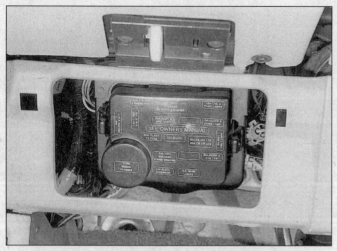

3.2 The removable cover over the fuse block contains fuse and circuit information for your vehicle

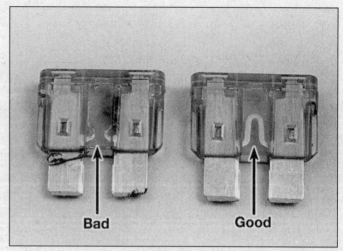

3.3 When a fuse blows, the element between the terminals burns - the fuse on the left is blown; the fuse on the right is good

The basic tools needed for electrical troubleshooting include a circuit tester or voltmeter (a 12-volt bulb with a set of test leads can also be used), a continuity tester, which includes a bulb, battery and set of test leads, and a jumper wire, preferably with a circuit breaker incorporated, which can be used to bypass electrical components.

Voltage checks should be performed if a circuit is not functioning properly. Connect one lead of a circuit tester to either the negative battery terminal or a known good ground. Connect the other lead to a connector in the circuit being tested, preferably nearest to the battery or fuse. If the bulb of the tester lights up, voltage is present, which means that the part of the circuit between the connector and the battery is problem free. Continue checking the rest of the circuit in the same fashion. When you reach a point at which no voltage is present, the problem lies between that point and the last test point with voltage. Most of the time the problem can be traced to a loose connection. **Note:** *Keep in mind that some circuits receive voltage only when the ignition key is in the Accessory or Run position.*

One method of finding shorts in a circuit is to remove the fuse and connect a test light or voltmeter in its place to the fuse terminals. There should be no voltage present in the circuit. Move the wiring harness from side-to-side while watching the test light. If the bulb goes on, there is a short to ground somewhere in that area, probably where the insulation has rubbed through. The same test can be performed on each component in the circuit, even a switch.

Perform a ground test to check whether a component is properly grounded. Disconnect the battery and connect one lead of a self-powered test light, known as a -continuity tester," to a known good ground. Connect the other lead to the wire or ground connection being tested. If the bulb goes on, the ground is good. If the bulb does not go on,

the ground is not good.

A continuity check determines if there are any breaks in a circuit- if it is conducting electricity properly. With the circuit off (no power in the circuit), a self-powered continuity tester can be used to check the circuit. Connect the test leads to both ends of the circuit (or to the -power" end and a good ground), and if the test light comes on the circuit is passing current properly. If the light doesn't come on, there is a break somewhere in the circuit. The same procedure can be used to test a switch, by connecting the continuity tester to the power in and power out sides of the switch. With the switch turned on, the test light should come on.

When diagnosing for possible open circuits, it is often difficult to locate them by sight because oxidation or terminal misalignment are hidden by the connectors. Merely wiggling a connector on a sensor or in the wiring harness may correct the open circuit condition. Remember this if an open circuit is indicated when troubleshooting a circuit. Intermittent problems may also be caused by oxidized or loose connections.

Electrical troubleshooting is simple if you keep in mind that all electrical circuits are basically electricity running from the battery, through the wires, switches, relays, fuses and fusible links to each electrical component (light bulb, motor, etc.) and back to ground, from which it is passed back to the battery. Any electrical problem is an interruption in the flow of electricity to and from the battery.

3   Fuses - general information

Refer to illustrations 3.2 and 3.3

The electrical circuits of the vehicle are protected by a combination of fuses, circuit breakers and fusible links. The fuse block is located under the dash on the driver's side.

Each fuse protects one or more circuits.

The protected circuit is identified on the fuse panel cover (**see illustration**). On these models, miniaturized fuses are installed in the fuse block. These compact fuses, with blade terminal design, allow fingertip removal and installation.

If an electrical component fails, always check the fuse first. The best way to check a fuse is with a test light. Check for power at the exposed terminal tips of each fuse. If power is present on one side of the fuse but not the other, the fuse is blown. A blown fuse can also be confirmed by visually inspecting it (**see illustration**).

Be sure to replace blown fuses with the correct type. Fuses of different ratings are physically interchangeable, but only fuses of the proper rating should be used. Replacing a fuse with one of a higher or lower value than specified is not recommended. Each electrical circuit needs a specific amount of protection. The amperage value of each fuse is molded into the fuse body. **Caution:** *Never bypass a fuse with pieces of metal or foil. Serious damage to the electrical system could result.*

If the replacement fuse immediately fails, do not replace it again until the cause of the problem is isolated and corrected.

4   Fusible links - replacement

Some circuits are protected by fusible links. These links are used in circuits which are not ordinarily fused, such as the ignition circuit. If a circuit protected by a fusible link becomes inoperative, inspect for a blown fusible link.

Although fusible links appear to be of heavier gauge than the wire they are protecting, their appearance is due to thicker insulation. All fusible links are several wire gauges smaller than the wire they are designed to protect. The location of the fusible links on your particular vehicle can be determined by referring to the wiring diagrams at the end of this Chapter.

Fusible links cannot be repaired. If you must replace one, make sure that the new fusible link is a duplicate of the one removed with respect to gauge, length and insulation. Replacement fusible links have insulation that is flame proof. Do not fabricate a fusible link from ordinary wire - the insulation may not be flame proof. **Warning:** *Do not mistake a resistor wire for a fusible link. The resistor wire is generally longer and is identified by a "Resistor - don't cut or splice" warning.*

Charging system fusible link

1   To replace the fusible link in the charging system, proceed as follows:

a) *Disconnect the negative cable at the battery.*

b) *Disconnect the fusible link from the wiring harness or the fusible link eyelet terminal from the battery terminal of the starter relay (on some vehicle applications, the fusible link is looped outside the wire harness).*

c) *Cut the damaged fusible link and the splices from the wires to which it is attached. Disconnect the feed wire part of the wiring and cut out the damaged portion as closely as possible behind the splice in the harness. If the fusible link wire insulation is burned or opened, disconnect the feed as close as possible behind the splice in the harness. If the damaged fusible link is between two splices (the weld points in the harness), cut out the damaged portion as close as possible to the weld points.*

d) *Strip the insulation back approximately 1/2-inch.*

e) *Splice and solder the new fusible link to the wires from which the old link was cut. Use rosin core solder at each end of the new link to obtain a good solder joint.*

f) *Wrap the splices completely with vinyl electrical tape around the soldered joint. No wires should be exposed.*

g) *Securely connect the eyelet terminals (if any) to the battery stud on the starter relay.*

h) *Install the repaired wiring as before, using existing clips, if provided.*

i) *Connect the battery ground cable.*

j) *Test the circuit for proper operation.*

All other fusible links

2   To service any other blown fusible link, use the following procedure:

a) *Determine which circuit is damaged, its location and the cause of the open fusible link. If the damaged fusible link is one of three fed by a common 10 or 12 gauge feed wire, determine the specific affected circuit.*

b) *Disconnect the negative cable at the battery.*

c) *Cut the damaged fusible link from the wiring harness and discard it. If the fusible link is one of three circuits fed by a single wire, cut it out of the harness at each splice and discard it.*

d) *Identify and procure the proper fusible link and butt connectors for attaching the fusible link to the harness.*

3   To service any fusible link in a three-link group with one feed:

a) *After cutting the open link out of the harness, cut each of the remaining undamaged fusible links close to the feed wire weld.*

b) *Strip approximately 1/2-inch of insulation from the detached ends of the two good fusible links. Insert two wire ends into one end of a Ford butt connector and carefully push one stripped end of the replacement fusible link into the same end of the butt connector and crimp all three firmly together.* **Note:** *Be very careful when fitting the three fusible links into the butt connector - the internal diameter is a snug fit for three wires. Be sure to use a proper crimping tool. Pliers, side cutters, etc. will not apply the proper crimp to retain the wires.*

c) *After crimping the butt connector to the three fusible links, cut the weld portion*

from the feed wire and strip about 1/2-inch of insulation from the cut end. Insert the stripped end into the open end of the butt connector and crimp very firmly.

d) *To attach the remaining end of the replacement fusible link, strip about 1/2-inch of insulation from the wire end of the circuit from which the blown fusible link was removed and firmly crimp a butt connector to the stripped wire. Insert the end of the replacement link into the other end of the butt connector and crimp firmly.*

---

5   Circuit breakers - general information

Circuit breakers protect components such as power windows, power door locks and headlights. Some circuit breakers are located in the fuse box.

Because a circuit breaker resets itself automatically, an electrical overload in a circuit breaker protected system will cause the circuit to fail momentarily, then come back on. If the circuit does not come back on, check it immediately. Once the condition is corrected, the circuit breaker will resume its normal function.

---

6   Turn signal and hazard flashers - check and replacement

Refer to illustration 6.1

Turn signal flasher

1   The turn signal flasher, a small canister shaped unit located in the fuse block **(see illustration)**, flashes the turn signals.

2   When the flasher unit is functioning properly, an audible click can be heard during its operation. If the turn signals fail on one side or the other and the flasher unit does not make its characteristic clicking sound, a faulty turn signal bulb is indicated.

3   If both turn signals fail to blink, the problem many be due to a blown fuse, a faulty flasher unit, a broken switch or a loose or open connection. If a quick check of the fuse box indicates that the turn signal fuse has blown, check the wiring for a short before installing a new fuse.

4   To replace the flasher, simply pull it out of the fuse block or wiring harness.

5   Make sure that the replacement unit is identical to the original. Compare the old one to the new one before installing it.

6   Installation is the reverse of removal.

Hazard flasher

7   The hazard flasher, a small canister shaped unit located in the fuse block, flashes all four turn signals simultaneously when activated.

8   The hazard flasher is checked in a fash-

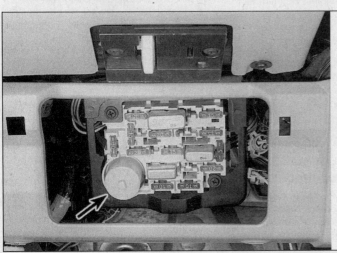

6.1 The turn signal flasher is located at the lower corner of the fuse panel (arrow)

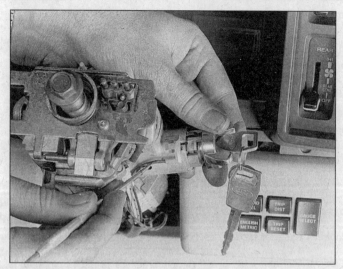

7.7 With the ignition key lock cylinder in the Run position, push in on the release lever to release the cylinder

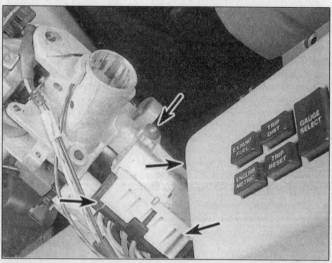

8.7 Unplug the ignition switch connector (lower arrows) and unscrew the break-off head bolts (upper arrows)

ion similar to the turn signal flasher (see Steps 2 and 3).

9   To replace the hazard flasher, pull it from the back of fuse block.

10   Make sure the replacement unit is identical to the one it replaces. Compare the old one to the new one before installing it.

11   Installation is the reverse of removal.

---

7   Ignition key lock cylinder - replacement

1986 through 1991 models

Refer to illustration 7.7

1   Disconnect the negative cable at the battery. Place the cable out of the way so it cannot accidentally come in contact with the negative terminal of the battery, as this would once again allow power into the electrical system of the vehicle.

2   Remove the steering wheel (Chapter 10).

3   If your vehicle is equipped with a tilt column, remove the steering column collar.

4   Remove the four retaining screws and detach the column cover.

5   Locate the key warning buzzer electrical lead (the insulated single wire coming out of the key lock cylinder housing), trace it back to the pigtail connector near the multi-terminal connector for the ignition switch and unplug it.

6   Turn the lock key to the Run position.

7   Place a 1/8-inch punch in the hole in the casting surrounding the lock cylinder. Depress the punch while pulling out on the lock cylinder to remove it from the column housing **(see illustration)**.

8   Install the lock cylinder by turning it to the Run position and depressing the retaining pin. Insert the lock cylinder into the lock cylinder housing. Make sure the cylinder is completely seated and aligned in the interlocking washer before turning the key to the

Off position. This will permit the retaining pin to extend into the hole.

9   Turn the lock to ensure that the operation is correct in all positions.

10   The remainder of installation is the reverse of removal.

1992 and later models

Warning: *Some models covered by this manual are equipped with Supplemental Restraint Systems (SRS), more commonly known as airbags. Always disable the airbag system before working in the vicinity of any airbag system component to avoid the possibility of accidental deployment of the airbag(s), which could cause personal injury (see Section 24).*

11   Disconnect the negative cable at the battery.

12   Insert the ignition key and turn it to the Run position.

13   Place a 1/8-inch punch in the hole in the trim shroud under the lock cylinder. Depress the retaining pin inside with the punch while pulling out on the lock cylinder. Remove the lock cylinder from the column housing.

14   Install the lock cylinder by turning it to the RUN position and depressing the retaining pin. Insert the lock cylinder into the lock cylinder housing. Make sure the lock cylinder is completely seated and aligned with the interlocking washer before turning the key to the Off position. This will permit the retaining pin to extend into the hole.

15   Turn the key to all positions to make sure all mechanical operations are correct.

---

8   Ignition switch - replacement

1986 through 1991 models

Refer to illustration 8.7

1   Disconnect the negative cable at the battery. Place the cable out of the way so it cannot accidentally come in contact with the

negative terminal of the battery, as this would once again allow power into the electrical system of the vehicle.

2   Remove the steering wheel (Chapter 10).

3   On tilt column models, grasp the upper extension shroud and remove it by squeezing it at the top and bottom and popping it free of the retaining plate.

4   Remove the screws and detach the steering column bottom cover.

5   Remove the screws and detach the panel to the right of the steering column.

6   Remove the key lock cylinder (Section 7).

7   Unplug the ignition switch electrical connector and use a hammer and chisel to unscrew the break-off head bolts one turn counterclockwise **(see illustration)**.

8   Use pliers to unscrew the bolts and remove them.

9   Disengage the ignition switch from the actuator pin.

10   Make sure the actuator pin slot in the new ignition switch is in the Run Position. **Note:** *A new replacement switch assembly will be set in the On (Run) position.*

11   Place the new switch in position and install the break-off head bolts. Tighten each bolt until the head breaks off.

12   Plug the electrical connector into the switch.

13   Install the steering column shroud and any other panels which were removed.

14   Install the ignition key lock cylinder.

15   Install the steering wheel.

16   Connect the battery negative cable.

1992 and later models

17   The ignition switch is part of the steering column bearing carrier. It is attached to the steering column in such a way that the steering column must be removed from the vehicle to remove the ignition switch. This procedure should be entrusted to a dealer service department or other qualified repair shop.

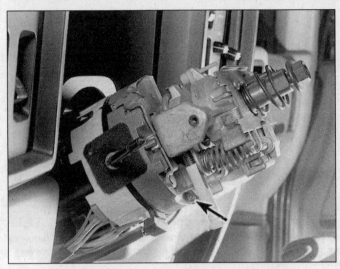

9.6 Remove the turn signal switch retaining screws (lower screw shown)

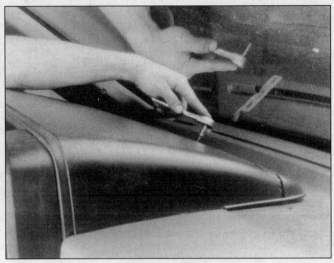

10.2a Remove the screws securing the instrument cluster panel . . .

9   Turn signal switch - removal and installation

1986 through 1991 models

Refer to illustration 9.6

1   Disconnect the negative cable at the battery. Place the cable out of the way so it cannot accidentally come in contact with the negative terminal of the battery, as this would once again allow power into the electrical system of the vehicle.
2   Remove the steering wheel.
3   Remove the steering column cover.
4   Remove the ignition key lock cylinder (Section 7).
5   Remove the control lever by grasping it securely and pulling it directly out of the switch.
6   Remove the two retaining screws, pull the switch out, unplug the connector and withdraw the switch assembly from the steering column **(see illustration)**.
7   Plug in the electrical connector, place the new switch in position and install the screws. Tighten the screws securely.
8   Install the steering column cover.
9   Install the ignition key lock cylinder.
10   Install the switch lever by aligning the key on the lever with the keyway in the switch and pushing the lever in until it engages.
11   Install the steering wheel.
12   Connect the negative battery cable.

1992 and later models

13   The multi-purpose switch (includes the turn signal switch) is attached to the steering column in such a way that the steering column must be removed from the vehicle to remove the switch. This procedure should be entrusted to a dealer service department or other qualified repair shop.

10   Headlight switch and rheostat - removal and installation

1986 through 1991 models

Refer to illustrations 10.2a, 10.2b, 10.4, 10.5 and 10.9

Headlight switch

1   Disconnect the negative cable at the battery. Place the cable out of the way so it cannot accidentally come in contact with the negative terminal of the battery, as this would once again allow power into the electrical system of the vehicle.
2   Remove the five instrument cluster panel retaining screws, pry up on the panel and remove it **(see illustrations)**.
3   Remove the three switch control pod retaining screws.
4   Disengage the clip at the corner and

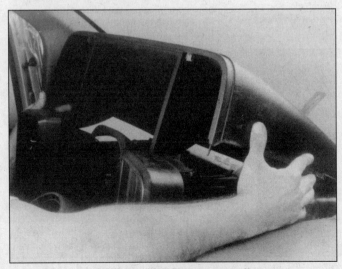

10.2b . . . then lift the cover off

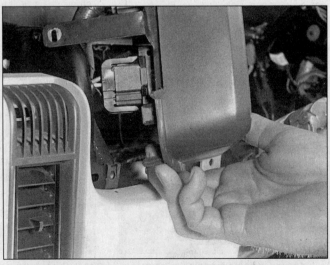

10.4 Disengage the clip and rotate the control pod out of the instrument panel for access to the switch

10.5 The rear of the control pod shows the electrical connectors at the back of the switches

10.9 Push the clip away from the switch with your thumb and unplug the connector

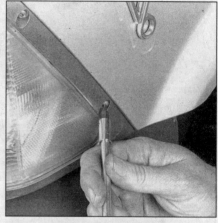

11.2a A torx-head screwdriver is required to remove the headlight door screws

rotate the pod out of the instrument panel for access to the switch **(see illustration)**.
5   Unplug the connector, remove the retaining screws and lift the switch from the pod **(see illustration)**.
6   Installation is the reverse of removal.

Rheostat

7   Remove the fuse panel cover to gain access to the rheostat.
8   Remove the two retaining screws and lower the rheostat from the instrument panel.
9   Unplug the electrical connector and remove the rheostat from the vehicle **(see illustration)**.
10   Installation is the reverse of removal.

1992 and later models

Warning: *Some models covered by this manual are equipped with Supplemental Restraint Systems (SRS), more commonly known as airbags. Always disable the airbag system before working in the vicinity of any airbag system component to avoid the possibility of accidental deployment of the airbag(s), which could cause personal injury (see Section 24).*
11   Disconnect the negative cable at

the battery.
12   Insert a small screwdriver into the headlight switch knob slot, depress the spring clip and remove the knob from the shaft.
13   If the vehicle is equipped with Auto lamp, pull the Auto lamp control knob off the switch shaft.
14   Remove the three steering column cover screws, pry up the panel and remove it from the four clips.
15   Remove the ash tray drawer, then remove the two center panel attaching screws. Carefully pull the center panel rearward and unsnap the retainers, then let it hang down.
16   Remove the left finish panel from the four clips.
17   Remove the cluster finish panel five attaching screws and carefully remove the panel from the two clips. Remove the cluster finish panel.
18   Remove the two switch attachment screws and pull the switch assembly partially out of the instrument panel. Disconnect the electrical connector from the switch and remove the switch from the vehicle.
19   Installation is the reverse of removal.

11 Headlights - removal and installation

Sealed beam headlights (1986 through 1991)

Refer to illustrations 11.2a, 11.2b, 11.3, 11.4, 11.5 and 11.8
1   Disconnect the negative cable at the battery. Place the cable out of the way so it cannot accidentally come in contact with the negative terminal of the battery, as this would once again allow power into the electrical system of the vehicle.
2   Remove the headlight door retaining screws and clips **(see illustrations)**.
3   Grasp the headlight door securely, push down on the bumper cover and remove the door **(see illustration)**.
4   Remove the headlight retaining ring screws **(see illustration)**, taking care not to disturb the adjustment screws.
5   Remove the retaining ring and pull the headlight out sufficiently to allow the connector to be unplugged **(see illustration)**.
6   Remove the headlight.

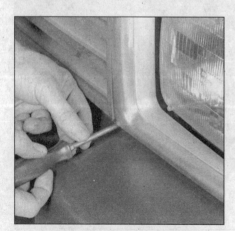

11.2b On some models the lower retaining screw is hidden by the bumper cover

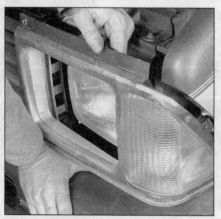

11.3 Push down on the bumper cover and lift out the headlight door

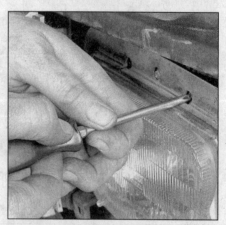

11.4 Use a Phillips screwdriver to remove the retaining ring screws - do not disturb the adjusting screws

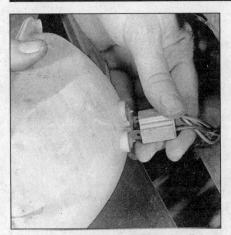

11.5 Support the headlight and pull the connector straight off

7   To install, plug the connector securely into the headlight, place the headlight in position and install the retaining ring and screws. Tighten the screws securely.
8   Place the headlight door in position, align the clips with their mounting holes and push the door firmly into place (see illustration).
9   Install the headlight screws and connect the battery negative cable.

Halogen bulb-type headlights (1992 and later models)

Warning: *Halogen gas-filled bulbs are under pressure and may shatter if the surface is scratched or the bulb is dropped. Wear eye protection and handle the bulbs carefully, grasping only the base whenever possible. Do not touch the surface of the bulb with your fingers because the oil from your skin could cause the bulb to overheat and fail prematurely. If you do touch the bulb surface, thoroughly clean it with rubbing alcohol.*
10   Disconnect the negative cable at the battery.
11   Remove the headlight door retaining screws and remove the door.
12   Remove the headlight retaining screws. **Note:** *Do not disturb the adjusting screw setting.*
13   Pull the headlight lens assembly forward and disconnect the electrical connector from bulb.
14   Use a screwdriver and unhook the bulb retaining wire.
15   Remove the headlight bulb from the headlight lens assembly.
16   Install the new bulb and install the retaining wire to the bulb. Make sure the bulb is secured correctly.
17   Connect the electrical connector onto the headlight bulb socket.
18   Move the headlight lens assembly into position and install the headlight retaining screws.
19   Install the headlight door and retaining screws and tighten securely.
20   Connect the battery negative cable.

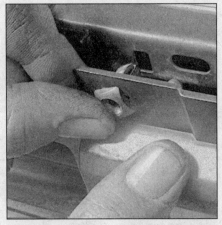

11.8 Line up the headlight door clip with the holes and push in to seat them

12   Headlights - adjustment

Refer to illustration 12.3
Note: *It is important that the headlights be aimed correctly. If adjusted incorrectly they could blind the driver of an oncoming vehicle and cause a serious accident or seriously reduce your ability to see the road. The head-lights should be checked for proper aim every 12 months and any time a new headlight is installed or front end body work is performed. It should be emphasized that the following procedure is only an interim step which will provide temporary adjustment until the head-lights can be adjusted by a properly equipped shop.*
1   On sealed beam headlights, there are two spring loaded adjusting screws, one on the top controlling up-and-down movement and one on the side controlling left-and-right movement.
2   On halogen bulb-type headlights, there are two spring loaded adjusting screws, one on the top front controlling up-and-down movement and one long adjuster on the top controlling left-and-right movement.
3   There are several methods of adjusting the headlights. The simplest method requires a blank wall 25 feet in front of the vehicle and a level floor (see illustration).
4   Position masking tape vertically on the wall in reference to the vehicle centerline and the centerlines of both headlights.
5   Position a horizontal tape line in reference to the centerline of all the headlights. **Note:** *It may be easier to position the tape on the wall with the vehicle parked only a few inches away.*

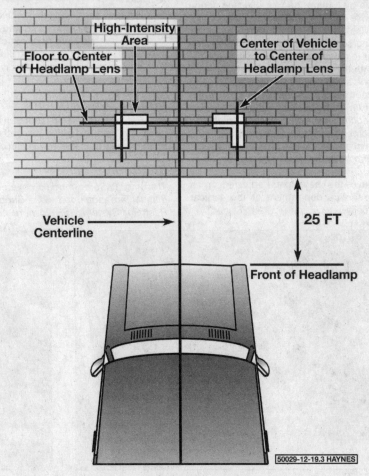

12.3 Headlight aiming details

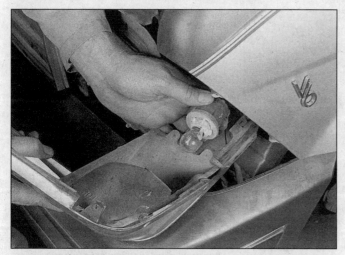

13.2 Pull the bulb holder out for access to the bulb

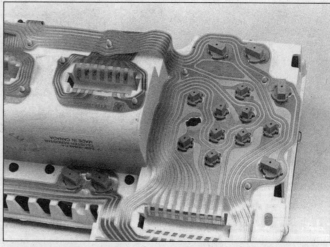

13.6 The bulb holders are located on the back of the instrument cluster and are removed by turning them about 1/4 turn

6   Adjustment should be made with the vehicle sitting level, the gas tank half-full and no unusually heavy load in the vehicle.

7   Starting with the low beam adjustment, position the high intensity zone so it is two inches below the horizontal line and two inches to the side of the headlight vertical line, away from oncoming traffic. Adjustment is made by turning the top adjusting screw clockwise to raise the beam and counterclockwise to lower the beam. The other adjusting screw should be used in the same manner to move the beam left or right.

8   With the high beams on, the high intensity zone should be vertically centered with the exact center just below the horizontal line. **Note:** *It may not be possible to position the headlight aim exactly for both high and low beams. If a compromise must be made, keep in mind that the low beams are the most used and have the greatest effect on driver safety.*

9   Have the headlights adjusted by a dealer service department at the earliest opportunity.

13 Bulb replacement

Refer to illustrations 13.2, 13.6, 13.13 and 13.14

Parking, turn and front side lamps

1   Remove the headlight door assembly **(see illustrations 11.2a, 11.2b and 11.3)**.

2   Hold the headlight door out of the way and pull the bulb holder out **(see illustration)**.

3   Push the bulb in and turn it counterclockwise to remove it.

Engine compartment lamp

4   Remove the lamp protective shield, grasp the bulb and pull it out of the socket. Push the new bulb securely in place.

Instrument cluster

Warning 1: *Some models covered by this manual are equipped with Supplemental Restraint Systems (SRS), more commonly known as airbags. Always disable the airbag system before working in the vicinity of any airbag system component to avoid the possibility of accidental deployment of the airbag(s), which could cause personal injury (see Section 24).*

Warning 2: *The bulbs on electronic display instrument clusters are under pressure and may shatter if scratched or dropped. Wear eye protection and handle the bulbs carefully, grasping only the base whenever possible. Do not touch the surface of the bulb with your fingers because the oil from your skin could cause the bulb to overheat and fail prematurely. If you do touch the bulb surface, thoroughly clean it with rubbing alcohol.*

5   Remove the instrument cluster (Section 18).

6   The instrument cluster bulbs can be replaced after rotating the holders and lifting them out **(see illustration)**.

Dome lamp

7   Use a screwdriver to pry the lens cover off and replace the bulb by grasping it

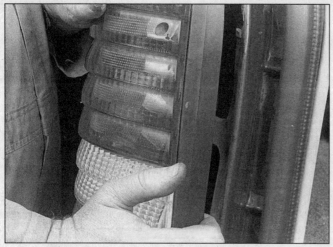

13.13 Pull the tail light lens away from the body - the sealant may make this job difficult

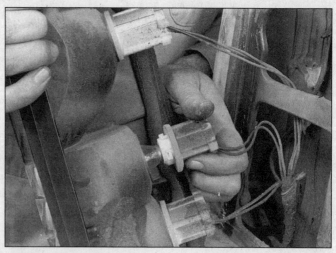

13.14 Rotate the bulb holders counterclockwise to remove them

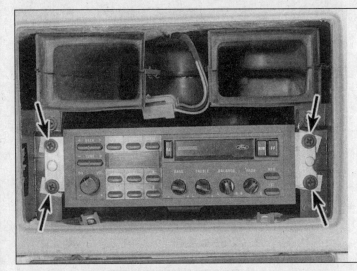

15.3 Radio retaining screw locations - 1986 through 1991 models (arrows)

securely and pulling it from the lamp assembly.

Map reading lamp

8   To replace the bulb in the center section, snap the lens carefully out of the lamp body and then remove the bulb.
9   To replace the bulb in the movable lamp gimble, depress the legs of the bulb cover with a thin tool and remove it. Snap the bulb gently from the socket.

Cargo lamp

10   Remove the two lens screws, lift off the lens and remove the bulb by depressing it and rotating it in a counterclockwise direction.

Brake and tail lamp

11   Open the rear liftgate or doors.
12   Remove the four tail light lens retaining screws.
13   Grasp the inner edge of the lens and rotate it away from the body to break it loose from the sealant **(see illustration)**.
14   Turn the bulb holders counterclockwise to remove them **(see illustration)**.
15   Turn the bulbs counterclockwise to remove them from the holders.

License plate lamp

16   On liftgate models, remove the two screws, lift off the bulb cover and pull the bulb from the socket.
17   On dual door models, remove the license plate assembly and pull the bulb out of the socket.

Hi-mount stoplight

18   Remove the two screws securing the rear hi-mount lamp to the liftgate.
19   Pull the hi-mount lamp away from the liftgate and disconnect the rear lamp assembly wiring connector.
20   Remove the bulb holder retaining screw and detach the bulb holder from the lamp assembly.
21   Pull the failed bulb(s) straight out of the holder and replace as required.
22   Installation is the reverse of removal.

14   Horn - check and replacement

Check

1   If the horn does not sound, first make sure the mounting bolt is securely tightened. Connect one jumper wire from the horn mounting bracket bolt to the negative battery terminal and another jumper wire from the horn terminal to the positive battery terminal.
2   If the horn does not sound and there is no sparking at the battery terminal, disconnect the jumper wires. Turn the adjusting screw counterclockwise between 1/4 and 3/8 turn and then use pliers to lock it from turning by squeezing the housing extrusions. If the horn does not sound when the jumper wires are reconnected, replace it with a new one.

Replacement

3   Disconnect the negative cable at the battery.
4   Unplug the horn electrical connector.
5   Remove the mounting bolt(s) and lift the horn from the vehicle.
6   Installation is the reverse of removal.

15   Radio and speakers - removal and installation

Refer to illustration 15.3

1   Disconnect the negative cable at the battery.

Radio (1986 through 1991 models)

2   Carefully pry off the finish panel.
3   Remove the retaining screws **(see illustration)**.
4   Pull the radio from the instrument panel, disconnect the antenna and electrical connectors and remove it from the vehicle.
5   Installation is the reverse of removal.

Radio (1992 and later models)

Warning: *Some models covered by this manual are equipped with Supplemental Restraint Systems (SRS), more commonly known as airbags. Always disable the airbag system before working in the vicinity of any airbag system component to avoid the possibility of accidental deployment of the airbag(s), which could cause personal injury (see Section 24).*

6   Remove the ash tray drawer, then remove the two center panel attaching screws.
7   Carefully pull the center panel rearward and unsnap the retainers.
8   Pull the center panel out and disconnect the electrical connectors from it.
9   Insert a special radio removal tool (available at most auto parts stores) or pieces of bent wire into the radio face plate. Press in about 1-inch to release the clips. Pull the radio out using the special tool.
10   Disconnect the antenna and the electrical connectors and remove it from the vehicle.

Speakers

Instrument panel mounted

11   Remove the instrument panel pad (Section 17).
12   Remove the retaining screws and lift the speaker out. Unplug the electrical connector and remove the speaker.
13   Installation is the reverse of removal.

Body side mounted

14   Pry the speaker grille off and remove the speaker retaining screws.
15   Disengage the speaker from the mounting plate, disconnect the wiring assembly and remove the speaker.
16   Installation is the reverse of removal.

Rear door mounted

17   On liftgate models, remove the trim panel (Chapter 11). Remove the mounting screws, withdraw the speaker and unplug the electrical connector and remove the speaker from the vehicle.
18   Installation is the reverse of removal.
19   On dual door models, remove the four screws and detach the speaker enclosure and bracket.
20   Remove the bracket assembly, unplug the wiring connector at the enclosure and remove the speaker.
21   Installation is the reverse or removal.

16   Radio antenna - removal and installation

Refer to illustration 16.1

Antenna mast

1   Unscrew the antenna mast from the base **(see illustration)**.
2   Installation is the reverse of removal.

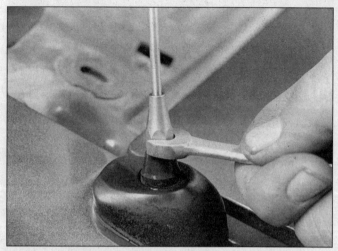

16.1 Unscrew the antenna mast from the base

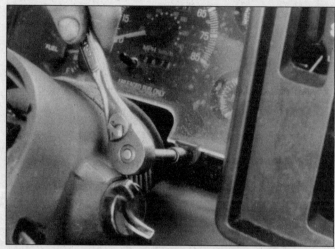

18.3a Remove the screws at the bottom of the instrument cluster, just above the steering column . . .

Cable

Warning: *Some models covered by this manual are equipped with Supplemental Restraint Systems (SRS), more commonly known as airbags. Always disable the airbag system before working in the vicinity of any airbag system component to avoid the possibility of accidental deployment of the airbag(s), which could cause personal injury (see Section 24).*

3   Remove the radio and disconnect the cable.
4   Disengage the cable from the clips and retainers along the top of the heater assembly.
5   Carefully pry off the radio antenna base cap, remove the mounting screws and pull the base and cable assembly out of the fender.
6   Installation is the reverse of removal.

17 Instrument panel - removal and installation

1986 through 1991 models

1   Disconnect the negative cable at the battery.
2   Remove the instrument cluster retaining screws and tap up at the corners to disengage the cluster from the control pods.
3   Remove the upper panel retaining screws **(see illustration 10.2a)**.
4   Remove the retaining screws and remove the cluster panel.
5   Lift the upper instrument panel out of the dash.
6   Remove the lower instrument panel trim panel.
7   Remove the two left side and one right side instrument panel cowl side attachment screws.
8   Remove the lower center instrument panel brace screws.
9   Remove the steering column-to-instrument panel bolts.
10   Disconnect the wiring connectors and

lift the instrument panel from the vehicle.
11   To install, place the panel in position on the cowl side brackets and align the centering stud into the cowl top. Line up the panel darts in the grooves on the cowl side brackets.
12   Connect the wiring connectors.
13   Install the steering column-to-instrument panel bolts.
14   Install the cowl top and side attachment screws.
15   Install the lower center brace screws.
16   Install the center brace-to-dash screws.
17   Install the upper instrument panel and cluster panel.
18   Install the lower instrument trim panel.

1992 and later models

Warning: *Some models covered by this manual are equipped with Supplemental Restraint Systems (SRS), more commonly known as airbags. Always disable the airbag system before working in the vicinity of any airbag system component to avoid the possibility of accidental deployment of the airbag(s), which could cause personal injury (see Section 24).*

19   Disconnect the negative cable at the battery.
20   Within the engine compartment, loosen the screw and disconnect the electrical wire connectors on the instrument panel.
21   Remove the left and right windshield inside moldings.
22   Remove the defroster grille.
23   Remove the glove box door assembly.
24   Remove the three steering column cover screws, pry up the panel and remove it from the four clips.
25   Remove the four steering column lower reinforcement screws and remove it.
26   Remove the right and left cowl side trim panels.
27   Remove the two nuts attaching the hood release handle assembly from the left cowl. Move the hood release handle assembly out of the way. Reinstall the electrical wiring harness onto the studs and reinstall

the two nuts.
28   Disconnect the electrical connectors and any vacuum connectors (if equipped) on the left cowl. **Note:** *The number of connections vary, depending on optional equipment.*
29   Disconnect the brake light switch wiring connector and the clutch interlock switch (manual transmission models).
30   Remove the pinch bolt from the steering column-to-extension shaft. Compress the extension shaft toward the engine and separate it from the column U-joint.
31   Disconnect the transmission shift cable from the steering column.
32   On the right side of the steering column, disconnect the electrical connectors for the windshield wiper intermittent control module (if equipped) and the wiring for the low oil relay.
33   On the right side cowl, disconnect the electrical wiring and radio antenna.
34   Disconnect the heater control cable and the air conditioner vacuum line connector (if equipped).
35   Remove the five screws attaching the top of the instrument panel to the cowl top.
36   Remove the one nut attaching the lower right side of the instrument panel to the cowl side.
37   Remove the two rearward nuts and loosen the two forward nuts located to the right side of the steering column attaching the instrument panel to the center mounting bracket.
38   **Note:** *The following steps require the aid of an assistant. The instrument panel is not heavy - just bulky.*
39   Support the instrument panel and remove the three bolts and one nut attaching the instrument panel to the left cowl side.
40   Carefully pull the instrument panel rearward and disconnect any remaining electrical connectors or other components that may still be attached.
41   Carefully remove the instrument panel through the driver's side door.
42   Installation is the reverse of removal.

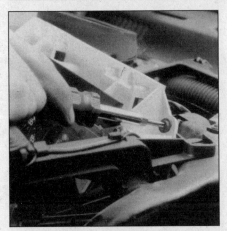

18.3b . . . and the two screws at the top of the cluster

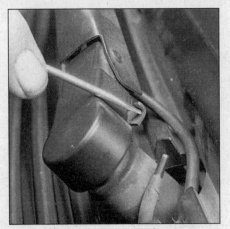

21.1 Lift the release lever out with a small screwdriver

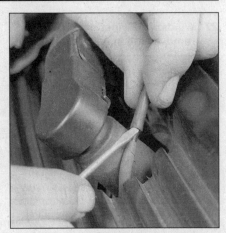

21.2 Push the washer hose off with a screwdriver

18 Instrument cluster - removal and installation

1986 through 1991 models

Refer to illustrations 18.3a and 18.3b

1 Disconnect the negative cable at the battery.

2 Remove the instrument cluster panel assembly (Section 17).

3 Remove the instrument cluster retaining screws **(see illustrations)**.

4 On conventional clusters, reach behind the cluster and disconnect the speedometer cable.

5 Pull the cluster out and unplug the electrical connectors.

6 Lift the cluster from the instrument panel.

7 Installation is the reverse of removal. On conventional clusters, apply speedometer cable lubricant to the drive hole of the speedometer.

1992 and later models

Warning: *Some models covered by this manual are equipped with Supplemental Restraint Systems (SRS), more commonly known as airbags. Always disable the airbag system before working in the vicinity of any airbag system component to avoid the possibility of accidental deployment of the airbag(s), which could cause personal injury (see Section 24).*

8 Disconnect the negative cable at the battery.

9 Remove the three steering column cover screws, pry up the panel and remove it from the four clips.

10 Remove the ash tray drawer, then remove the two center panel attaching screws. Carefully pull the center panel rearward and unsnap the retainers. Carefully let it hang down.

11 Remove the left finish panel from the four clips.

12 Remove the cluster finish panel five attaching screws and carefully remove the panel from the two clips. Remove the cluster finish panel.

13 Remove the four screws securing the instrument cluster to the instrument panel.

14 Disconnect the transmission indicator cable from the instrument cluster.

15 Partially pull the instrument cluster out and disconnect the electrical connectors.

16 Lift the cluster out of the instrument panel.

17 Installation is the reverse of removal. Make sure the instrument cluster engages the locating pin on the instrument panel prior to installing the screws.

19 Speedometer cable - replacement

Warning: *Some models covered by this manual are equipped with Supplemental Restraint Systems (SRS), more commonly known as airbags. Always disable the airbag system before working in the vicinity of any airbag system component to avoid the possibility of accidental deployment of the airbag(s), which could cause personal injury (see Section 24).*

1 Disconnect the negative cable at the battery.

2 Disconnect the speedometer cable at the transmission.

3 Detach the cable from the fasteners in the engine compartment and pull it up to provide enough slack to allow disconnection from the speedometer.

4 Remove the instrument cluster screws and disconnect the speedometer cable from the back of the cluster.

5 Remove the cable from the vehicle.

6 Installation is the reverse of removal.

20 Power door lock switch and actuator motor - removal and installation

Switch

1 Disconnect the negative cable from the battery and then carefully pry the switch trim bezel off with a thin screwdriver.

2 Remove the wiring connector retaining nuts from the back of the bezel and then pry the switch off the bezel.

3 Installation is the reverse of removal.

Actuator

4 Disconnect the negative cable at the battery.

5 Remove the door trim panel (Chapter 11).

6 Detach the actuator motor link from the door latch.

7 Use a drill with a suitable size drill bit to drill out the actuator motor retaining rivet.

8 Unplug the electrical connector and lift the motor and bracket from the door.

9 To install, plug in the electrical connector, place the actuator motor in position and install a new rivet. The rivet must be installed securely or the motor and bracket will move during operation, causing rattles and binding.

10 Connect the actuator motor link to the door latch.

11 Connect the battery cable and check the actuator for proper operation.

12 Install the door trim panel.

21 Windshield wiper arm - removal and installation

Refer to illustrations 21.1 and 21.2

1 Lift the wiper arm up several inches, then use a small screwdriver to pry the release lever out **(see illustration)**. While holding the lever out, allow the wiper arm to go down to lock the lever.

2 Disconnect the washer fluid hose by pushing it off with the screwdriver **(see illustration)**.

3 Press downward on the wiper arm and wiggle to lever it off.

4 Installation is the reverse of removal.

22   Windshield wiper motor - removal and installation

Refer to illustrations 22.4 and 22.5

1   Disconnect the negative cable at the battery.
2   Remove the wiper arms (Section 21).
3   Remove the cowl ventilator grille (Chapter 11).
4   Remove the clip from the wiper motor and disconnect the wiper actuating arms **(see illustration)**.
5   Unplug the electrical connector, remove the retaining nuts and lower the wiper motor into the engine compartment **(see illustration)**.
6   Installation is the reverse of removal.

23   Windshield washer motor, seal and pump - removal and installation

1   Remove and drain the washer reservoir (Chapter 3).
2   Use a small screwdriver and pry out the retaining ring.
3   Use pliers to grip the wall around the electrical terminal and pull the motor, seal and impeller assembly from the reservoir.
4   Clean all foreign material from the pump chamber before installing the pump assembly into the reservoir.
5   Lubricate the outside diameter of the seal with a powdered graphite to prevent the seal from sticking to the reservoir.
6   Align the small projection on the motor end with the slot in the reservoir and assemble so the seal seats against the bottom of the motor cavity.
7   Use a 1-inch (12 point preferably) socket to hand press the retaining ring securely against the motor. **Caution:** *Do not operate the pump until the reservoir has been filled with fluid.*
8   Replace the reservoir. Plug in the electrical connector and insert the hose.
9   Fill the reservoir.
10   Check for leaks.

24   Airbag system - general information

All models are equipped with a Supplemental Restraint System (SRS), more commonly known as an airbag. This system is designed to protect the driver and (on models equipped with a passenger's side airbag) the front seat passenger from serious injury in the event of a head-on or frontal collision. It consists of an airbag module in the center of the

22.4 Detach the wiper actuating arm clip by prying up with a small screwdriver

22.5 Wiper motor retaining nuts (arrows)

steering wheel and, on models so equipped, the right side of the instrument panel, a crash sensor mounted at the front of the vehicle, a crash sensor mounted behind each kick panel and a diagnostic monitor located inside the passenger compartment, under the instrument panel, to the left of the steering column.

Airbag module

Driver's side

The airbag inflator module contains a housing incorporating the cushion (airbag) and inflator unit, mounted in the center of the steering wheel The inflator assembly is mounted on the back of the housing over a hole through which gas is expelled, inflating the bag almost instantaneously when an electrical signal is sent from the system. A sliding contact assembly on the steering column under the module carries this signal to the module.

This sliding contact assembly can transmit an electrical signal regardless of steering wheel position. The igniter in the airbag converts the electrical signal to heat and ignites the sodium azide/copper oxide powder, producing a small explosion which inflates the bag.

Passenger's side

The airbag is mounted above the glove compartment and designated by the letters SRS (Supplemental Restraint System). It consists of an inflator containing an igniter, a bag assembly, a reaction housing and a trim cover.

The airbag is considerably larger than the steering wheel-mounted unit and is supported by the steel reaction housing. The trim cover is textured and painted to match the instrument panel and has a molded seam which splits when the bag inflates.

Sensors

The system has three sensors: a forward crash sensor at the front of the vehicle (mounted to the radiator support) and a crash sensor mounted behind each kick panel.

The sensors are basically pressure sensitive switches that complete an electrical circuit during an impact of sufficient G force.

Electronic diagnostic monitor

The electronic diagnostic monitor supplies the current to the airbag system in the event of the collision, even if battery power is cut off. It checks this system every time the vehicle is started, causing the "AIR BAG" light to go on then off, if the system is operating properly. If there is a fault in the system, the light will go on and stay on, flash, or the dash will make a beeping sound. If this happens, the vehicle should be taken to your dealer immediately for service.

Disabling the system

Whenever working in the vicinity of the steering wheel, steering column or near other components of the airbag system, the system should be disarmed. To do this, perform the following steps:

a)   *Turn the ignition switch to Off.*
b)   *Detach the cable from the negative battery terminal then detach the positive cable. Wait 2 minutes for the backup power supply to be depleted.*

Enabling the system

a)   *Turn the ignition switch to the Off position.*
b)   *Connect the positive battery cable first, then connect the negative cable.*

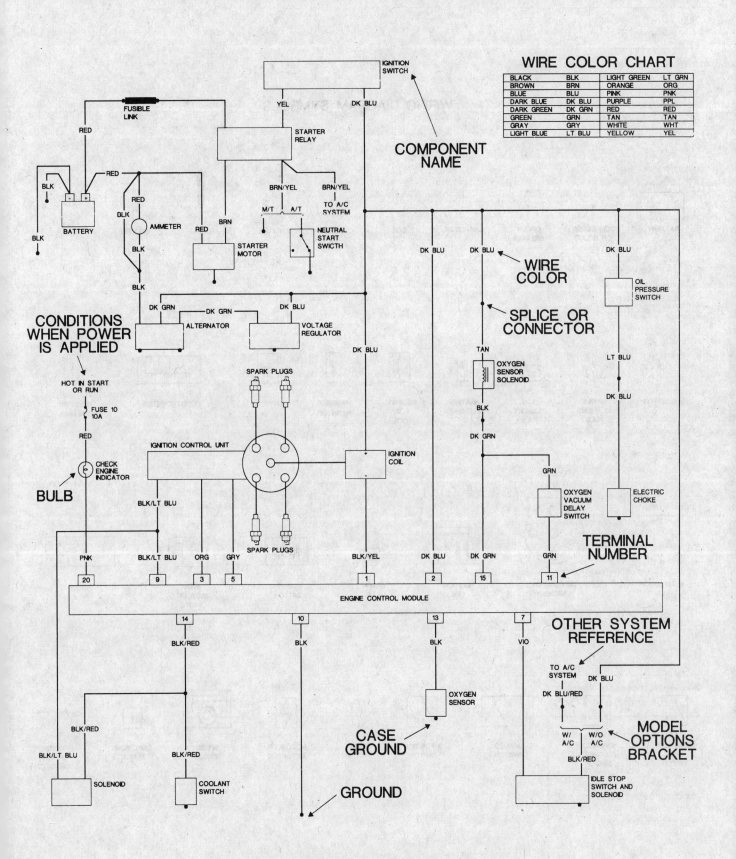

Sample diagram - how to read and interpret wiring diagrams

WIRING DIAGRAM SYMBOLS

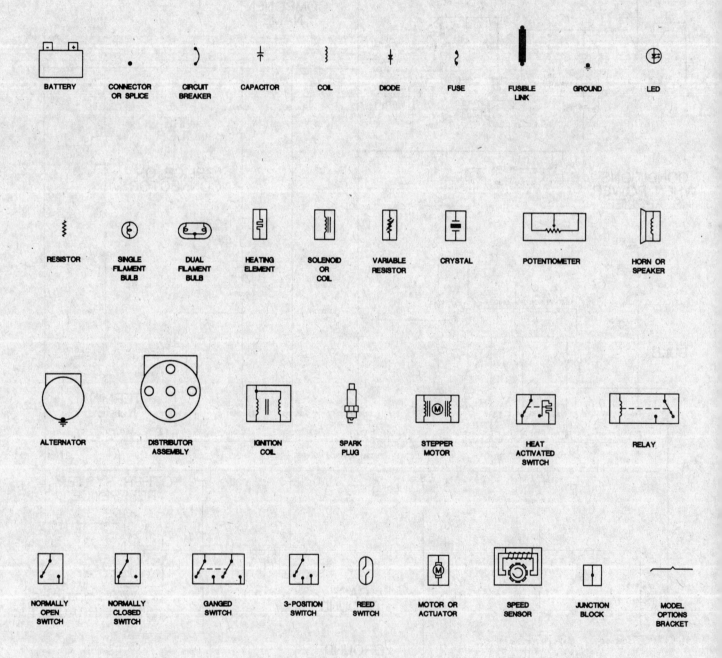

Common wiring diagram symbols

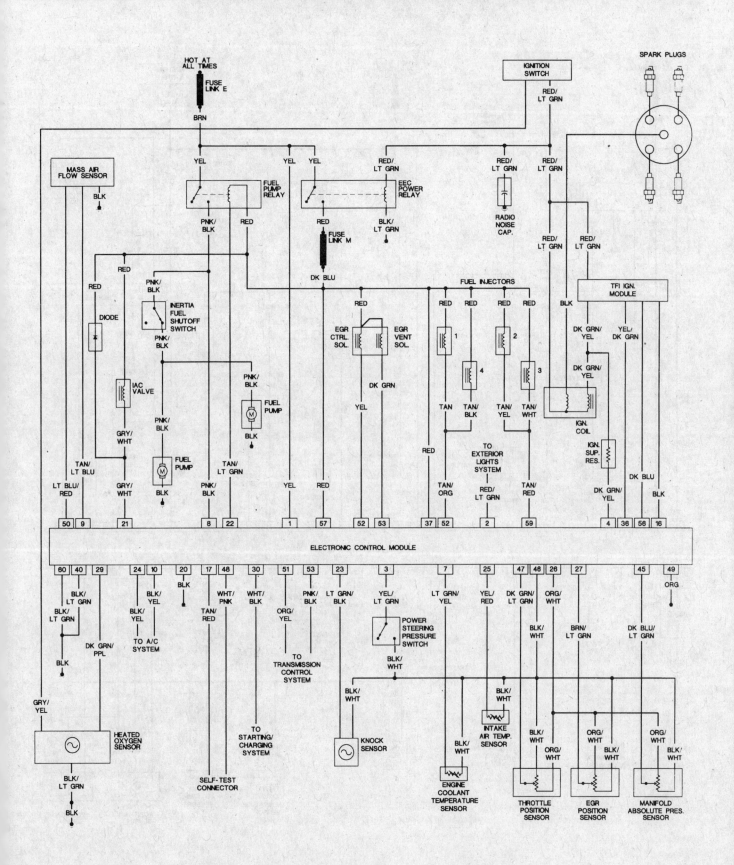

Engine control wiring - 1986 and 1987 2.3L four-cylinder engine

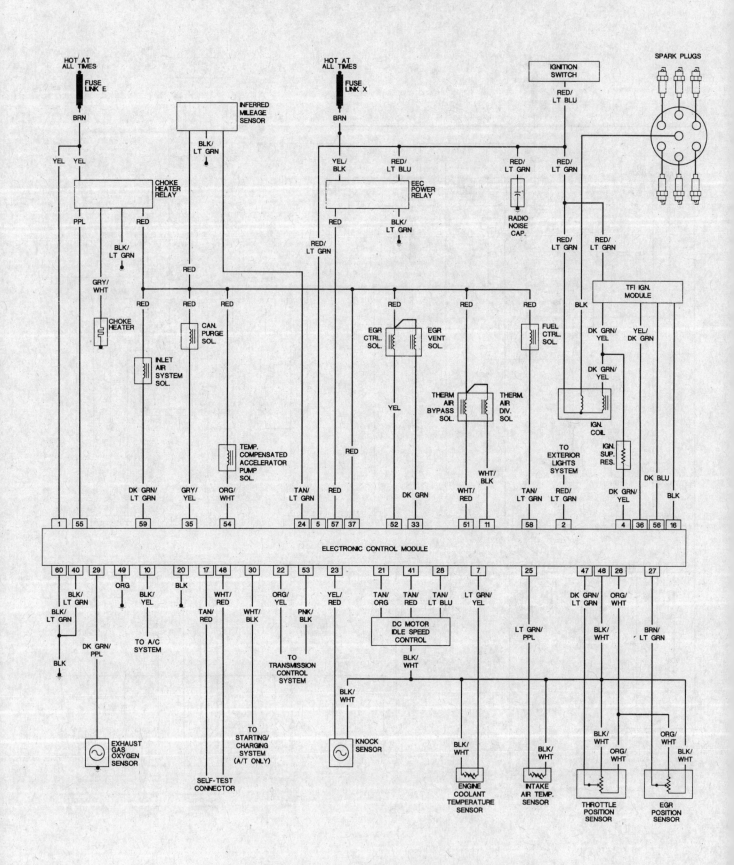

Engine control wiring - 1986 and 1987 2.8L V6 engine

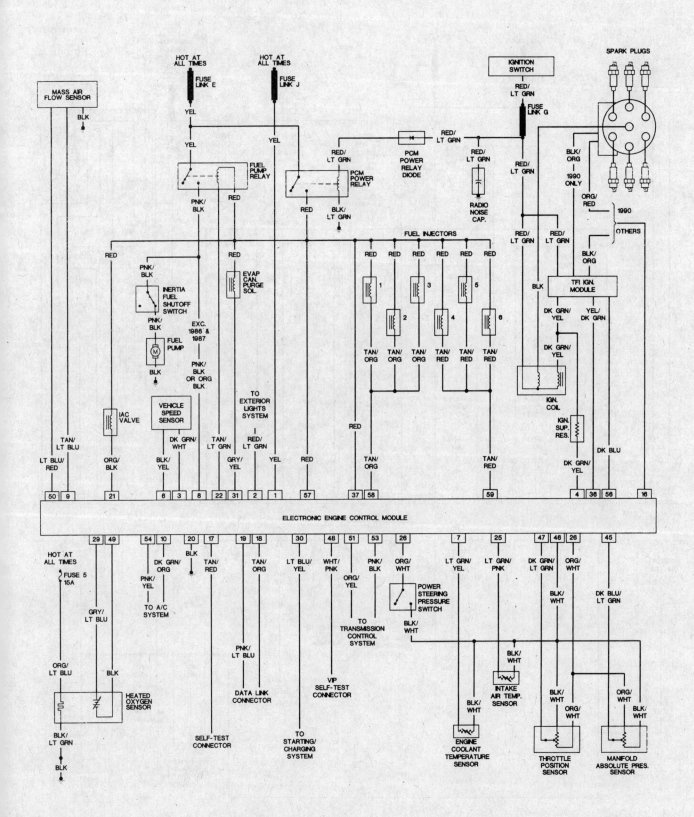

Engine control wiring - 1986 through 1991 3.0L V6 engine

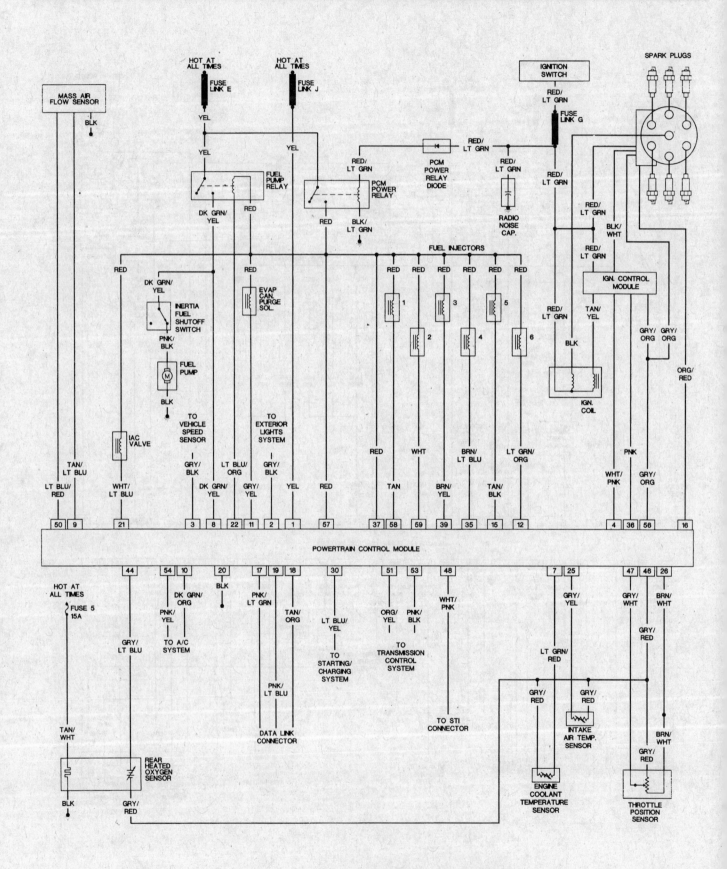

Engine control wiring - 1992 through 1995 3.0L V6 engine

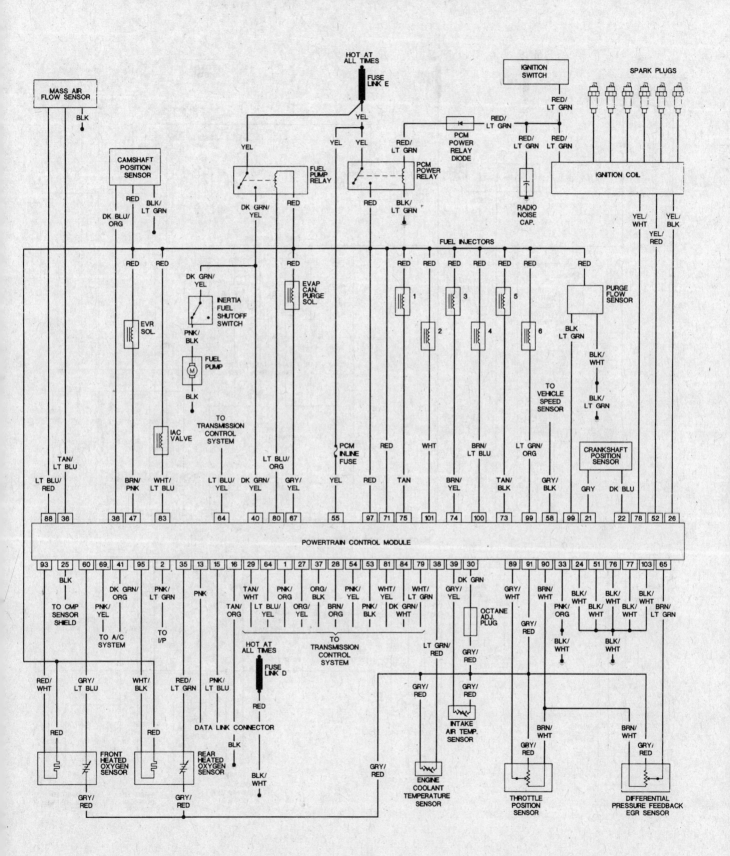

Engine control wiring - 1997 3.0L V6 engine

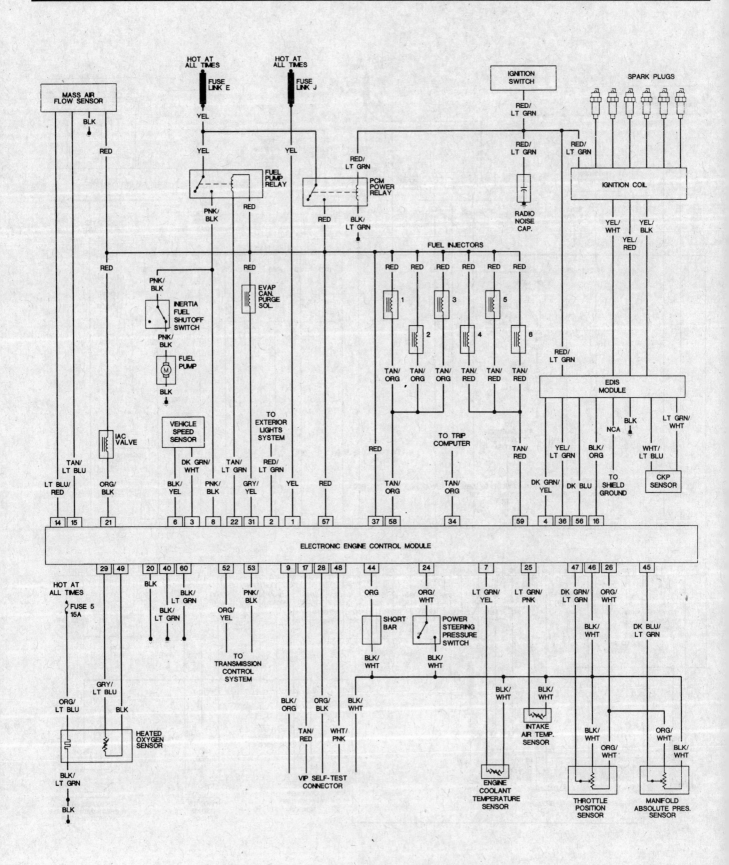

Engine control wiring - 1990 and 1991 4.0L V6 engine

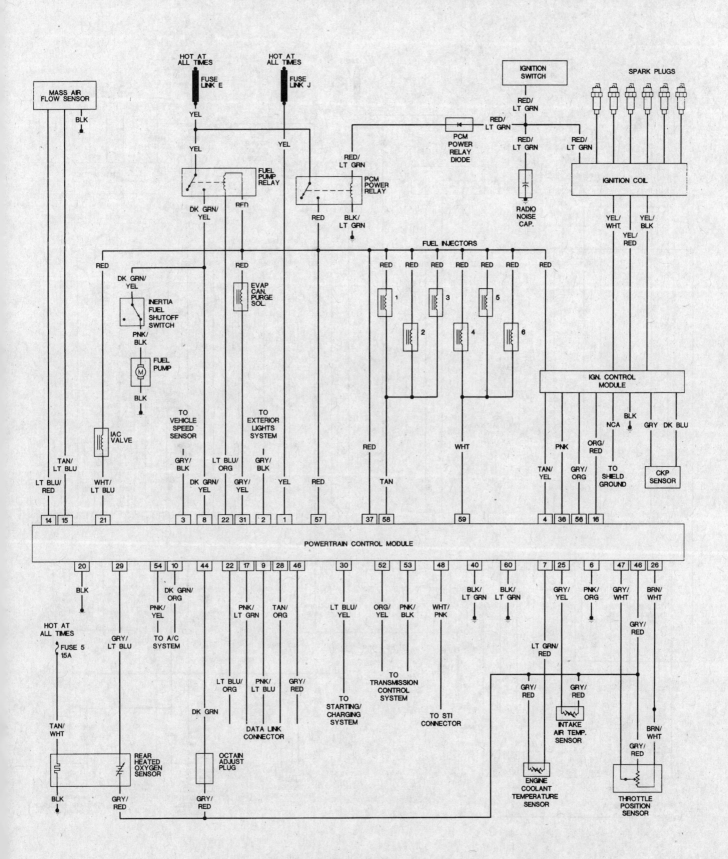

Engine control wiring - 1992 through 1995 4.0L V6 engine

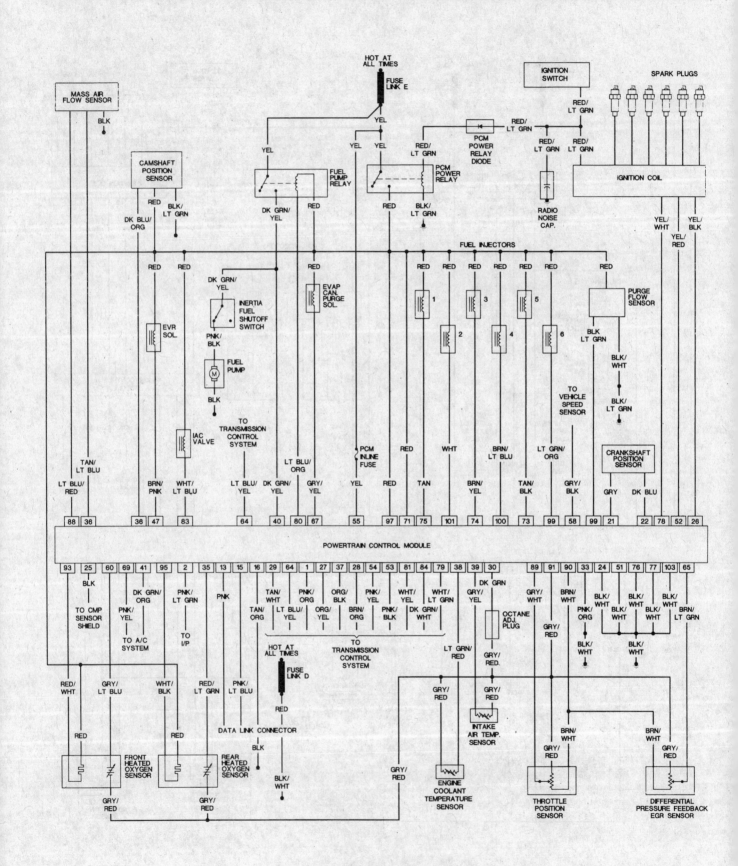

Engine control wiring - 1997 4.0L V6 engine

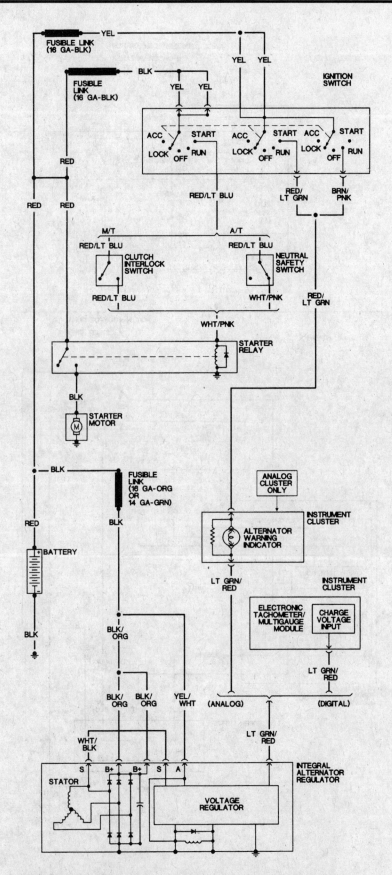

Starting and charging systems - 1986 through 1989 models

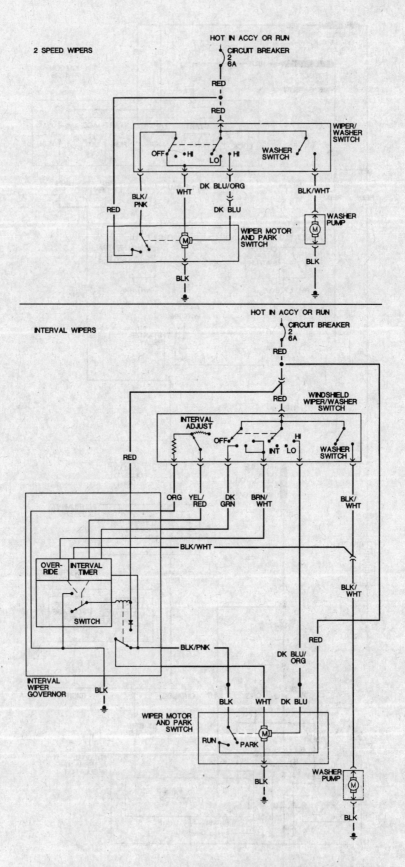

Windshield wiper/washer systems - 1986 through 1989 models

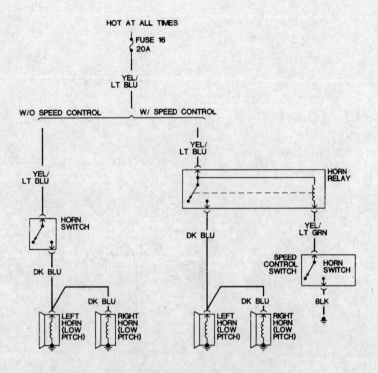

Horn wiring diagram - 1986 through 1989 models

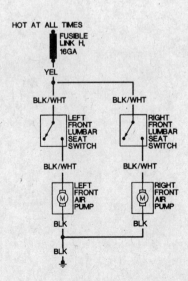

Seat lumbar support wiring diagram - 1986 through 1989 models

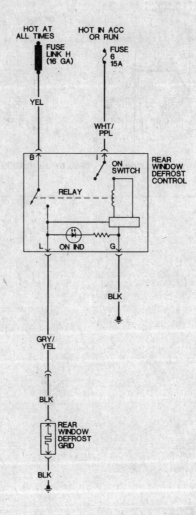

Rear window defroster wiring diagram - 1986 through 1989 models

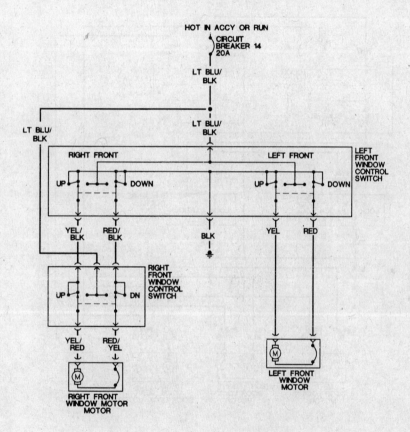

Power window wiring diagram - 1986 through 1989 models

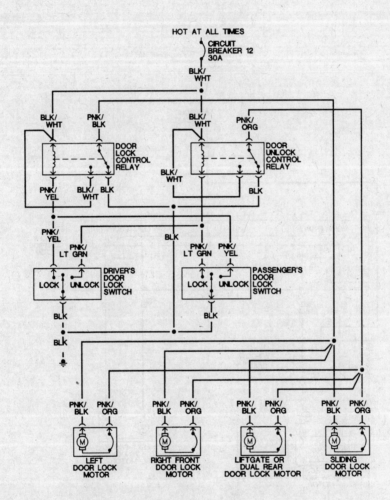

Power door lock wiring diagram - 1986 through 1989 models

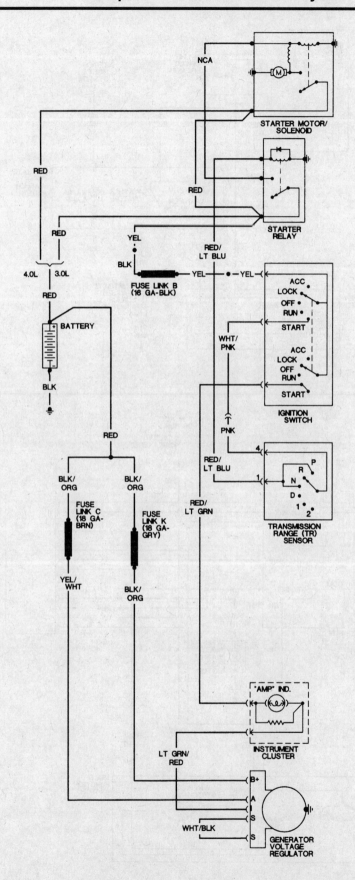

Starting and charging systems - 1990 through 1997 models

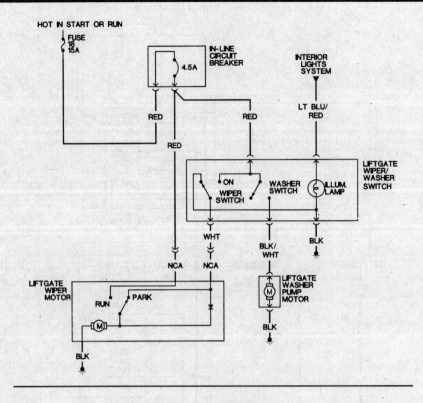

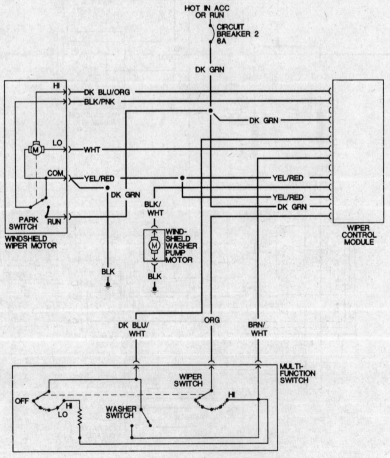

Windshield wiper/washer systems - 1990 through 1997 models

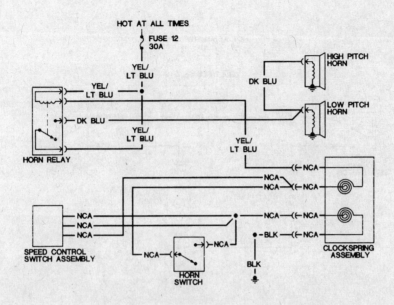

Horn wiring diagram - 1990 through 1997 models

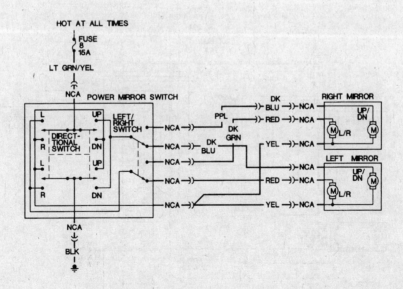

Power mirror wiring diagram - 1990 through 1997 models

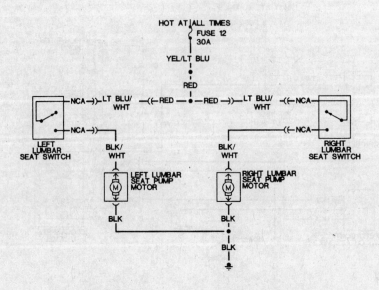

Seat lumbar support wiring diagram - 1990 through 1997 models

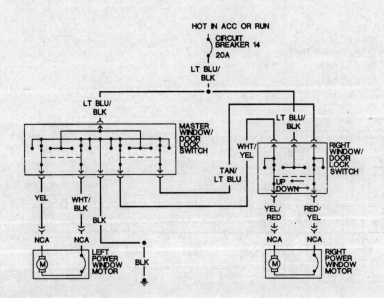

Power window wiring diagram - 1990 through 1997 models

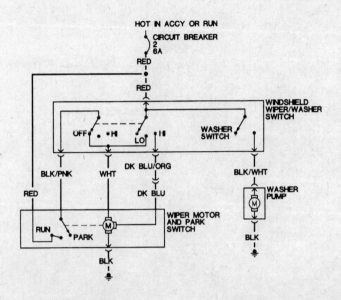

Windshield wiper/washer systems - 1990 through 1997 models

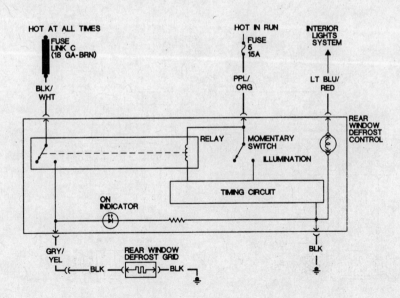

Rear window defroster wiring diagram - 1990 through 1997 models

Notes

Index

Haynes Automotive Manuals

NOTE: If you do not see a listing for your vehicle, consult your local Haynes dealer for the latest product information.

HAYNES XTREME CUSTOMIZING
11101	Sport Compact Customizing
11102	Sport Compact Performance
11110	In-car Entertainment
11150	Sport Utility Vehicle Customizing
11213	Acura
11255	GM Full-size Pick-ups
11314	Ford Focus
11315	Full-size Ford Pick-ups
11373	Honda Civic

ACURA
12020	Integra '86 thru '89 & Legend '86 thru '90
12021	Integra '90 thru '93 & Legend '91 thru '95

AMC
	Jeep CJ - see JEEP (50020)
14020	Concord/Hornet/Gremlin/Spirit '70 thru '83
14025	(Renault) Alliance & Encore '83 thru '87

AUDI
15020	4000 all models '80 thru '87
15025	5000 all models '77 thru '83
15026	5000 all models '84 thru '88

AUSTIN
	Healey Sprite - see MG Midget (66015)

BMW
18020	3/5 Series '82 thru '92
18021	5 Series including Z3 models '92 thru '98
18022	3-Series, E46 chassis '99 thru '05, Z4 models '03 thru '05
18025	320i all 4 cyl models '75 thru '83
18050	1500 thru 2002 except Turbo '59 thru '77

BUICK
19010	Buick Century '97 thru '05
	Century (front-wheel drive) - see GM (38005)
19020	Buick, Oldsmobile & Pontiac Full-size (Front wheel drive) '85 thru '05
19025	Buick Oldsmobile & Pontiac Full-size (Rear wheel drive) '70 thru '90
19030	Mid-size Regal & Century '74 thru '87
	Regal - see GENERAL MOTORS (38010)
	Skyhawk - see GM (38030)
	Skylark - see GM (38020, 38025)
	Somerset - see GENERAL MOTORS (38025)

CADILLAC
21030	Cadillac Rear Wheel Drive '70 thru '93
	Cimarron, Eldorado & Seville - see GM (38015, 38030, 38031)

CHEVROLET
10305	Chevrolet Engine Overhaul Manual
24010	Astro & GMC Safari Mini-vans '85 thru '03
24015	Camaro V8 all models '70 thru '81
24016	Camaro all models '82 thru '92, Cavalier - see GM (38015), Celebrity - see GM (38005)
24017	Camaro & Firebird '93 thru '02
24020	Chevelle, Malibu, El Camino '69 thru '87
24024	Chevette & Pontiac T1000 '76 thru '87
	Citation - see GENERAL MOTORS (38020)
24027	Colorado & GMC Canyon '04 thru '06
24032	Corsica/Beretta all models '87 thru '96
24040	Corvette all V8 models '68 thru '82
24041	Corvette all models '84 thru '96
24045	Full-size Sedans Caprice, Impala, Biscayne, Bel Air & Wagons '69 thru '90
24046	Impala SS & Caprice and Buick Roadmaster '91 thru '96
	Lumina '90 thru '94 - see GM (38010)
24048	Lumina & Monte Carlo '95 thru '05
	Lumina APV - see GM (38035)
24050	Luv Pick-up all 2WD & 4WD '72 thru '82
	Malibu - see GM (38026)
24055	Monte Carlo all models '70 thru '88
	Monte Carlo '95 thru '01 - see LUMINA
24059	Nova all V8 models '69 thru '79
24060	Nova/Geo Prizm '85 thru '92
24064	Pick-ups '67 thru '87 - Chevrolet & GMC, all V8 & in-line 6 cyl, 2WD & 4WD '67 thru '87; Suburbans, Blazers & Jimmys '67 thru '91
24065	Pick-ups '88 thru '98 - Chevrolet & GMC, all full-size models '88 thru '98; C/K Classic '99 & '00; Blazer & Jimmy '92 thru '94; Suburban '92 thru '99; Tahoe & Yukon '95 thru '99
24066	Pick-ups '99 thru '06 - Chevrolet Silverado & GMC Sierra '99 thru '06; Suburban/Tahoe/Yukon/Yukon XL/Avalanche '00 thru '06
24070	S-10 & S-15 Pick-ups '82 thru '93
24071	S-10, Sonoma & Jimmy '94 thru '04
24072	Chevrolet TrailBlazer & TrailBlazer EXT, GMC Envoy & Envoy XL, Oldsmobile Bravada '02 thru '06
24075	Sprint '85 thru '88, Geo Metro '89 thru '01
24080	Vans - Chevrolet & GMC '68 thru '96
24081	Chevrolet Express & GMC Savana Full-size Vans '96 thru '05

CHRYSLER
10310	Chrysler Engine Overhaul Manual
25015	Chrysler Cirrus, Dodge Stratus, Plymouth Breeze, '95 thru '00
25020	Full-size Front-Wheel Drive '88 thru '93
	K-Cars - see DODGE Aries (30008)
	Laser - see DODGE Daytona (30030)
25025	Chrysler LHS, Concorde & New Yorker, Dodge Intrepid, Eagle Vision, '93 thru '97
25026	Chrysler LHS, Concorde, 300M, Dodge Intrepid '98 thru '03
25027	Chrysler 300, Dodge Charger & Magnum '05 thru '07
25030	Chrysler/Plym. Mid-size '82 thru '95
	Rear-wheel Drive - see DODGE (30050)
25035	PT Cruiser all models '01 thru '03
25040	Chrysler Sebring/Dodge Avenger '95 thru '05, Dodge Stratus '01 thru '05

DATSUN
28005	200SX all models '80 thru '83
28007	B-210 all models '73 thru '78
28009	210 all models '78 thru '82
28012	240Z, 260Z & 280Z Coupe '70 thru '78
28014	280ZX Coupe & 2+2 '79 thru '83
	300ZX - see NISSAN (72010)
28018	510 & PL521 Pick-up '68 thru '73
28020	510 all models '78 thru '81
28022	620 Series Pick-up all models '73 thru '79
	720 Series Pick-up - see NISSAN (72030)
28025	810/Maxima all gas models, '77 thru '84

DODGE
	400 & 600 - see CHRYSLER (25030)
30008	Aries & Plymouth Reliant '81 thru '89
30010	Caravan & Ply. Voyager '84 thru '95
30011	Caravan & Ply. Voyager '96 thru '02
30012	Challenger/Plymouth Saporro '78 thru '83
	Challenger '67-'76 - see DART (30025)
30013	Caravan, Chrysler Voyager, Town & Country '03 thru '06
30016	Colt/Plymouth Champ '78 thru '87
30020	Dakota Pick-ups all models '87 thru '96
30021	Durango '98 & '99, Dakota '97 thru '99
30022	Dodge Durango models '00 thru '03 Dodge Dakota models '00 thru '04
30023	Dodge Durango '04 thru '06, Dakota '05 and '06
30025	Dart, Challenger/Plymouth Barracuda & Valiant 6 cyl models '67 thru '76
30030	Daytona & Chrysler Laser '84 thru '89
	Intrepid - see Chrysler (25025, 25026)
30034	Dodge & Plymouth Neon '95 thru '99
30035	Omni & Plymouth Horizon '78 thru '90
30036	Dodge and Plymouth Neon '00 thru '05
30040	Pick-ups all full-size models '74 thru '93
30041	Pick-ups all full-size models '94 thru '01
30042	Dodge Full-size Pick-ups '02 thru '05
30045	Ram 50/D50 Pick-ups & Raider and Plymouth Arrow Pick-ups '79 thru '93
30050	Dodge/Ply./Chrysler RWD '71 thru '89
30055	Shadow/Plymouth Sundance '87 thru '94
30060	Spirit & Plymouth Acclaim '89 thru '95
30065	Vans - Dodge & Plymouth '71 thru '03

EAGLE
	Talon - see MITSUBISHI (68030, 68031)
	Vision - see CHRYSLER (25025)

FIAT
34010	124 Sport Coupe & Spider '68 thru '78
34025	X1/9 all models '74 thru '80

FORD
10355	Ford Automatic Transmission Overhaul
10320	Ford Engine Overhaul Manual
36004	Aerostar Mini-vans '86 thru '97
36006	Aspire - see Ford Festiva (36030)
36006	Contour/Mercury Mystique '95 thru '00
36008	Courier Pick-up all models '72 thru '82
36012	Crown Victoria & Mercury Grand Marquis '88 thru '06
36016	Escort/Mercury Lynx '81 thru '90
36020	Escort/Mercury Tracer '91 thru '00
	Expedition - see FORD Pick-up (36059)
36022	Ford Escape & Mazda Tribute '01 thru '03
36024	Explorer & Mazda Navajo '91 thru '01
36025	Ford Explorer & Mercury Mountaineer '02 thru '06
36028	Fairmont & Mercury Zephyr '78 thru '83
36030	Festiva & Aspire '88 thru '97
36032	Fiesta all models '77 thru '80
36034	Focus all models '00 thru '05
36036	Ford & Mercury Full-size '75 thru '87
36044	Ford & Mercury Mid-size '75 thru '86
36048	Mustang V8 all models '64-1/2 thru '73
36049	Mustang II 4 cyl, V6 & V8 '74 thru '78
36050	Mustang & Mercury Capri '79 thru '86
36051	Mustang all models '94 thru '04
36052	Mustang '05 thru '07
36054	Pick-ups and Bronco '73 thru '79
36058	Pick-ups and Bronco '80 thru '96
36059	F-150 & Expedition '97 thru '03, F-250 '97 thru '99 & Lincoln Navigator '98 thru '02
36060	Super Duty Pick-ups, Excursion '99 thru '06
36061	F-150 full-size '04 thru '06
36062	Pinto & Mercury Bobcat '75 thru '80
36066	Probe all models '89 thru '92
36070	Ranger/Bronco II gas models '83 thru '92
36071	Ford Ranger '93 thru '05 & Mazda Pick-ups '94 thru '05
36074	Taurus & Mercury Sable '86 thru '95
36075	Taurus & Mercury Sable '96 thru '01
36078	Tempo & Mercury Topaz '84 thru '94
36082	Thunderbird/Mercury Cougar '83 thru '88
36086	Thunderbird/Mercury Cougar '89 thru '97
36090	Vans all V8 Econoline models '69 thru '91
36094	Vans full size '92 thru '05
36097	Windstar Mini-van '95 thru '03

GENERAL MOTORS
10360	GM Automatic Transmission Overhaul
38005	Buick Century, Chevrolet Celebrity, Olds Cutlass Ciera & Pontiac 6000 '82 thru '96
38010	Buick Regal, Chevrolet Lumina, Oldsmobile Cutlass Supreme & Pontiac Grand Prix front wheel drive '88 thru '05
38015	Buick Skyhawk, Cadillac Cimarron, Chevrolet Cavalier, Oldsmobile Firenza Pontiac J-2000 & Sunbird '82 thru '94
38016	Chevrolet Cavalier/Pontiac Sunfire '95 thru '04
38017	Chevrolet Cobalt & Pontiac G5 '05 thru '07
38020	Buick Skylark, Chevrolet Citation, Olds Omega, Pontiac Phoenix '80 thru '85
38025	Buick Skylark & Somerset, Olds Achieva, Calais & Pontiac Grand Am '85 thru '98
38026	Chevrolet Malibu, Olds Alero & Cutlass, Pontiac Grand Am '97 thru '03
38027	Chevrolet Malibu '04 thru '07
38030	Cadillac Eldorado & Oldsmobile Toronado '71 thru '85, Seville '80 thru '85, Buick Riviera '79 thru '85
38031	Cadillac Eldorado & Seville '86 thru '91, DeVille & Buick Riviera '86 thru '93, Fleetwood & Olds Toronado '86 thru '92
38032	DeVille '94 thru '05, Seville '92 thru '04
38035	Chevrolet Lumina APV, Oldsmobile Silhouette & Pontiac Trans Sport '90 thru '96
38036	Chevrolet Venture, Olds Silhouette, Pontiac Trans Sport & Montana '97 thru '05
	General Motors Full-size Rear-wheel Drive - see BUICK (19025)

GEO
	Metro - see CHEVROLET Sprint (24075)
	Prizm - see CHEVROLET (24060) or TOYOTA (92036)
40030	Storm all models '90 thru '93
	Tracker - see SUZUKI Samurai (90010)

GMC
	Vans & Pick-ups - see CHEVROLET

HONDA
42010	Accord CVCC all models '76 thru '83
42011	Accord all models '84 thru '89
42012	Accord all models '90 thru '93
42013	Accord all models '94 thru '97
42014	Accord all models '98 thru '02
42015	Honda Accord models '03 thru '05
42020	Civic 1200 all models '73 thru '79
42021	Civic 1300 & 1500 CVCC '80 thru '83
42022	Civic 1500 CVCC all models '75 thru '79
42023	Civic all models '84 thru '91
42024	Civic & del Sol '92 thru '95
42025	Civic '96 thru '00, CR-V '97 thru '01, Acura Integra '94 thru '00
	Passport - see ISUZU Rodeo (47017)
42026	Civic '01 thru '04, CR-V '02 thru '04
42035	Honda Odyssey models '99 thru '04
42037	Honda Pilot '03 thru '07, Acura MDX '01 thru '07
42040	Prelude CVCC all models '79 thru '89

HYUNDAI
43010	Elantra all models '96 thru '01
43015	Excel & Accent all models '86 thru '98

ISUZU
	Hombre - see CHEVROLET S-10 (24071)
47017	Rodeo '91 thru '02, Amigo '89 thru '02, Honda Passport '95 thru '02
47020	Trooper '84 thru '91, Pick-up '81 thru '93

JAGUAR
49010	XJ6 all 6 cyl models '68 thru '86
49011	XJ6 all models '88 thru '94
49015	XJ12 & XJS all 12 cyl models '72 thru '85

JEEP
50010	Cherokee, Comanche & Wagoneer Limited all models '84 thru '01
50020	CJ all models '49 thru '86
50025	Grand Cherokee all models '93 thru '04
50029	Grand Wagoneer & Pick-up '72 thru '91
50030	Wrangler all models '87 thru '03
50035	Liberty '02 thru '04

KIA
54070	Sephia '94 thru '01, Spectra '00 thru '04

LEXUS
	ES 300 - see TOYOTA Camry (92007)

LINCOLN
	Navigator - see FORD Pick-up (36059)
59010	Rear Wheel Drive all models '70 thru '05

MAZDA
61010	GLC (rear wheel drive) '77 thru '83
61011	GLC (front wheel drive) '81 thru '85
61015	323 & Protegé '90 thru '03
61016	MX-5 Miata '90 thru '97
61020	MPV all models '89 thru '94
	Navajo - see FORD Explorer (36024)
61030	Pick-ups '72 thru '93
	Pick-ups '94 on - see Ford (36071)
61035	RX-7 all models '79 thru '85
61036	RX-7 all models '86 thru '91
61040	626 (rear wheel drive) '79 thru '82
61041	626 & MX-6 (front wheel drive) '83 thru '92
61042	626 '93 thru '01, & MX-6/Ford Probe '93 thru '01

MERCEDES-BENZ
63012	123 Series Diesel '76 thru '85
63015	190 Series 4-cyl gas models, '84 thru '88
63020	230, 250 & 280 6 cyl sohc '68 thru '72
63025	280 123 Series gas models '77 thru '81
63030	350 & 450 all models '71 thru '80

MERCURY
64200	Villager & Nissan Quest '93 thru '01
	All other titles, see FORD listing.

MG
66010	MGB Roadster & GT Coupe '62 thru '80
66015	MG Midget & Austin Healey Sprite Roadster '58 thru '80

MITSUBISHI
68020	Cordia, Tredia, Galant, Precis & Mirage '83 thru '93
68030	Eclipse, Eagle Talon & Plymouth Laser '90 thru '94
68031	Eclipse '95 thru '01, Eagle Talon '95 thru '98
68035	Mitsubishi Galant '94 thru '03
68040	Pick-up '83 thru '96, Montero '83 thru '93

NISSAN
72010	300ZX all models incl. Turbo '84 thru '89
72015	Altima all models '93 thru '04
72020	Maxima all models '85 thru '92
72021	Maxima all models '93 thru '01
72030	Pick-ups '80 thru '97, Pathfinder '87 thru '95
72031	Frontier Pick-up '98 thru '04, Xterra '00 thru '04, Pathfinder '96 thru '04
72040	Pulsar all models '83 thru '86
72050	Sentra all models '82 thru '94
72051	Sentra & 200SX all models '95 thru '04
72060	Stanza all models '82 thru '90

OLDSMOBILE
73015	Cutlass '74 thru '88
	For other OLDSMOBILE titles, see BUICK, CHEVROLET or GM listings.

PLYMOUTH
	For PLYMOUTH titles, see DODGE.

PONTIAC
79008	Fiero all models '84 thru '88
79018	Firebird V8 models except Turbo '70 thru '81
79019	Firebird all models '82 thru '92
79040	Mid-size Rear-wheel Drive '70 thru '87
	For other PONTIAC titles, see BUICK, CHEVROLET or GM listings.

PORSCHE
80020	911 Coupe & Targa models '65 thru '89
80025	914 all 4 cyl models '69 thru '76
80030	924 all models incl. Turbo '76 thru '82
80035	944 all models incl. Turbo '83 thru '89

RENAULT
	Alliance, Encore - see AMC (14020)

SAAB
84010	900 including Turbo '79 thru '88

SATURN
87010	Saturn all models '91 thru '02
87011	Saturn Ion '03 thru '07
87020	Saturn all L-series models '00 thru '04

SUBARU
89002	1100, 1300, 1400 & 1600 '71 thru '79
89003	1600 & 1800 2WD & 4WD '80 thru '94
89100	Legacy all models '90 thru '99
89101	Legacy & Forester '00 thru '06

SUZUKI
90010	Samurai/Sidekick/Geo Tracker '86 thru '01

TOYOTA
92005	Camry all models '83 thru '91
92006	Camry all models '92 thru '96
92007	Camry/Avalon/Solara/Lexus ES 300 '97 thru '01
92008	Toyota Camry, Avalon and Solara & Lexus ES 300/330 all models '02 thru '05
92015	Celica Rear Wheel Drive '71 thru '85
92020	Celica Front Wheel Drive '86 thru '99
92025	Celica Supra all models '79 thru '92
92030	Corolla all models '75 thru '79
92032	Corolla rear wheel drive models '80 thru '87
92035	Corolla front wheel drive models '84 thru '92
92036	Corolla & Geo Prizm '93 thru '02
92037	Corolla models '03 thru '05
92040	Corolla Tercel all models '80 thru '82
92045	Corona all models '74 thru '82
92050	Cressida all models '78 thru '82
92055	Land Cruiser FJ40/43/45/55 '68 thru '82
92056	Land Cruiser FJ60/62/80/FZJ80 '80 thru '96
92065	MR2 all models '85 thru '87
92070	Pick-up all models '69 thru '78
92075	Pick-up all models '79 thru '95
92076	Tacoma '95 thru '04, 4Runner '96 thru '02, T100 '93 thru '98
92078	Tundra '00 thru '05, Sequoia '01 thru '04
92079	Previa all models '91 thru '95
92081	Prius '01 thru '08
92082	RAV4 all models '96 thru '05
92085	Tercel all models '87 thru '94
92090	Sienna all models '98 thru '02
92095	Highlander & Lexus RX-330 '99 thru '06

TRIUMPH
94007	Spitfire all models '62 thru '81
94010	TR7 all models '75 thru '81

VW
96008	Beetle & Karmann Ghia '54 thru '79
96009	New Beetle '98 thru '05
96016	Rabbit, Jetta, Scirocco, & Pick-up gas models '75 thru '92 & Convertible '80 thru '92
96017	Golf, GTI & Jetta '93 thru '98, Cabrio '95 thru '98
96018	Golf, GTI, Jetta & Cabrio '99 thru '02
96020	Rabbit, Jetta, Pick-up diesel '77 thru '84
96023	Passat '98 thru '01, Audi A4 '96 thru '01
96030	Transporter 1600 all models '68 thru '79
96035	Transporter 1700, 1800, 2000 '72 thru '79
96040	Type 3 1500 & 1600 '63 thru '73
96045	Vanagon air-cooled models '80 thru '83

VOLVO
97010	120, 130 Series & 1800 Sports '61 thru '73
97015	140 Series all models '66 thru '74
97020	240 Series all models '76 thru '93
97040	740 & 760 Series all models '82 thru '88

TECHBOOK MANUALS
10205	Automotive Computer Codes
10206	OBD-II & Electronic Engine Management Systems
10210	Automotive Emissions Control Manual
10215	Fuel Injection Manual, 1978 thru 1985
10220	Fuel Injection Manual, 1986 thru 1999
10225	Holley Carburetor Manual
10230	Rochester Carburetor Manual
10240	Weber/Zenith/Stromberg/SU Carburetor
10305	Chevrolet Engine Overhaul Manual
10310	Chrysler Engine Overhaul Manual
10320	Ford Engine Overhaul Manual
10330	GM and Ford Diesel Engine Repair
10333	Building Engine Power Manual
10340	Small Engine Repair Manual
10345	Suspension, Steering & Driveline
10355	Ford Automatic Transmission Overhaul
10360	GM Automatic Transmission Overhaul
10405	Automotive Body Repair & Painting
10410	Automotive Brake Manual
10415	Automotive Detailing Manual
10420	Automotive Electrical Manual
10425	Automotive Heating & Air Conditioning
10430	Automotive Reference Dictionary
10435	Automotive Tools Manual
10440	Used Car Buying Guide
10445	Welding Manual
10450	ATV Basics
10452	Scooters, Automatic Transmission 50cc to 250cc

SPANISH MANUALS
98903	Reparación de Carrocería & Pintura
98904	Carburadores para los modelos Holley & Rochester
98905	Códigos Automotrices de la Computadora
98910	Frenos Automotriz
98913	Electricidad Automotriz
98915	Inyección de Combustible 1986 al 1999
99040	Chevrolet & GMC Camionetas '67 al '87
99041	Chevrolet & GMC Camionetas '88 al '98
99042	Chevrolet & GMC Camionetas Cerradas '68 al '95
99055	Dodge Caravan/Ply. Voyager '84 al '95
99075	Ford Camionetas y Bronco '80 al '94
99077	Ford Camionetas Cerradas '69 al '91
99088	Ford Modelos de Tamaño Mediano '75 al '86
99091	Ford Taurus & Mercury Sable '86 al '95
99095	GM Modelos de Tamaño Grande '70 al '90
99100	GM Modelos de Tamaño Mediano '70 al '88
99106	Jeep Cherokee, Wagoneer & Comanche '84 al '00
99110	Nissan Camionetas '80 al '96, Pathfinder '87 al '95
99118	Nissan Sentra '82 al '94
99125	Toyota Camionetas y 4-Runner '79 al '95

Over 100 Haynes motorcycle manuals also available

10-07